OXFORD MATHEMATICAL MONOGRAPHS

Series Editors

OXFORD MATHEMATICAL MONOGRAPHS

The Computational Complexity of Differential and Integral Equations

An information-based approach

ARTHUR G. WERSCHULZ

Division of Science and Mathematics
Fordham University / College at Lincoln Center
New York

and

Department of Computer Science
Columbia University
New York

OXFORD NEW YORK TOKYO
OXFORD UNIVERSITY PRESS
1991

Oxford University Press, Walton Street, Oxford OX2 6DP
Oxford New York Toronto
Delhi Bombay Calcutta Madras Karachi
Petaling Jaya Singapore Hong Kong Tokyo
Nairobi Dar es Salaam Cape Town
Melbourne Auckland
and associated companies in
Berlin Ibadan

Oxford is a trade mark of Oxford University Press

Published in the United States
by Oxford University Press, New York

A catalogue record for this book is available from the British Library

Library of Congress Cataloging in Publication Data
Werschulz, Arthur G.
The computational complexity of differential and integral
equations : an information-based approach / Arthur G. Werschulz.
(Oxford mathematical monographs)
Includes bibliographical references and indexes.
1. Differential equations—Numerical solutions. 2. Integral
equations—Numerical solutions. 3. Computational complexity.
I. Title. II. Series.
QA372.W47   1991   515'.35—dc20   91–17226
ISBN 0–19–853589–9 (h'bk.)

Typeset from TEX Data Supplied
Printed in Great Britain by
St Edmundsbury Press Ltd
Bury St Edmunds, Suffolk

ב"ה
כ' סיון תש"נ
12–13 June 1990

Dedicated to the memory of my father
Harry Wolff Werschulz
אהרן זאב בן אריה ע"ה
on the nineteenth anniversary of his passing
כ' סיון תשל"א — 12–13 June 1971

ת.נ.ש.ל.ב.ע.

Acknowledgements

I would like to thank all those who, in one way or another, have helped make this work possible.

First, I would like to thank all those who have participated in the information-based complexity group of the Columbia University Computer Science Department, for providing a stimulating research environment. My special thanks go to J. F. Traub and H. Woźniakowski, who have been both my mentors and my colleagues during the past fifteen years. I would also like to thank T. E. Boult, B. Z. Kacewicz, and G. W. Wasilkowski for their comments on various portions of the manuscript.

In addition, I would like to thank my friend and colleague A. K. Aziz of the University of Maryland, from whom I learned to really appreciate the beauty and utility of finite element methods.

Next, I would like to thank those who extended me the financial support necessary for this project. Fordham University granted me a Faculty Fellowship for the 1988-89 academic year. Moreover, the Columbia University Computer Science Department made me a Visiting Lecturer for that year. In addition, the National Science Foundation supported me for two summers under Grant Number CCR-8901664.

Finally, I would like to thank my wife Patricia and my sons Aaron and Nathaniel for their patience, understanding, support, and love.

Contents

1

Introduction

Most problems of scientific computation must be solved with uncertainty. Prime examples include the solution of differential equations (DE) and integral equations (IE). We typically have only *partial* and *contaminated* information about such problems, this information being *priced*.

We motivate these ideas by considering the solution of the Poisson equation. Let $\Omega \subset \mathbb{R}^d$. We want to find a function $u \colon \overline{\Omega} \to \mathbb{R}$, such that $-\Delta u = f$ in Ω, subject to homogeneous Dirichlet boundary conditions $u = 0$ on $\partial \Omega$. Of course, the function f is defined at an uncountable number of points, whereas a digital computer can only handle a finite set of numbers. Hence we cannot enter "the function f" into our computer. We typically (although often implicitly) assume that the only way we can input f is by sampling f at a finite set of points in $\overline{\Omega}$, and then entering these function values into the computer. In other words, the only information an algorithm can have about f consists of $f(x_1), \ldots, f(x_n)$ at points $x_1, \ldots, x_n \in \overline{\Omega}$. We call this information "standard information," since it is so commonly used.

We see that our information is *partial*. That is, since we only know the value of f at a finite set of points, there is an infinite set of functions $\tilde{f}$ which share the same information as f, i.e. such that $\tilde{f}(x_i) = f(x_i)$ for $1 \leq i \leq n$. Moreover, this information is usually *contaminated* by roundoff error, measurement error, and the like. Finally, there is a *price* associated with computing each $f(x_i)$. If we obtain $f(x_i)$ by a computation, this price might be the number of operations used in its computation, whereas if we get $f(x_i)$ by experimental measurement, the price might be the cost of running the experiment.

Because only partial and/or contaminated information is available, we see that there must be uncertainty in the computed solution provided by any algorithm using this information. That is, any computed solution must have positive error ε. Thus it is natural to ask how we can guarantee an error no greater than ε at minimal cost. More precisely, we wish to answer the following questions:

(i) *What is the complexity of the problem?* For $\varepsilon > 0$, we wish to compute an ε-approximation, i.e. an approximation whose error is at most ε. What is the minimal cost of doing this?

(ii) *How can we find optimal complexity algorithms?* We wish to obtain an algorithm whose error is at most ε at minimal cost. Note that this includes the selection of optimal information.

(iii) *To what extent are the usual algorithms optimal?* That is, we wish to find out whether algorithms commonly used in practice have almost minimal cost.

Such questions belong to the area of research known as *information-based complexity*.

We illustrate these ideas with the Poisson example $-\Delta u = f$ mentioned above. Assume that only standard information about the function f is available. Suppose that f is in the unit ball of the Sobolev space[1] $H^r(\Omega)$. Let us measure the error of an algorithm by the maximum error in the $L_2(\Omega)$-norm over all such f. Then:

(i) The problem complexity grows as $\varepsilon^{-d/r}$ as $\varepsilon \to 0$.

(ii) The standard implementation of the finite element method (FEM) of degree k as a finite element method with quadrature (FEMQ) is an almost optimal algorithm if $k \geq r$. Note that this method samples f on a quasi-uniform set of points. Hence, quasi-uniform sampling is nearly optimal for this problem.

(iii) If $k < r$, then the FEMQ is no longer optimal. Moreover, the asymptotic penalty for using an FEMQ whose degree is too low is unbounded.

Note that (i) states that the complexity of our Poisson problem is exponential in the dimension d. This means that the Poisson problem is intractable for problems of high dimension. Note that high-dimensional Poisson problems are not of merely academic interest. Common examples of such problems include high-dimensional random walks and simultaneous Brownian motion of many non-interacting particles. Another example of a high-dimensional problem is given by Schrödinger's equation from quantum mechanics, the solution of a p-particle problem in three dimensions being a function of $d = 3p$ variables. For instance, if we wanted to find the wave function of the electron cloud in an uranium atom, we would have $p = 92$ electrons, so the dimension of the problem is $d = 276$.

This "curse of dimensionality" is not restricted to problems of high dimension. Suppose, for example, that we wish to solve a three-dimensional problem with $r = 1$. Then the complexity of computing an ε-approximation is proportional to ε^{-3}. If we need four-place accuracy ($\varepsilon = 10^{-4}$), this means that we must use approximately 10^{12} operations. On a megaflop machine, this is about two weeks!

Is there some way of overcoming the intractability of the Poisson problem? Our first idea might be to choose a different algorithm. This ap-

---

[1]Functions belonging to the space $H^r(\Omega)$ have r derivatives in a "weak" sense; see §A.2 in the Appendix for details.

proach is doomed. The complexity (being the minimal cost of computing an approximation) is a problem invariant, independent of any particular algorithm. So, changing the algorithm won't help at all, since the *minimal* cost of solving our problem is too big. Another approach might be to change the problem being solved. For instance, we might demand that f be analytic, or we might only require an approximate solution at a few points, rather than everywhere in the region. We choose not to follow that path; we shall assume that the problem presented to us is the one that we must solve.

A means by which we might be able to circumvent intractability is to change the setting. Complexity result (i) holds for the the *worst case setting*. That is, the error and cost of an algorithm are defined by their suprema over all possible f. Suppose that we no longer require that an ε-approximation have error at most ε for all f. If we know a probability measure on the space of all admissible f, then two possibilities immediately suggest themselves:

- an *average case setting*, in which error and cost are given by expected values, and
- a *probabilistic setting*, in which we ignore the error over sets of small measure.

Later on, we shall see that there *are* reasonable measures on $H^r(\Omega)$ such that the Poisson problem is tractable in the average case and probabilistic settings.

The purpose of this book is to answer questions (i)–(iii) for certain classes of DE and IE that arise in scientific computation, namely:

- elliptic partial differential equations,
- elliptic systems,
- Fredholm integral equations of the second kind,
- Fredholm integral equations of the first kind, and
- initial-value problems for ordinary differential equations.

Again, our goals are to determine the complexity (minimal cost) of computing ε-approximations to these problems, as well as to find optimal algorithms for their approximate solution. We give results for the worst case, average, and probabilistic settings.

We briefly outline the contents of this book.

In Chapter 2, we use the simple example of a two-point boundary value problem to explain the main ideas of information-based complexity. This chapter is self-contained, requiring no results from the general theory. In Chapter 3, we give a brief abstract formulation of information-based complexity. We describe problem formulation, information, and algorithms. Moreover, we formally introduce the settings studied in this book.

Chapters 4–6 cover the worst case setting. This is the setting that is (at least tacitly) most often used in scientific computation. We give general theoretical results for this setting in Chapter 4. In Chapter 5, we look at the complexity of boundary value problems for linear elliptic partial differential equations (PDE). We give conditions that are necessary and sufficient for the finite element method (FEM) of given degree to be a nearly optimal complexity algorithm. Moreover, we consider both unrestricted linear information and standard information for these problems. (The "pure" FEM uses the former information, whereas the FEMQ uses the latter.) As a result, we can quantify the penalty for using standard information instead of unrestricted linear information.

In Chapter 6, we examine other problems in the worst case setting. For linear elliptic systems and Fredholm integral equations of the second kind, our results are similar to those for scalar linear elliptic PDE (except that we use a least-squares FEM or FEMQ for elliptic systems). Next, we turn to ill-posed problems. Ill-posedness arises mainly because the "solution operator" defining the problem is unbounded (for example, the Fredholm problem of the first kind) We find that any algorithm for such a problem must have infinite error. Hence, the complexity of ill-posed problems is infinite, no matter how large we choose our desired error. However, suppose that we change our error criterion from the absolute error criterion that we have been using to a residual error criterion. For the other problems listed above, this causes no appreciable difference. But if we make this change for the Fredholm problem of the first kind, the problem now becomes solvable, and we describe optimal algorithms for its approximate solution. Finally, we discuss the complexity of initial value problems for ordinary differential equations.

In Chapter 7, we investigate the average case and probabilistic settings. First, we study elliptic PDE and systems, as well as the Fredholm IE of the second kind. For certain measures (including the classical Wiener measure), we find that these problems are now tractable. Furthermore, we study the optimality of the FEM and FEMQ in these new settings, determining conditions under which they are optimal. Finally, we discuss ill-posed linear problems, showing that their average case and probabilistic complexities are finite; moreover, we describe nearly-optimal algorithms for their solution.

In Chapter 8, we discuss the asymptotic and randomized settings. Note that in the previous settings, our goal was to find a single *fixed* choice of information and algorithm that would minimize the cost of computing an ε-approximation over the set of problem elements. In the *asymptotic* setting, we instead wish to find a *sequence* of information and algorithms with the fastest possible speed of convergence. Using known relationships between the various settings, we show that the same information and algorithms that were optimal in the worst case setting are optimal in the asymptotic setting. Moreover, in our previous work, the choice of information and algorithm was *deterministic*. In the *randomized* setting, we allow *nondeterministic* information and algorithms (such as Monte Carlo methods). Our hope is that randomization may help us to beat intractability, as it does for the problem of high-dimensional numerical quadrature. Our result is that this is not the case. Randomization does not help us if we want to approximate the solution of an elliptic PDE, elliptic system, or the Fredholm problem of the second kind throughout the region over which the problem is defined. Furthermore, ill-posed problems have infinite complexity in the randomized setting.

To make this book reasonably self-contained, we have included additional technical material in an Appendix. We also have a bibliography, listing about 140 references. Notes and Remarks are included at the end of many sections and subsections, containing material of historical or bibliographic interest, as well as additional commentary to (or extensions of) main results.

We close this introduction by noting that information-based complexity has become a major area of research over the past decade. The reader who is interested in learning more about this area should consult Traub *et al.* (1988) and Traub and Woźniakowski (1980) . These monographs are frequently referenced throughout this book; for brevity, we refer to them as IBC and GTOA (which are acronyms for their titles). Moreover, several survey articles have recently appeared. Packel and Traub (1987), Packel and Woźniakowski (1987), Traub and Woźniakowski (1984), and Woźniakowski (1986) all give an introductory overview of information-based complexity. In addition, Traub and Woźniakowski (1990) poses research questions for mathematicians.

Note that we follow a standard numbering scheme for chapters, sections, and subsections of this book (for example, §2.1 denotes the first section of the second chapter and §2.1.1 denotes the first subsection of §2.1). Theorems, corollaries, and lemmas are numbered separately and consecutively within the section or subsection in which they appear (for example, Theorem 4.5.3.2 is the second theorem in §4.5.3). References to the Appendix are in the form §A.2 (for the second section in the Appendix), Lemma A.2.2.1 (for the first lemma in the second subsection of §A.2), and so forth. Finally, the Notes and Remarks are numbered essentially in the same way as the theorems, corollaries, and lemmas, i.e. **NR 2.1.1:2** is the second entry in the Notes and Remarks for §2.1.1.

2

Example: a two-point boundary value problem

2.1 Introduction

We use a simple example of a two-point boundary value problem

$$(Lu)(x) \equiv -u''(x) + u(x) = f(x) \qquad \text{for } 0 < x < 1,$$
$$u'(0) = u'(1) = 0.$$

to illustrate major issues and concepts of information-based complexity.

First, we must describe our problem more carefully. That is, we specify

- the *problem formulation*,
- permissible *information*, and
- a *model of computation*.

Once these are all specified, we define the ε-complexity of our problem to be the minimal cost of computing an ε-approximation, i.e. an approximation whose error is at most ε.

Before proceeding, we establish some notational conventions used in the rest of this book. If X and Y are linear spaces, and $A \colon X \to Y$ is a mapping (which need not be linear), we will generally write Ax, instead of $A(x)$, for the image of x under A. We let $\mathbb{R}$, $\mathbb{Z}$, $\mathbb{N}$, and $\mathbb{P}$ respectively denote the sets of real numbers, integers, nonnegative integers, and positive integers. Given functions $f, g \colon \mathbb{N} \to \mathbb{R}$, we say that

$$f(n) = O\big(g(n)\big) \text{ as } n \to \infty \qquad \text{if} \qquad \overline{\lim_{n \to \infty}} \left| \frac{f(n)}{g(n)} \right| < \infty,$$

$$f(n) = \Omega\big(g(n)\big) \text{ as } n \to \infty \qquad \text{if} \qquad g(n) = O\big(f(n)\big) \text{ as } n \to \infty,$$

$$f(n) = \Theta\big(g(n)\big) \text{ as } n \to \infty \qquad \text{if} \qquad \left\{ \begin{array}{c} f(n) = O\big(g(n)\big) \\ \text{and} \\ f(n) = \Omega\big(g(n)\big) \end{array} \right\} \text{ as } n \to \infty,$$

$$f(n) = o\big(g(n)\big) \text{ as } n \to \infty \qquad \text{if} \qquad \lim_{n \to \infty} \frac{f(n)}{g(n)} = 0,$$

and

$$f(n) \sim g(n) \text{ as } n \to \infty \qquad \text{if} \qquad \lim_{n \to \infty} \frac{f(n)}{g(n)} = 1.$$

If $f, g \colon (0, \varepsilon_0) \to \mathbb{R}$ for some $\varepsilon_0 > 0$, then $f(\varepsilon) = O(g(\varepsilon))$ as $\varepsilon \to 0$, $f(\varepsilon) = \Omega(g(\varepsilon))$ as $\varepsilon \to 0$, etc., are defined analogously.

Notes and Remarks

NR 2.1:1 The $O-$, $o-$, and $\sim$-notations are standard. The Ω- and Θ- notations may be found in Knuth (1976).

2.1.1 Problem formulation

First, we recall the definition of two standard function spaces that we will use throughout this chapter. Let $I = (0, 1)$ denote the unit interval. Then $L_2(I)$ is the usual space of (Lebesgue) square-integrable real-valued functions, which is a Hilbert space under the inner product

$$\langle f, g \rangle_{L_2(I)} = \int_0^1 f(x) g(x) \, dx \qquad \forall f, g \in L_2(I)$$

and norm

$$\| f \|_{L_2(I)} = \sqrt{\langle f, f \rangle_{L_2(I)}} \qquad \forall f \in L_2(I).$$

We also let

$$H^1(I) = \{ v \in L_2(I) : v' \in L_2(I) \},$$

the prime denoting weak derivative. Define an inner product and norm on $H^1(I)$ by

$$\langle v, w \rangle_{H^1(I)} = \langle v, w \rangle_{L_2(I)} + \langle v', w' \rangle_{L_2(I)} \qquad \forall v, w \in H^1(I)$$

and

$$\| v \|_{H^1(I)} = \sqrt{\langle v, v \rangle_{H^1(I)}} \qquad \forall v \in H^1(I),$$

so that $H^1(I)$ is also a Hilbert space.

We are now ready to define our problem. Let

$$F = \{ f \in L_2(I) : \| f \|_{L_2(I)} \leq 1 \}$$

denote a set of *problem elements*. For $f \in F$, we wish to solve the problem

$$-u''(x) + u(x) = f(x) \qquad \text{for } 0 < x < 1,$$
$$u'(0) = u'(1) = 0.$$

Note that the Riesz representation theorem implies that that there exists a unique $u \in H^1(I)$ such that

$$\langle u, v \rangle_{H^1(I)} = \langle f, v \rangle_{L_2(I)} \qquad \forall\, v \in H^1(I).$$

Such an element u is called a *variational solution* to our problem. We get this formulation by multiplying our original differential equation by an arbitrary $v \in H^1(I)$, and then integrating by parts. We write $u = Sf$, where

$$S \colon F \to H^1(I)$$

is called a *solution operator*, since it maps each problem element f to its matching *solution element* Sf.

Notes and Remarks

NR 2.1.1:1 The space $H^1(I)$ is an example of a Sobolev space. Such spaces will be extensively used in this book. The reader who is unfamiliar with Sobolev spaces should consult §A.2 for further background.

NR 2.1.1:2 What is the connection between variational solutions and classical solutions? (By a classical solution, we mean a twice-continuously differentiable function satisfying the original boundary value problem.) Clearly, any classical solution is a variational solution. It is not too difficult to show that if f is continuous, then the variational solution is a classical solution of our model problem.

2.1.2 Information

To compute an approximation to Sf, we need information about the problem element f. Such information will consist of evaluating continuous linear functionals defined on $L_2(I)$. We assume that *any* such functional is allowed in this chapter (except in §2.6). That is, let Λ denote the class of permissible information operations, and let Λ^* denote the class of all continuous linear functionals on $L_2(I)$. Then $\Lambda = \Lambda^*$ unless stated otherwise.

2.1.3 Model of computation

We assume that the cost of evaluating a linear functional does not depend on the functional. We denote this cost by c, where $c > 0$. We also assume that we can do arithmetic operations and comparisons of real numbers exactly, with unit cost. Finally, since the problem elements and solution elements belong to real linear spaces, we assume that we can do addition and real scalar multiplication with unit cost.

2.2 Error, cost, and complexity

We wish to find the ε-complexity of our problem, i.e. the minimal cost of finding an ε-approximation. What linear functionals must we evaluate, and how should we combine them, to find an ε-approximation with minimal cost?

We make these notions more precise. First, to compute an approximation of the solution Sf, we must know something about the problem element $f \in F$. For the purposes of this chapter, we do this as follows. Choose functions $v_1, \ldots, v_n \in L_2(I)$. Then $\langle \cdot, v_1 \rangle_{L_2(I)}, \ldots, \langle \cdot, v_n \rangle_{L_2(I)}$ are continuous linear functionals on $L_2(I)$. We assume that we can evaluate *information*

$$Nf = \begin{bmatrix} \langle f, v_1 \rangle_{L_2(I)} \\ \vdots \\ \langle f, v_n \rangle_{L_2(I)} \end{bmatrix}$$

about any $f \in F$. Since N consists of n functionals, we say that N is information of *cardinality* n, and write card $N = n$.

Information Nf consists of linear functionals, which are given in advance, independent of any particular $f \in F$. Such information N is called *nonadaptive information*. However, we could have loosened these requirements. For instance, we could have allowed *adaptive* information, in which the v_i depend on f, or information of *varying cardinality*, for which n depends on f. As we will see later in this book, there is no advantage in doing this for our model problem, and so no generality is lost by considering only linear nonadaptive information of fixed cardinality.

Knowing $y = Nf$, we must use y to create an approximation $\phi(y)$ of Sf. The mapping $\phi \colon N(F) \to H^1(I)$ is said to be an *algorithm* using N.

The cost of computing $\phi(Nf)$ may be divided into two parts. First, we evaluate the information $y = Nf$, with *informational cost* denoted by $\mathrm{cost}(N, f)$. Of course, if card $N = n$, then $\mathrm{cost}(N, f) = c\,n$ for all $f \in F$. Next, once we have computed $y = Nf$, we can compute $\phi(y)$. We denote the *combinatory cost* of computing $\phi(y)$, given y, by $\mathrm{cost}(\phi, y)$.

We now define $e(\phi, N)$ and $\mathrm{cost}(\phi, N)$, the error and the cost of (ϕ, N). In this introductory chapter, they will be defined in a worst case setting, i.e.

$$e(\phi, N) = \sup_{f \in F} \| Sf - \phi(Nf) \|_{H^1(I)}$$

and

$$\mathrm{cost}(\phi, N) = \sup_{f \in F} \{ \mathrm{cost}(N, f) + \mathrm{cost}(\phi, Nf) \}.$$

Other settings will be discussed in later chapters.

We can now define the ε-*complexity* of our variational boundary value problem as

$$\mathrm{comp}(\varepsilon) = \inf\{\, \mathrm{cost}(\phi, N) : (\phi, N) \text{ such that } e(\phi, N) \leq \varepsilon \,\}.$$

Our basic goal is to find:

    (i) the ε-complexity $\mathrm{comp}(\varepsilon)$, and
    (ii) an algorithm ϕ_ε using information N_ε such that

$$e(\phi_e, N_\varepsilon) \leq \varepsilon \qquad \text{and} \qquad \mathrm{cost}(\phi_\varepsilon, N_\varepsilon) = \mathrm{comp}(\varepsilon).$$

We say that ϕ_ε is an *optimal ε-complexity algorithm*, or, for brevity, an *optimal complexity algorithm*.

2.3   A minimal error algorithm

To accomplish our basic goal, we first determine the minimal error achievable by an algorithm using information of given cardinality. Then, we show how such minimal error algorithms are related to optimal complexity algorithms.

    Let N be information of cardinality n. We wish to determine

$$r(N) = \inf_{\phi} e(\phi, N),$$

the minimal error among all algorithms using N. For reasons explained in §4.2, we call $r(N)$ the *radius of information*. Since $r(N)$ is the minimal error among all algorithms using N, we say that an algorithm ϕ^* using N is an *optimal error algorithm using* N if $e(\phi^*, N) = r(N)$.

THEOREM 2.3.1. *For any information* N,

$$r(N) = \sup_{h \in \ker N \cap F} \|Sh\|_{H^1(I)},$$

where $\ker N = \{\, h \in L_2(I) : Nh = 0 \,\}$ denotes the kernel of N. Moreover, if

$$Nf = \begin{bmatrix} \langle f, v_1 \rangle_{L_2(I)} \\ \vdots \\ \langle f, v_n \rangle_{L_2(I)} \end{bmatrix} \qquad \forall f \in F$$

for orthonormal $v_1, \ldots, v_n$, then the algorithm ϕ^* defined by

$$\phi^*(Nf) = \sum_{j=1}^{n} \langle f, v_j \rangle_{L_2(I)} S v_j \qquad \forall f \in F$$

is an optimal error algorithm using N.

PROOF. Let N be information of cardinality n. Without loss of generality, we assume that $v_1, \ldots, v_n$ are orthonormal in $L_2(I)$, since if not, we can delete any linearly dependent functions from $\{v_1, \ldots, v_n\}$ and then do Gram-Schmidt orthogonalization on the remaining linearly independent elements.

We first show that

$$r(N) \geq \sup_{h \in \ker N \cap F} \|Sh\|_{H^1(I)}.$$

Let ϕ be an algorithm using N. Let $h \in \ker N \cap F$. Setting $f = h$ and $\tilde{f} = -h$, we have $f, \tilde{f} \in \ker N \cap F$. Thus $\phi(Nf) = \phi(0) = \phi(N\tilde{f})$, and so

$$\begin{aligned} 2\|Sh\|_{H^1(I)} &= \|Sf - S\tilde{f}\|_{H^1(I)} \\ &\leq \|Sf - \phi(Nf)\|_{H^1(I)} + \|S\tilde{f} - \phi(N\tilde{f})\|_{H^1(I)} \\ &\leq 2\,e(\phi, N), \end{aligned}$$

the last inequality because $e(\phi, N)$ is defined by a worst case over F. Since $h \in \ker N \cap F$ and ϕ using N are arbitrary, the lower bound follows.

To complete the proof, we only need to show that

$$e(\phi^*, N) \leq \sup_{h \in \ker N \cap F} \|Sh\|_{H^1(I)}.$$

Let $f \in F$. Define

$$h = f - \sum_{j=1}^{n} \langle f, v_j \rangle_{L_2(I)} v_j \, .$$

Since $v_1, \ldots, v_n$ are orthonormal, we can check that

$$\|h\|_{L_2(I)}^2 = \|f\|_{L_2(I)}^2 - \sum_{j=1}^{n} \langle f, v_j \rangle_{L_2(I)}^2 \leq \|f\|_{L_2(I)}^2 \leq 1$$

and

$$\langle h, v_i \rangle_{L_2(I)} = 0 \qquad (1 \leq i \leq n).$$

Hence $h \in \ker N \cap F$, and so

$$\|Sf - \phi^*(Nf)\|_{H^1(I)} = \|Sh\|_{H^1(I)} \leq \sup_{h \in \ker N \cap F} \|Sh\|_{H^1(I)}.$$

Since $f \in F$ is arbitrary, the desired inequality follows immediately. $\quad\square$

Now that we know how to find the best algorithm using given information, we now try to find the best information of given cardinality. That is, for $n \in \mathbb{N}$, we wish to find the nth *minimal radius*

$$r(n) = \inf\{\, r(N) : \operatorname{card} N \leq n \,\}.$$

If N_n is information of cardinality at most n for which $r(N_n) = r(n)$, then N_n is said to be *nth optimal information*. An algorithm ϕ_n using N_n for which $e(\phi_n, N_n) = r(n)$ is an *nth minimal error algorithm*, since there is no algorithm using information of cardinality at most n whose error is smaller.

To find $r(n)$, we first note that S, when extended from F to $L_2(I)$ by linearity, is a compact injection of $L_2(I)$ into $H^1(I)$; see **NR 2.3:2** for a proof. Letting $S^* \colon H^1(I) \to L_2(I)$ denote the usual Hilbert space adjoint of S, we see that S^*S is a self-adjoint, compact, positive-definite operator on $L_2(I)$, whose positive square root we denote by $K = (S^*S)^{1/2}$.

Let $\kappa_1 \geq \kappa_2 \geq \ldots > 0$ be the eigenvalues of K, and let $y_1, y_2, \ldots \in L_2(I)$ denote the corresponding orthonormal eigenfunctions. Since our model problem is simple, we can determine the eigenpairs of K explicitly. Indeed, note that

$$\langle f, S^*v \rangle_{L_2(I)} = \langle Sf, v \rangle_{H^1(I)} = \langle f, v \rangle_{L_2(I)} \qquad \forall v \in H^1(I), f \in L_2(I),$$

and so $S^*v = v$ for any $v \in H^1(I)$. Thus for any $j \in \mathbb{P}$,

$$Ky_j = \kappa_j y_j \qquad \text{iff} \qquad Sy_j = \kappa_j^2 y_j,$$

which is true iff

$$-y_j'' + y_j = \kappa_j^{-2} y_j \qquad \text{in } (0,1),$$
$$y_j'(0) = y_j'(1) = 0.$$

Using standard techniques, we find that

$$\kappa_j = \frac{1}{\sqrt{1 + (j-1)^2 \pi^2}}$$

and

$$y_j(x) = \begin{cases} 1 & \text{for } j = 1, \\ \sqrt{2}\cos\left((j-1)\pi x\right) & \text{for } j \geq 2. \end{cases}$$

THEOREM 2.3.2. *For $n \in \mathbb{P}$, define information N_n by*

$$N_n f = \begin{bmatrix} \langle f, y_1 \rangle_{L_2(I)} \\ \vdots \\ \langle f, y_n \rangle_{L_2(I)} \end{bmatrix} \qquad \forall f \in F$$

and an algorithm ϕ_n using N_n by

$$\phi_n(N_n f) = \sum_{j=1}^{n} \langle f, y_j \rangle_{L_2(I)} \, \kappa_j^2 y_j \qquad \forall f \in F.$$

For $n = 0$, define $N_0 f \equiv 0$ and $\phi_0(N_0 f) \equiv 0$. Then for any $n \in \mathbb{N}$, N_n is *n*th optimal information and ϕ_n is an *n*th minimal error algorithm, and

$$e(\phi_n, N_n) = r(N_n) = r(n) = \kappa_{n+1} = \frac{1}{\sqrt{1 + n^2 \pi^2}}.$$

PROOF. Since $S y_j = \kappa_j^2 y_j$ for all $j \in \mathbb{N}$, we have

$$\phi_n(N_n f) = \sum_{j=1}^{n} \langle f, y_j \rangle_{L_2(I)} S y_j \qquad \forall f \in F,$$

and so ϕ_n is an optimal error algorithm using N_n by Theorem 2.3.1. It only remains to show that N_n is *n*th optimal information and that

$$r(N_n) = r(n) = \kappa_{n+1}.$$

Let N be information of cardinality at most n, say

$$N f = \begin{bmatrix} \langle f, v_1 \rangle_{L_2(I)} \\ \vdots \\ \langle f, v_k \rangle_{L_2(I)} \end{bmatrix} \qquad \forall f \in F,$$

for $v_1, \ldots, v_k \in L_2(I)$ and $k \le n$. Formally set $v_i = 0$ for $k + 1 \le i \le n$. From Theorem 2.3.1, we have

$$r(N) = \sup_{\substack{h \in \ker N \cap F}} \|S h\|_{H^1(I)} = \sup_{\substack{h \perp \{v_1, \ldots, v_n\} \\ \|h\|_{L_2(I)} \le 1}} \|K h\|_{L_2(I)},$$

and so

$$r(n) = \inf_{v_1, \ldots, v_n \in L_2(I)} \sup_{\substack{h \perp \{v_1, \ldots, v_n\} \\ \|h\|_{L_2(I)} \le 1}} \|K h\|_{L_2(I)}.$$

By the minimax principle (see, for example, Dunford and Schwartz (1963), p. 908) we then have

$$r(n) = \kappa_{n+1} = \sup_{\substack{h \perp \{y_1, \ldots, y_n\} \\ \|h\|_{L_2(I)} \le 1}} \|K h\|_{L_2(I)} = \sup_{\substack{h \in \ker N_n \cap F}} \|S h\|_{H^1(I)} = r(N_n),$$

as required. $\qquad\qquad\qquad\qquad\qquad\qquad\qquad\qquad\qquad\qquad\qquad\qquad\square$

Notes and Remarks

NR 2.3:1 The proof of Theorem 2.3.1 is an adaption of a more general technique. See Section 5.7 in Chapter 4 of IBC and Theorem 4.2 in Chapter 5 of GTOA.

NR 2.3:2 We sketch the proof that $S\colon L_2(I) \to H^1(I)$ is a compact injection, as well as prove a bound that will be useful in §2.5. Let

$$H^2(I) = \{\, v \in L_2(I) : v', v'' \in L_2(I) \,\},$$

the primes again denoting weak derivatives. Define an inner product and norm on $H^2(I)$ by

$$\langle v, w \rangle_{H^2(I)} = \langle v, w \rangle_{L_2(I)} + \langle v', w' \rangle_{L_2(I)} + \langle v'', w'' \rangle_{L_2(I)} \qquad \forall\, v, w \in H^1(I)$$

and

$$\|v\|_{H^2(I)} = \sqrt{\langle v, v \rangle_{H^2(I)}} \qquad \forall\, v \in H^1(I),$$

so that $H^2(I)$ is a Hilbert space.

For $f \in L_2(I)$, let $u = Sf \in H^1(I)$ be a variational solution of $Lu = f$, so that

$$\langle u, v \rangle_{H^1(I)} = \langle f, v \rangle_{L_2(I)} \qquad \forall\, v \in H^1(I).$$

We claim that

$$\|u''\|_{L_2(I)} \le 2\|f\|_{L_2(I)} \quad \text{and} \quad \|u\|_{H^2(I)} \le 5\|f\|_{L_2(I)} \qquad \forall\, v \in H^2(I).$$

To prove the first inequality, note that since the space $C(I)$ of continuous functions is dense in $L_2(I)$, we can assume without essential loss of generality that $f \in C(I)$. Hence, u is a classical solution of our problem, i.e. $-u'' + u = f$, from which we find

$$\|u''\|_{L_2(I)} \le \|u\|_{L_2(I)} + \|f\|_{L_2(I)}.$$

Moreover,

$$\|u\|^2_{H^1(I)} = \langle u, u \rangle_{H^1(I)} = \langle f, u \rangle_{L_2(I)} \le \|f\|_{L_2(I)} \|u\|_{L_2(I)} \le \|f\|_{L_2(I)} \|u\|_{H^1(I)},$$

which implies that

$$\|u\|_{L_2(I)} \le \|u\|_{H^1(I)} \le \|f\|_{L_2(I)}.$$

Hence,

$$\|u''\|_{L_2(I)} \le \|u\|_{L_2(I)} + \|f\|_{L_2(I)} \le 2\|f\|_{L_2(I)},$$

establishing the first inequality. Substituting the first inequality into

$$\|u\|^2_{H^2(I)} = \|u\|^2_{H^1(I)} + \|u''\|^2_{L_2(I)},$$

the second inequality follows, as claimed.

Note that S is an injection, since $Sf = 0$ implies that $\langle f, v \rangle_{L_2(I)} = 0$ for all $v \in H^1(I)$, implying that $f = 0$. So, S may be momentarily viewed as a bounded injection of $L_2(I)$ into $H^2(I)$. By the Rellich-Kondrasov compactness theorem (Lemma A.2.2.3), $H^2(I)$ is compactly embedded in $H^1(I)$. Thus $S\colon L_2(I) \to H^1(I)$ is a compact injection.

NR 2.3:3 The proof that N_n is nth optimal information is a streamlined version of that in Section 5.3 of Chapter 4 of IBC. The older proof does not directly use the minimax principle, but essentially re-proves it in the process of showing the optimality of N_n.

2.4 Complexity bounds

We now show how to use our results on minimal error algorithms to obtain sharp bounds on complexity.

THEOREM 2.4.1. *For $\varepsilon > 0$, let*

$$m(\varepsilon) = \begin{cases} \left\lceil \pi^{-1}\sqrt{\varepsilon^{-2}-1} \right\rceil & \text{if } 0 < \varepsilon \leq 1, \\ 0 & \text{if } \varepsilon \geq 1. \end{cases}$$

Define information N_ε by

$$N_\varepsilon f = \begin{bmatrix} \langle f, y_1 \rangle_{L_2(I)} \\ \vdots \\ \langle f, y_{m(\varepsilon)} \rangle_{L_2(I)} \end{bmatrix} \qquad \forall f \in F$$

and an algorithm ϕ_ε using N_ε by

$$\phi_\varepsilon(N_\varepsilon f) = \sum_{j=1}^{m(\varepsilon)} \langle f, y_j \rangle_{L_2(I)} \, \kappa_j^2 y_j \qquad \forall f \in F,$$

with $N_\varepsilon f \equiv 0$ and $\phi_\varepsilon(N_\varepsilon f) \equiv 0$ when $\varepsilon \geq 1$. Then

$$c\,m(\varepsilon) \leq \text{comp}(\varepsilon) \leq (c+2)m(\varepsilon) - 1.$$

Moreover,

$$e(\phi_\varepsilon, N_\varepsilon) \leq \varepsilon \qquad \text{and} \qquad \text{cost}(\phi_\varepsilon, N_\varepsilon) \leq (c+2)m(\varepsilon) - 1.$$

Hence, ϕ_ε is (almost) an optimal ε-complexity algorithm.

PROOF. It is easy to check that

$$m(\varepsilon) = \inf\{\, n \in \mathbb{N} : r(n) \leq \varepsilon \,\}.$$

We need only show that

$$\text{comp}(\varepsilon) \geq c\,m(\varepsilon)$$

and that

$$e(\phi_\varepsilon, N_\varepsilon) \leq \varepsilon \qquad \text{with} \qquad \text{cost}(\phi_\varepsilon, N_\varepsilon) \leq (c+2)m(\varepsilon) - 1.$$

To prove the first statement, let ϕ be any algorithm using information N of cardinality n such that $e(\phi, N) \leq \varepsilon$. Then

$$r(n) \leq r(N) \leq e(\phi, N) \leq \varepsilon,$$

and so $n \geq m(\varepsilon)$. Hence

$$\text{cost}(\phi, N) \geq c\,m(\varepsilon)$$

by the remarks in §2.2.

To prove the second statement, note that N_ε is $m(\varepsilon)$th optimal information and ϕ_ε is an nth minimal error algorithm by Theorem 2.3.2, and so

$$e(\phi_\varepsilon, N_\varepsilon) = r\big(m(\varepsilon)\big) \leq \varepsilon.$$

Moreover, we can compute $\phi_\varepsilon(N_\varepsilon f)$ for any $f \in F$ by evaluating $m(\varepsilon)$ inner products, and then performing $m(\varepsilon)$ scalar multiplications followed by $m(\varepsilon) - 1$ additions. Hence, $\mathrm{cost}(\phi_\varepsilon, N_\varepsilon) \leq (c+2)m(\varepsilon) - 1$.   □

Of course, we normally expect that $c \gg 1$, i.e. that evaluating an arbitrary continuous linear functional is *much* harder than doing an arithmetic operation. We then find that

$$\mathrm{comp}(\varepsilon) \simeq c\,m(\varepsilon) \sim \frac{c}{\pi\varepsilon} \qquad \text{as } \varepsilon \to 0$$

and

$$e(\phi_\varepsilon, N_\varepsilon) \leq \varepsilon \qquad \text{with} \qquad \mathrm{cost}(\phi_\varepsilon, N_\varepsilon) \simeq c\,m(\varepsilon) \sim \frac{c}{\pi\varepsilon} \qquad \text{as } \varepsilon \to 0,$$

explaining why we say that ϕ_ε is "almost" an optimal complexity algorithm.

Notes and Remarks

NR 2.4:1 The proof that ϕ_ε is almost optimal is an adaptation of a more general technique, see Section 5.8 of Chapter 2 of IBC.

2.5   Comparison with the finite element method

In the previous sections, we were able to find almost optimal complexity algorithms for our model problem. Although the overall procedure is the same for more complicated problems, certain details (such as determining the eigenvalues and eigenfunctions of K) will become far more difficult when we replace our model problem by more complicated ones. With this in mind, we should look at standard algorithms, to see how close they are to being optimal. To be specific, we will look at the finite element method using piecewise linear elements over a uniform mesh.

First, for $n \geq 2$, we define an n-dimensional subspace S_n of $H^1(I)$ as follows. Let

$$x_j = \frac{j-1}{n-1} \qquad (1 \leq j \leq n).$$

Then $v \in S_n$ iff v is continuous on I and v is a linear polynomial (i.e. of degree at most 1) on each subinterval $[x_j, x_{j+1}]$ for $1 \leq j \leq n-1$.

Let $v_1, \ldots, v_n$ be a basis for S_n satisfying

$$v_i(x_j) = \delta_{i,j} \qquad (1 \leq i, j \leq n).$$

Then *finite element information* N_n^{FE} of cardinality n is given by

$$N_n^{\mathrm{FE}} f = \begin{bmatrix} \langle f, v_1 \rangle_{L_2(I)} \\ \vdots \\ \langle f, v_n \rangle_{L_2(I)} \end{bmatrix} \qquad \forall f \in F.$$

Next, we define an algorithm ϕ_n^{FE} using N_n^{FE} as follows. First, define the Gram matrix

$$G = [g_{i,j}]_{1 \le i,j \le n} = [\langle v_j, v_i \rangle_{H^1(I)}]_{1 \le i,j \le n},$$

so that

$$g_{i,j} = \begin{cases} (n-1) + \frac{1}{3}(n-1)^{-1} & \text{if } i = j = 1 \text{ or } i = j = n, \\ 2(n-1) + \frac{2}{3}(n-1)^{-1} & \text{if } 2 \le i = j \le n-1, \\ -(n-1) + \frac{1}{6}(n-1)^{-1} & \text{if } |i-j| = 1, \\ 0 & \text{otherwise.} \end{cases}$$

Let $f \in F$. Then we wish to find $u_n \in \mathcal{S}_n$ satisfying

$$\langle u_n, v \rangle_{H^1(I)} = \langle f, v \rangle_{L_2(I)} \qquad \forall v \in H^1(I).$$

If we write

$$u_n(x) = \sum_{j=1}^{n} \alpha_j v_j(x),$$

then

$$Ga = b,$$

with

$$a = \begin{bmatrix} \alpha_1 \\ \vdots \\ \alpha_n \end{bmatrix} \quad \text{and} \quad b = \begin{bmatrix} \langle f, v_1 \rangle_{L_2(I)} \\ \vdots \\ \langle f, v_n \rangle_{L_2(I)} \end{bmatrix} = N_n^{\mathrm{FE}} f.$$

Since G is positive definite, we see that u_n is well-defined, and depends on f only through the information $N_n^{\mathrm{FE}} f$. We write $u_n = \phi_n^{\mathrm{FE}}(N_n^{\mathrm{FE}} f)$, and say that the algorithm ϕ_n^{FE} is the *finite element method* (FEM) using N_n^{FE}.

How good is the FEM?

THEOREM 2.5.1. *For $n \ge 2$,*

$$e(\phi_n^{\mathrm{FE}}, N_n^{\mathrm{FE}}) \le \sqrt{1 + \frac{1}{\pi^2} \frac{2}{\pi(n-1)}}.$$

PROOF. Let $f \in F$, and let $e = Sf - \phi_n^{\text{FE}}(N_n^{\text{FE}}f)$. We also let $\tilde{u} \in \mathcal{S}_n$ be the interpolant of u satisfying

$$\tilde{u}(x_j) = u(x_j) \qquad (1 \le j \le n),$$

and let $\tilde{e} = Sf - \tilde{u}$ denote the error in the interpolant. Since

$$\langle e, v \rangle_{H^1(I)} = 0 \qquad \forall v \in \mathcal{S}_n,$$

we have

$$\|e\|_{H^1(I)} = \inf_{v \in \mathcal{S}_n} \|Sf - v\|_{H^1(I)} \le \|\tilde{e}\|_{H^1(I)}.$$

Choose any subinterval $[x_j, x_{j+1}]$. Since $\tilde{e}(x_j) = \tilde{e}(x_{j+1}) = 0$, we expand $\tilde{e}$ in a sine series to find that

$$\int_{x_j}^{x_{j+1}} \tilde{e}(x)^2 \, dx \le \left[\frac{1}{\pi(n-1)} \right]^2 \int_{x_j}^{x_{j+1}} \tilde{e}'(x)^2 \, dx \le \frac{1}{\pi^2} \int_{x_j}^{x_{j+1}} \tilde{e}'(x)^2 \, dx$$

and

$$\int_{x_j}^{x_{j+1}} \tilde{e}'(x)^2 \, dx \le \left[\frac{1}{\pi(n-1)} \right]^2 \int_{x_j}^{x_{j+1}} \tilde{u}''(x)^2 \, dx,$$

the latter since $\tilde{u}'' = 0$ in (x_j, x_{j+1}). Hence,

$$\int_{x_j}^{x_{j+1}} [\tilde{e}(x)^2 + \tilde{e}'(x)^2] \, dx \le \left(1 + \frac{1}{\pi^2} \right) \int_{x_j}^{x_{j+1}} \tilde{e}'(x)^2 \, dx$$

$$\le \left(1 + \frac{1}{\pi^2} \right) \left[\frac{1}{\pi(n-1)} \right]^2 \int_{x_j}^{x_{j+1}} u''(x)^2 \, dx .$$

Summing over all the subintervals and taking square roots, we have

$$\|\tilde{e}\|_{H^1(I)} \le \sqrt{1 + \frac{1}{\pi^2}} \, \frac{1}{\pi(n-1)} \, \|u''\|_{L_2(I)}.$$

From **NR 2.3:2**, we have

$$\|u''\|_{L_2(I)} \le 2\|f\|_{L_2(I)} \le 2,$$

the latter because $f \in F$ implies that $\|f\|_{L_2(I)} \le 1$. Combining these results, we find that

$$\|Sf - \phi_n^{\text{FE}}(N_n^{\text{FE}}f)\|_{H^1(I)} = \|e\|_{H^1(I)} \le \|\tilde{e}\|_{H^1(I)} \le \sqrt{1 + \frac{1}{\pi^2}} \, \frac{2}{\pi(n-1)} .$$

Taking the supremum over all $f \in F$, the result follows.   □

Using Theorems 2.3.2 and 2.5.1, we find that

$$\frac{e(\phi_n^{\text{FE}}, N_n^{\text{FE}})}{r(n)} \le 2\sqrt{1 + \frac{1}{\pi^2}} \doteq 2.099.$$

Thus the error of the FEM is at most about twice the minimal error.

What is the cost of using the FEM to get an ε-approximation? That is, we wish to find

$$\text{cost}^{\text{FE}}(\varepsilon) = \inf\{ \, \text{cost}(\phi_n^{\text{FE}}, N_n^{\text{FE}}) : n \ge 2 \text{ and } e(\phi_n^{\text{FE}}, N_n^{\text{FE}}) \le \varepsilon \, \}.$$

THEOREM 2.5.2. *For any $\varepsilon > 0$,*

$$\mathrm{cost}^{\mathrm{FE}}(\varepsilon) \le (c+8)\left(\left\lceil\sqrt{1 + \frac{1}{\pi^2}\frac{2}{\pi\varepsilon}}\,\right\rceil + 1\right) - 6.$$

PROOF. Let

$$n = \left\lceil\sqrt{1 + \frac{1}{\pi^2}\frac{2}{\pi\varepsilon}}\,\right\rceil + 1.$$

Then $e(\phi_n^{\mathrm{FE}}, N_n^{\mathrm{FE}}) \le \varepsilon$ by Theorem 2.5.1, which implies that $\mathrm{cost}^{\mathrm{FE}}(\varepsilon) \le \mathrm{cost}(\phi_n^{\mathrm{FE}}, N_n^{\mathrm{FE}})$. We need only determine an upper bound on the cost of ϕ_n^{FE}. Now to evaluate $\phi_n^{\mathrm{FE}}(N_n^{\mathrm{FE}}f)$ for $f \in F$, we first evaluate $y = N_n^{\mathrm{FE}}f$ with cost cn. Given y, we can compute $\phi_n^{\mathrm{FE}}(y)$ by first solving the tridiagonal linear system $Ga = b$, and then constructing $\phi_n^{\mathrm{FE}}(y) = \sum_{j=1}^{n} a_j v_j$. The tridiagonal system can be solved using $6(n-1)+1$ arithmetic operations, and the construction of $\phi_n^{\mathrm{FE}}(y)$ can be done using $2n-1$ more arithmetic operations. Hence the combinatory cost of ϕ_n^{FE} is at most $8n-6$ arithmetic operations, showing that $\mathrm{cost}(\phi_n^{\mathrm{FE}}, N_n^{\mathrm{FE}}) \le (c+8)n - 6$, as required.   □

Note that if $\varepsilon \le \frac{1}{2}$, then

$$\frac{\mathrm{cost}^{\mathrm{FE}}(\varepsilon)}{\mathrm{comp}(\varepsilon)} \le \frac{\pi}{\sqrt{3}}\frac{c+2}{c} + \sqrt{\frac{8}{3}}\sqrt{1+\frac{1}{\pi^2}}\frac{c+8}{c} \doteq 1.814\frac{c+2}{c} + 1.714\frac{c+8}{c}.$$

Thus, for large c, the cost of using the FEM to find an ε-approximation is about 3.5 as much as the minimal cost of computing an ε-approximation. Moreover,

$$\varlimsup_{\varepsilon\to 0}\frac{\mathrm{cost}^{\mathrm{FE}}(\varepsilon)}{\mathrm{comp}(\varepsilon)} \le 2\sqrt{1+\frac{1}{\pi^2}}\frac{c+8}{c} \doteq 2.099\frac{c+8}{c}.$$

Thus for small ε and large c, we see that the cost of using the FEM is at most about twice the minimal cost.

Notes and Remarks

NR 2.5:1 The presentation of the finite element method as an algorithm using finite element information is somewhat non-standard, since standard finite element texts are not written from an information-based approach. See Werschulz (1982, 1986, 1987b) for further information.

NR 2.5:2 The best-approximation properties of the FEM are well-known; for example, see Babuška and Aziz (1972), Ciarlet (1978), Oden and Reddy (1976), and Strang and Fix (1973a).

NR 2.5:3 The idea of using a sine series to estimate the interpolation error $\tilde{e}$ is found in Chapter 1 of Strang and Fix (1973a). This idea does not generalize well to problems in more than one dimension unless the problem and the grid are special; see Fix and Strang (1969) for details.

2.6 Standard information

So far in this chapter, we have assumed that any continuous linear functional is permissible information, i.e. that $\Lambda = \Lambda^*$. In particular, this allows us to compute nth optimal information and finite element information of cardinality n, since each of these consist of n inner products.

In practice, the assumption $\Lambda = \Lambda^*$ does not generally hold. More often, we assume only that we may evaluate a function at any point in its domain. For instance, we might decide to replace the integrals appearing in the FEM by numerical quadratures, i.e. we might use rules of the form

$$\int_0^1 f(x)v_j(x)\,dx \sim \sum_{i=1}^m \alpha_{ij} f(x_i),$$

with the scalars α_{ij} being independent of the problem element f, but depending on the basis function v_j.

Hence, we wish to consider information for which the class Λ of permissible information is the class Λ^{std} of *standard information* defined by

$$\lambda \in \Lambda^{\mathrm{std}} \qquad \text{iff} \qquad \exists\, x = x_\lambda \in I \text{ such that } \forall f \in F,\ \lambda(f) = f(x).$$

Then (nonadaptive) standard information N of cardinality n has the form

$$Nf = \begin{bmatrix} f(x_1) \\ \vdots \\ f(x_n) \end{bmatrix} \qquad \forall f \in F.$$

Since standard information is not continuous on $L_2(I)$, it would not be surprising to find that standard information is not good for problem elements in the unit ball of $L_2(I)$. To obtain stronger (and hence more interesting) results, we restrict our class F, so that standard information will be continuous. Perhaps the most natural restriction is to take

$$F = \{\, f \in C(I) : \|f\|_{\sup} \le 1 \,\},$$

where $C(I)$ is the space of continuous functions on the interval $I = [0,1]$, normed by the sup norm

$$\|f\|_{\sup} = \sup_{x \in I} |f(x)| \qquad \forall f \in C(I).$$

Once we have thus restricted our class of problem elements, we can now define our basic concepts (algorithm, error of an algorithm, and radius of information) as we did before.

In this section, we show that our restriction of F to the unit ball of $C(I)$ is not strong enough. There is no algorithm using standard information whose error is less than unity.

THEOREM 2.6.1. *Let N be any standard information of cardinality n, and let ϕ^0 be the* zero algorithm *using N, which is defined to be*

$$\phi^0(Nf) = 0 \qquad \forall f \in F.$$

Then ϕ^0 is an optimal error algorithm using N, and

$$e(\phi^0, N) = r(N) = 1.$$

PROOF. We first show that $r(N) \geq 1$. Let $x_1, \ldots x_n$ be the evaluation points for N. It is no essential loss of generality to only consider the case where $0 = x_1 < x_2 < \cdots < x_{n-1} < x_n = 1$, so that $n \geq 2$. Let $\delta_0 = \min_{1 \leq i \leq n-1}(x_{i+1} - x_i)$, and take $\delta \in (0, \frac{1}{2}\delta_0)$. Setting $\delta_1 = \delta/(n-1)$, we define a function $h \in C(I)$ by

$$h(x) = \begin{cases} \delta_1^{-1}|x - x_i| & \text{if } |x - x_i| \leq \delta_1 \text{ for some } i \in \{1, \ldots, n\}, \\ 1 & \text{otherwise.} \end{cases}$$

Then $h \in \ker N \cap F$, and so

$$r(N) \geq \|Sh\|_{H^1(I)}.$$

Choose $v \equiv 1$, so that $v \in H^1(I)$, with $\|v\|_{H^1(I)} \leq 1$. Then

$$\|Sh\|_{H^1(I)} = \|Sh\|_{H^1(I)}\|v\|_{H^1(I)} \geq \langle Sh, v\rangle_{H^1(I)}$$

$$= \langle h, v\rangle_{L_2(I)} = \int_0^1 h(x)\,dx = 1 - \delta_1(n-1) = 1 - \delta.$$

Since $\delta > 0$ is arbitrary, we find that $\|Sh\|_{H^1(I)} \geq 1$, and so $r(N) \geq 1$, as required.

To show that $e(\phi^0, N) = 1$, let $f \in F$. Then

$$\|Sf\|_{H^1(I)}^2 = \langle Sf, Sf\rangle_{H^1(I)} = \langle f, Sf\rangle_{L_2(I)} \leq \|f\|_{L_2(I)}\|Sf\|_{H^1(I)}.$$

Since $f \in F$ implies that $\|f\|_{L_2(I)} \leq \|f\|_{\sup} \leq 1$, this yields

$$\|Sf\|_{H^1(I)} \leq \|f\|_{L_2(I)} \leq 1,$$

as required. □

Thus the naive zero algorithm is optimal if the set of problem elements is the unit ball in $C(I)$ and we use standard information. Moreover, since $r(N) = 1$ for any standard information N, we see that we cannot find ε-approximations for $\varepsilon < 1$. This tells us that if we want to find such ε-approximations using standard information, then we must know more about the regularity of problem elements. That is, we will have to restrict the class F even further. We will discuss this topic further in Chapter 5.

2.7   Final remarks

In this chapter, we have developed almost optimal complexity algorithms for a simple model problem. In the remainder of this book, we will develop almost optimal complexity algorithms for important classes of differential and integral equations. Moreover, we will often be interested in finding out whether the "standard" methods for these problems are nearly optimal.

As with our model problem, we will need to specify the problem formulation (i.e. the class F of problem elements and the solution operator S), the permissible information N, and the model of computation. Moreover, we will need to specify a setting. Since an optimal complexity algorithm for a given problem may depend on the setting, we will examine these problems in a setting-by-setting basis.

3
General formulation

3.1 Introduction

Our main goal in this book is to find approximate solutions of differential and integral equations at minimal cost. Such approximations are given by *optimal complexity algorithms*. Instead of starting from scratch when finding optimal complexity algorithms for each new class of problems, it is more efficient to first recall general results from the theory of information-based complexity, which we can then apply to each class of problems.

As in the previous chapter, our key concepts are the *problem formulation*, the permissible *information* that may be used to solve the problem, and the *model of computation* that tells us which operations are permissible, and at what cost. We also introduce several *settings*, which allow us to measure the error and cost of algorithms. In particular, we are interested in the *worst case* setting (error and cost are measured at a hardest problem element), the *average case* setting (error and cost are defined on the average), and the *probabilistic* setting (error and cost in the worst case, but disregarding a set of problem elements of given measure).

Note that the information and algorithms that we have dealt with so far have all been *deterministic*. We shall see later that if we allow only deterministic algorithms, then many of the problems we wish to solve are intractable for problems of high dimension. It is well-known that *non-deterministic* techniques that make a random choice of information and algorithm are sometimes more effective than deterministic methods. We will therefore study complexity under a *randomized setting*, hoping that randomization will help us beat intractability.

Although our basic goal is to find an approximation with given error at minimal cost, we sometimes wish to find a sequence of approximations that converges as quickly as possible to the solution at each problem element. This leads us to consider the *asymptotic setting*.

In this chapter, we describe assumptions and terminology common to all settings. The remainder of this book will be organized on a setting-by-setting basis. For each setting, we first describe general results for that setting; we then apply these results to find optimal complexity algorithms for specific problems in that setting.

Notes and Remarks

NR 3.1:1 For further discussion of the reasons underlying the assumptions of this chapter, the reader is referred to IBC.

3.2 Problem formulation

Let $S\colon F \to G$ be a mapping from a set F to a normed linear space G. We call F the set of *problem elements*, and say that S is a *solution operator*. The element Sf is called the *solution element* corresponding to the problem element $f \in F$. Our goal will be to approximate solution elements Sf.

The class F of problem elements often has a special form, being given by *restriction*. That is, there exists a linear space F_1, a normed linear space X, and a linear surjective *restriction operator* $T\colon F_1 \to X$ such that

$$F = \{ f \in F_1 : \|Tf\| \le 1 \}.$$

This is essentially equivalent to requiring that F be *convex* ($f_1, f_2 \in F$ and $0 \le t \le 1$ implies that $tf_1 + (1 - t)f_2 \in F$) and *balanced* ($f \in F$ implies that $-f \in F$), see **NR 3.2:1**.

EXAMPLE.   For the problem considered in Chapter 2, we had $G = H^1(I)$, $F_1 = L_2(I)$, and F was the unit ball of F_1, so that the restriction operator T was the identity mapping on $L_2(I)$. The solution operator S was given by

$$\langle Sf, v \rangle_{H^1(I)} = \langle f, v \rangle_{L_2(I)} \qquad \forall f \in H^1(I), v \in L_2(I). \qquad \square$$

Notes and Remarks

NR 3.2:1 Clearly, if F is given by restriction, then F is convex and balanced. The proof that F being convex and balanced essentially implies that F is given by restriction may be found on pp. 32–33 of GTOA.

3.3 Information

If we are going to approximate Sf, we must know something about the problem element f. Typically, we find out each piece of information about f by evaluating some $\lambda(f)$, where $\lambda\colon F \to \mathbb{R}$ is a functional on F. Let Λ be a class of *permissible information operations*, i.e. $\lambda(f)$ is computable for each $f \in F$.

Usually, we will be concerned with *linear information*, i.e. F is a subset of a linear space F_1, and Λ is a class of linear functionals on F_1. We will be particularly interested in the following classes of linear information:

(i) *Arbitrary* linear information: Here, any linear functional on F_1 is permissible. We let Λ' denote the class of arbitrary linear information.

(ii) *Continuous* linear information: Here, F_1 is a *normed* linear space, and any *continuous* linear functional on F_1 is permissible. We let Λ^* denote the class of continuous linear information.

(iii) *Standard* information: Here, F_1 is a space of real-valued functions defined over some multidimensional domain Ω. For $f \in F$ and $x \in \Omega$, the evaluation of f (or, possibly, a derivative of f) at x is permissible. We let Λ^{std} denote the class of standard information.

EXAMPLE (*continued*). Consider the model problem of Chapter 2. In most of the chapter, we had $\Lambda = \Lambda^*$, i.e, we assumed that any continuous linear functional was permissible. In §2.6, we assumed that $\Lambda = \Lambda^{\mathrm{std}}$, i.e. that only standard information was available. $\square$

The simplest kind of information is nonadaptive information. Let $n \in \mathbb{N}$, and let $\lambda_1, \ldots, \lambda_n \in \Lambda$. Then information N is said to be *nonadaptive information* if

$$Nf = \begin{bmatrix} \lambda_1(f) \\ \vdots \\ \lambda_n(f) \end{bmatrix} \qquad \forall f \in F.$$

The number n is said to be the *cardinality* of N; we write $n = \mathrm{card}\, N$. Hence, nonadaptive information of cardinality n is an operator $N \colon F \to \mathbb{R}^n$.

When information N is nonadaptive, the number n of samples and the choice of the linear functionals making up Nf do not depend on the specific problem element f. Alternatively, we can consider information for which this no longer holds. Information N is said to be *adaptive* if

$$Nf = \begin{bmatrix} \lambda_1(f) \\ \lambda_2(f; y_1) \\ \vdots \\ \lambda_{n(f)}(f; y_1, \ldots, y_{n(f)-1}) \end{bmatrix} \qquad \forall f \in F$$

where $y_1 = \lambda_1(f)$ and $y_i = \lambda_i(f; y_1, \ldots, y_{i-1})$ for $2 \leq i \leq n(f)$. That is, the choice of the ith information functional depends on the $i-1$ previously computed values. Of course, since we can use only permissible information operations, we must require that $\lambda_i(\cdot; y_1, \ldots, y_{i-1}) \in \Lambda$ for any $y_1, \ldots, y_{i-1} \in \mathbb{R}$.

We call $n(f)$ the *cardinality* of N at f; it measures the number of information evaluations in Nf. We determine $n(f)$ dynamically, by choosing new functionals and evaluating information until some condition is satisfied. For example, we might keep on going until some computable estimate of the error is sufficiently small. We formalize this idea by letting $\mathrm{ter}_i \colon \mathbb{R}^i \to \{0, 1\}$ denote the ith *termination function* for our information. Then we stop as soon as we reach an index i for which $\mathrm{ter}_i(y_1, \ldots, y_i) = 1$. Hence, the cardinality $n(f)$ at f is given by

$$n(f) = \min\{\, i \in \mathbb{N} : \mathrm{ter}_i(y_1, \ldots, y_i) = 1\,\},$$

with the convention that $\min \emptyset = \infty$.

Clearly, adaptive information is a generalization of nonadaptive information. Furthermore, nonadaptive information N can be computed in parallel with linear speedup, by splitting the task of evaluating the functionals in N equally among the processors. However, adaptive information is inherently sequential, since the choice of which functional to compute depends on the previously computed results. It will turn out that for many of the problems we study, nonadaptive information is as powerful as adaptive information (or nearly so).

EXAMPLE (*continued*). For the sample problem in Chapter 2, we considered only nonadaptive information. Our information would have become adaptive in either of the following cases:

   (i) The number n of evaluations is allowed to vary.
 (ii) The choice of which information to evaluate is allowed to vary. More precisely, this would have happened as follows:
     (a) For continuous linear information, the functions $v_1, \ldots, v_n$ making up Nf are determined dynamically, depending on 'the problem element f.
     (b) For standard information (see §2.6), the points $x_1, \ldots, x_n$ of evaluation are determined dynamically, depending on the problem element f.

As mentioned in Chapter 2, there was no need to consider adaptive information, because nonadaptive information is as powerful as adaptive information for this problem (see Chapter 4).   □

3.4 Model of computation

Next, we specify which operations are permissible, and how much they cost. This is our *model of computation*, which is given by the following three postulates:

   (i) Let $c(\lambda, f)$ denote the cost of computing $\lambda(f)$ for $\lambda \in \Lambda$, $f \in F$. We assume that $c(\lambda, f)$ is independent of λ and f, i.e. there exists $c > 0$ for which $c(\lambda, f) \equiv c$.
 (ii) Let Ω denote a set of combinatory operations, including computation of the sum of any two elements in G and the scalar multiplication between any real number and any element of G, the usual arithmetic operations on and comparisons between the real numbers, and evaluation of certain elementary functions. We assume that each combinatory operation of Ω may be exactly computed with unit cost.

(iii) Unless specified otherwise, only one operation can be done at a time, so that the cost of a set of operations is the sum of the individual costs.

3.5   Algorithms, their errors, and their costs

Now that we have defined what information is assumed known about a problem element, as well as the kinds of operations that can be performed, we define our class of admissible algorithms. Let N be information. An (idealized) *algorithm* ϕ using N is any mapping

$$\phi \colon N(F) \to G.$$

That is, ϕ combines the information Nf about a problem element f to produce an element $\phi(Nf) \in G$ approximating Sf.

How do we assess the quality of an algorithm? We are mainly interested in its error and its cost. Normally, there is little disagreement on how the error and cost at a particular problem element should be defined. Indeed, we usually measure the error of ϕ using N at a particular $f \in F$ by $\|Sf - \phi(Nf)\|$. To determine the cost of ϕ using N at f, note that we compute $\phi(Nf)$ in two stages. The first stage is the computation of the information $y = Nf$. We denote the cost of this stage by $\mathrm{cost}(N, f)$, and call it the *informational cost*. From our description of information and our model of computation, we have

$$\mathrm{cost}(N, f) \begin{cases} \geq cn(f) & \text{for adaptive information } N, \\ = cn & \text{for nonadaptive information } N \text{ of cardinality } n. \end{cases}$$

The second stage is the computation of $\phi(y)$, given $y = Nf$. Its cost is denoted $\mathrm{cost}(\phi, y)$, and is called the *combinatory cost* of computing $\phi(y)$. From our model of computation, we see that

$$\mathrm{cost}(\phi, y) = \begin{cases} k & \text{if computing } \phi(y) \text{ requires } k \text{ combinatory operations from } \Omega, \\ \infty & \text{if computing } \phi(y) \text{ uses at least one combinatory operation not in } \Omega. \end{cases}$$

The *cost* of computing $\phi(Nf)$ is then defined to be the sum of the informational and combinatory costs, i.e.

$$\mathrm{cost}(\phi, N, f) = \mathrm{cost}(N, f) + \mathrm{cost}(\phi, Nf) \qquad \forall f \in F.$$

Now that we know how to define error and cost at a *particular* problem element, what is a reasonable way of determining the error and cost over *all*

problem elements? We describe several *settings*, which allow us to determine this "global" error and cost from our more basic "local" definitions.

In the *worst case setting*, the error and cost of an algorithm ϕ using
information N are respectively defined to be

$$e(\phi, N) = \sup_{f \in F} \|Sf - \phi(Nf)\|$$

and

$$\mathrm{cost}(\phi, N) = \sup_{f \in F} \mathrm{cost}(\phi, N, f).$$

This is the most basic setting, and is most familiar to those involved in
scientific computing. We need to guarantee that the error never exceeds a
certain threshold, and that the cost never rises above a certain amount.

In the *average case setting*, the error and cost of an algorithm ϕ using
information N are respectively defined to be

$$e(\phi, N) = \left(\int_F \|Sf - \phi(Nf)\|^2 \, \mu(df) \right)^{1/2}$$

and

$$\mathrm{cost}(\phi, N) = \int_F \mathrm{cost}(\phi, N, f) \, \mu(df).$$

Here, μ is a probability measure on F. Here, we are willing to live with an
approximation that is within a certain level of accuracy, and whose cost is
not too expensive, on the average.

Of course, the definitions of error and cost in the average case setting
appear to require the measurability of the mapping $f \mapsto \|Sf - \phi(Nf)\|^2$ and
of $\mathrm{cost}(\phi, N, \cdot)$ over F. As we shall see in Chapter 7, these measurability
requirements can be waived under rather general conditions. Note that
we define error in an L_2 average case sense, and cost in an L_1 average
case sense. These can be generalized to the case where error and cost are
respectively measured in an L_p and L_q sense; we choose $p = 2$ and $q = 1$
only for convenience.

In the *probabilistic setting*, the error and cost of an algorithm ϕ using
information N are respectively defined to be

$$e(\phi, N, \delta) = \inf_A \sup_{f \in F - A} \|Sf - \phi(Nf)\|,$$

the infimum being over all measurable sets A such that $\mu(A) \leq \delta$, and

$$\mathrm{cost}(\phi, N) = \sup_{f \in F} \mathrm{cost}(\phi, N, f).$$

As in the average case setting, μ is a probability measure on F. Furthermore, the error depends on an additional parameter $\delta \in [0, 1]$. In this
setting, we are willing to live with an approximation that is good outside
a set of (presumably small) measure δ; however, cost is measured in the
worst case. When it will cause no confusion, we suppress the dependence
of the probabilistic error on δ in the rest of this chapter.

EXAMPLE (*continued*). In Chapter 2, we dealt with the worst case setting exclusively. For the nth minimal error algorithm ϕ_n, we found that

$$cn \leq \text{cost}(\phi_n, N_n) \leq (c+2)n - 1$$

and

$$e(\phi_n, N_n) = \frac{1}{\sqrt{1 + n^2\pi^2}}.$$

For the FEM ϕ_n^{FE}, we found that

$$cn \leq \text{cost}(\phi_n^{\text{FE}}, N_n^{\text{FE}}) \leq (c+8)n - 6$$

and

$$\frac{1}{\sqrt{1 + n^2\pi^2}} \leq e(\phi_n^{\text{FE}}, N_n^{\text{FE}}) \leq \sqrt{1 + \frac{1}{\pi^2}\frac{2}{\pi(n-1)}}. \qquad \square$$

3.6 Complexity

Let us say that an algorithm ϕ using information N produces an *ε-approximation* if $e(\phi, N) \leq \varepsilon$. Then the ε-complexity of our problem is the minimal cost of computing an ε-approximation. That is, the *ε-complexity* (or the *complexity* for short) of our problem is defined to be

$$\text{comp}(\varepsilon) = \inf\{\,\text{cost}(\phi, N) : \phi \text{ and } N \text{ such that } e(\phi, N) \leq \varepsilon\,\},$$

with the convention that $\inf \emptyset = \infty$. An algorithm ϕ_ε using information N_ε is said to be an *ε-optimal complexity algorithm* (briefly, an *optimal complexity algorithm*) if

$$e(\phi_\varepsilon, N_\varepsilon) \leq \varepsilon \quad \text{and} \quad \text{cost}(\phi_\varepsilon, N_\varepsilon) = \text{comp}(\varepsilon).$$

Unfortunately, it is usually very difficult to find optimal complexity algorithms for problems of practical interest. More realistically, we will be happy to find an ε-approximation at almost minimal cost. That is, we hope that there exists a constant d (not too much greater than one) such that for $\varepsilon > 0$, we can find an algorithm ϕ_ε using information N_ε such that

$$e(\phi_\varepsilon, N_\varepsilon) \leq \varepsilon \quad \text{and} \quad \text{cost}(\phi_e, N_\varepsilon) \leq d \cdot \text{comp}(\varepsilon).$$

Such an algorithm is said to be an *almost optimal complexity algorithm*.

EXAMPLE (*continued*). In Chapter 2, we found that comp(ε) $= \Theta(\varepsilon^{-1})$ as
$\varepsilon \to 0$. The algorithms $\phi_{m(\varepsilon)}$ and ϕ_n^{FE} (with $m(\varepsilon)$ and n as respectively
defined in Theorems 2.4.1 and 2.5.2) were almost optimal complexity algo-
rithms. $\square$

Notes and Remarks

NR 3.6:1 Strictly speaking, we should respectively refer to ϕ_ε and N_ε as an ε-*optimal*
complexity algorithm and ε-*optimal complexity information*, since we are concerned with
the choice of optimal information and algorithm. However for brevity, we often refer
merely to the algorithm ϕ_ε as an ε-optimal complexity algorithm. From the context,
it will generally be clear that the information used by this algorithm is ε-optimal complexity information.

3.7   Randomized setting

The settings above have all assumed that the information N and algorithm ϕ that will be used to approximate Sf for $f \in F$ are deterministic.
Another possibility is to allow a random (or nondeterministic) choice of
information and algorithm.

We use a problem of multivariate numerical quadrature to illustrate the
randomized setting. Let $I_d = [0,1]^d$ be the d-dimensional unit cube and let
$C(I_d)$ denote the continuous functions on I_d. Our set F of problem elements
will be the unit ball of $C(I_d)$. Our solution operator is $Sf = \int_{I_d} f(x)\,dx$ for
$f \in F$. We assume that only standard information is permissible.

First, note that the worst case (deterministic) ε-complexity of this problem is infinite for *any* $\varepsilon > 0$, no matter how small; the proof is similar to
that of Theorem 2.6.1.

We now look at randomized algorithms for this problem. Let $T = I_d^n$
and let ρ be the uniform distribution on I_d. The *Monte Carlo algorithm*
for approximating Sf starts by randomly choosing $t = (t_1, \dots, t_n) \in T$. It
then computes

$$\phi(N_t f) = \frac{1}{n} \sum_{j=1}^{n} f(t_j).$$

Note that we are using standard information of the form

$$N_t f = \begin{bmatrix} f(t_1) \\ \vdots \\ f(t_n) \end{bmatrix},$$

where $t = (t_1, \dots, t_n) \in T$ is chosen randomly according to the distribution ρ. More generally, let us consider algorithms of the form $\phi_t \colon N_t(F) \to$
$\mathbb{R}$. That is, we allow a nondeterministic choice of algorithm, as well as of

information. (The Monte Carlo algorithm uses randomization only in its choice of information, and not in its combination of the information.) Then the error and cost of an algorithm in the *randomized setting* are given by

$$\left(\sup_{f \in F} \int_T \| Sf - \phi_t(N_t f) \|^2 \, \rho(dt) \right)^{1/2}$$

and

$$\sup_{f \in F} \int_T \mathrm{cost}(\phi_t, N_t) \, \rho(dt).$$

Complexity and optimal complexity algorithms are now defined as in §3.6.

Bakhvalov (1959) has shown that the the complexity of this quadrature problem is $\Theta(\varepsilon^{-2})$ as $\varepsilon \to 0$ in the randomized setting. Hence this problem is solvable in the randomized setting, but is not solvable in the deterministic setting. Furthermore, it is straightforward to check that the Monte Carlo method described above has error $n^{-1/2}$ and cost $cn + n$. So, the classical Monte Carlo method is almost optimal in the randomized setting.

We will examine the randomized setting in more detail in Chapter 8.

3.8 Asymptotic setting

So far, we have been interested in finding an algorithm producing an ε-approximation with minimal cost over a class of problem elements. Another approach that is commonly used in scientific computation is an *asymptotic setting*. That is, we consider sequences $\{\phi_n(N_n f)\}_{n \in \mathbb{N}}$ of approximations to Sf, where N_n is information of cardinality n. We then look for such a sequence with the fastest possible speed of convergence, i.e. we want $\| Sf - \phi_n(N_n f) \|$ to converge to zero as quickly as possible, either

(i) for all $f \in F$, or
(ii) for almost every $f \in F$ (with respect to some measure μ on F.)

Such a sequence is said to be optimal in the *asymptotic setting*. We develop this setting more completely in Chapter 8.

EXAMPLE (*continued*). It is easy to see that for the problem of Chapter 2,

$$\| Sf - \phi_n(N_n f) \|_{H^1(I)} = \Theta(n^{-1}) \qquad \text{as } n \to \infty$$

and

$$\| Sf - \phi_n^{\mathrm{FE}}(N_n^{\mathrm{FE}} f) \|_{H^1(I)} = \Theta(n^{-1}) \qquad \text{as } n \to \infty$$

for all $f \in F$. Are these algorithm sequences optimal in the asymptotic setting, or are there sequences for which the exponent 1 can be replaced by a larger exponent? From the results of Chapter 8, we will find that the exponent 1 cannot be improved for any algorithm, and so these algorithms are optimal. □

4

The worst case
setting: general results

4.1 Introduction

In this chapter, we describe general results for the worst case setting. Recall
that we are given a set F of problem elements, as well as a normed linear
space G with norm $\|\cdot\|$. We wish to approximate the solution operator
$S\colon F \to G$. We measure error and cost in a worst case sense. That is, if ϕ
is an algorithm using information N, then the error and cost of (ϕ, N) are
respectively given by

$$e(\phi, N) = \sup_{f \in F} \|Sf - \phi(Nf)\|$$

and

$$\mathrm{cost}(\phi, N) = \sup_{f \in F} \mathrm{cost}(\phi, N, f).$$

We give an overview of this chapter. In §4.2, we describe relations
between the radius (minimal error) of information and a quantity that is
often easier to compute, namely, the diameter of information. In §4.3,
we use these results to determine complexity bounds in the worst case.
The results of §4.2 and §4.3 are specialized to the important class of linear
problems in §4.4. Finally in §4.5, we consider complexity under the *residual
error criterion*, which is often useful when the solution operator is given as
the inverse of a known linear "problem operator."

Notes and Remarks

NR 4.1:1 For proof and discussion of the results contained in §§4.2–4.4, see Chapters 1–
5 of GTOA and Chapter 4 of IBC.

4.2 Radius and diameter

For information N, let us try to find an algorithm ϕ^* using N with error as small as possible, i.e. with

$$e(\phi^*, N) = \inf_{\phi \text{ using } N} e(\phi, N).$$

Such an algorithm is called an *optimal error algorithm.*

To help us find optimal (or almost optimal) error algorithms, we need the concepts of radius and diameter of information. First, for a subset $A \subset G$, recall that the *radius* and *diameter* of A are respectively defined as

$$\text{rad}(A) = \inf_{g \in G} \sup_{a \in A} \|g - a\|$$

and

$$\text{diam}(A) = \sup_{a_1, a_2 \in A} \|a_1 - a_2\|.$$

Of course, we have the elementary inequality

$$\text{rad}(A) \leq \text{diam}(A) \leq 2\,\text{rad}(A).$$

Now let $f \in F$ be a problem element, and let $y = Nf$ be the information about f provided by N. Then

$$N^{-1}(y) = \{\, \tilde{f} \in F : N\tilde{f} = y \,\}$$

denotes the set of problem elements that cannot be distinguished from f by the information N. If we define the *local radius of information* to be

$$r(N, y) = \text{rad}\left(SN^{-1}(y)\right),$$

then we see that $r(N, y)$ is the radius of the set of solution elements $S\tilde{f}$ for problem elements $\tilde{f}$ that cannot be distinguished from f. The (*global*) *radius of information* is defined to be

$$r(N) = \sup_{y \in N(F)} r(N, y),$$

i.e. the global radius is the local radius for a worst y. Analogously, we may define the *local diameter of information* by

$$d(N, y) = \text{diam}\left(SN^{-1}(y)\right)$$

and the (*global*) *diameter of information* by

$$d(N) = \sup_{y \in N(F)} d(N, y).$$

It is easy to see that

$$r(N) = \sup_{f \in F} \inf_{g \in G} \sup_{\substack{\tilde{f} \in F \\ N\tilde{f}=Nf}} \|S\tilde{f} - g\|,$$

$$d(N) = \sup_{f \in F} \sup_{\substack{\tilde{f} \in F \\ N\tilde{f}=Nf}} \|S\tilde{f} - Sf\|,$$

$$r(N, y) \le d(N, y) \le 2\, r(N, y) \qquad \forall\, y \in N(F), \text{ and}$$

$$r(N) \le d(N) \le 2\, r(N).$$

Usually, the diameter $d(N)$ is easier to compute than the radius $r(N)$. Moreover, for many of the problems we will look at in this book, $d(N) = 2\, r(N)$.

Recall that we are trying to find optimal error algorithms. First, note that for an algorithm ϕ using information N, we may define its *local error at* $y \in N(F)$ to be

$$e(\phi, N, y) = \sup_{f \in N^{-1}(y)} \|Sf - \phi(Nf)\|.$$

Then the (global) error of an algorithm ϕ using N is related to its local error by

$$e(\phi, N) = \sup_{y \in N(F)} e(\phi, N, y).$$

We can determine whether a given algorithm has (almost) optimal error by using the following result.

THEOREM 4.2.1. *The (global or local) radius of information is the minimal (global or local) error among all algorithms using that information, i.e. for any information* N,

$$r(N, y) = \inf_{\phi \text{ using } N} e(\phi, N, y) \qquad \forall\, y \in N(F)$$

and

$$r(N) = \inf_{\phi \text{ using } N} e(\phi, N). \qquad\qquad \square$$

Let us say that an algorithm ϕ^* using information N is a *strongly optimal error algorithm* if

$$e(\phi^*, N, y) = \inf_{\phi \text{ using } N} e(\phi, N, y) \qquad \forall\, y \in N(F).$$

Then we see that

(i) ϕ^* is a strongly optimal error algorithm iff $e(\phi^*, N, y) = r(N, y)$ for all $y \in N(F)$, and

(ii) ϕ^* is an optimal error algorithm iff $e(\phi^*, N) = r(N)$.

In general, it will be difficult to find optimal error algorithms. However, if we are willing to sacrifice a factor of two, we can achieve nearly optimal results with an interpolatory algorithm. Let $y = Nf$ be information about the problem element f. Knowing y, we find an element $\tilde{f} \in F$ such that $N\tilde{f} = y$, i.e. $\tilde{f}$ *interpolates* the information y. Then the algorithm

$$\phi^I(y) = S\tilde{f}$$

is said to be an *interpolatory algorithm* using N. If ϕ^I is an interpolatory algorithm using N, then

$$e(\phi^I, N, y) \le d(N, y) \le 2\, r(N, y) \qquad \forall\, y \in N(F),$$

and so

$$e(\phi^I, N) \le d(N) \le 2\, r(N).$$

Thus an interpolatory algorithm is an almost (strongly) optimal error algorithm.

4.3 Complexity

We now show how results on optimal error algorithms may be used to find ε-complexity bounds for any $\varepsilon > 0$ in the worst case setting.

First, let ϕ be an algorithm using information N. Since we are using a worst case setting,

$$\text{cost}(\phi, N) = \sup_{f \in F} \{\text{cost}(N, f) + \text{cost}(\phi, Nf)\},$$

where $\text{cost}(N, f)$ is the informational cost of N at f, and $\text{cost}(\phi, Nf)$ is the combinatory cost of ϕ at Nf. Recall that $n(f)$ denotes the cardinality of N at $f \in F$, so that Nf uses $n(f)$ information functionals. Since each information functional has cost c under our model of computation,

$$\text{cost}(\phi, N) \ge \sup_{f \in F} \text{cost}(N, f) \ge c \sup_{f \in F} n(f).$$

Letting

$$\text{card}\, N = \sup_{f \in F} n(f)$$

denote the (worst case) *cardinality* of the information N, we then see that

$$\mathrm{cost}(\phi, N) \geq c \text{ card } N$$

for any algorithm ϕ using information N. In particular, note that if N is *nonadaptive* information of cardinality n as defined in §3.3, then $n(f) = n$ for any problem element $f \in F$. Thus card $N = n$ and $\mathrm{cost}(N, f) = cn$ for any problem element f when N is nonadaptive.

Let $\varepsilon > 0$. How many information operations are needed to get an ε-approximation? The answer is given by the ε-*cardinality number*, which is defined to be

$$m(\varepsilon) = \inf\{\, n \in \mathbb{N} : r(N) \leq \varepsilon \text{ for some } N \text{ with } \text{card } N \leq n \,\}.$$

If no such n exists, we set $m(\varepsilon) = \infty$.

We are now able to give a lower bound on ε-complexity. If $m(\varepsilon) = \infty$, then there is no information N of finite cardinality for which $r(N) \leq \varepsilon$, so that there is no algorithm yielding an ε-approximation for our problem. Thus $\mathrm{comp}(\varepsilon) = \infty$ when $m(\varepsilon) = \infty$. Suppose now that $m(\varepsilon) < \infty$. Let ϕ be an algorithm using information N such that $e(\phi, N) \leq \varepsilon$. Then card $N \geq m(\varepsilon)$, so that

$$\mathrm{cost}(\phi, N) \geq c\, m(\varepsilon).$$

Taking the infimum over all (ϕ, N) such that $e(\phi, N) \leq \varepsilon$, we find that

$$\mathrm{comp}(\varepsilon) \geq c\, m(\varepsilon).$$

We now seek an upper bound on complexity that is close to this lower bound. Suppose that there exists information N_ε and an algorithm ϕ_ε using N_ε such that

    (i)   card $N_\varepsilon = m(\varepsilon)$,
    (ii) $r(N_\varepsilon) = \varepsilon$,
    (iii)   the cost of selecting the functionals in N_ε can be neglected, i.e.

$$\sup_{f \in F} \mathrm{cost}(N_\varepsilon, f) \simeq c \text{ card } N_\varepsilon,$$

    (iv) $e(\phi_\varepsilon, N_\varepsilon) = r(N_\varepsilon)$, and
    (v)   the combinatory cost of ϕ_ε is dominated by the informational cost, i.e.

$$\mathrm{cost}(\phi_e, y) \ll c \text{ card } N_\varepsilon \qquad \forall\, y \in N_\varepsilon(F).$$

Then

$$\mathrm{cost}(\phi_\varepsilon, N_\varepsilon) \simeq c\, m(\varepsilon).$$

From the previous lower bound, we conclude that

$$\mathrm{comp}(\varepsilon) \simeq c\, m(\varepsilon),$$

and that ϕ_ε is an almost optimal complexity algorithm.

In summary, we have

THEOREM 4.3.1. *For any $\varepsilon > 0$,*

$$\mathrm{comp}(\varepsilon) \geq c\, m(\varepsilon).$$

Moreover, if there exists information N_ε satisfying (i)–(iii), *and an algorithm ϕ_ε using N_ε satisfying* (iv) *and* (v), *then ϕ_ε is an almost optimal complexity algorithm, and*

$$\mathrm{comp}(\varepsilon) \simeq c\, m(\varepsilon). \qquad \square$$

The assumptions (i)–(v) hold for many important problems. In particular, they hold for many of the problems involving differential and integral equations that we study in this book.

Notes and Remarks

NR 4.3:1 Assumption (iii) holds automatically for nonadaptive information.

4.4 Linear problems

In this section, we will restrict our attention to linear problems. The class of linear problems is general enough to include many of the problems we study in this book. Moreover, it has enough structure that we can find general results that are more concrete than the ones we have seen so far. Hence, it is worthwhile to spend some time noting results that hold specifically for linear problems.

We briefly outline the contents of this section. First, we give the definition of a linear problem. Next, we describe why adaption is no more powerful than nonadaption for linear problems. We then discuss optimal information for linear problems. After that, we ask to what extent linear algorithms can have optimal error for linear problems. Finally, these results are used to determine complexity bounds for linear problems, as well to find almost optimal complexity algorithms.

4.4.1 Definition

A problem is said to be *linear* if the following three conditions are satisfied:

   (i) The set F of problem elements is generated by a *linear* restriction operator. That is,

$$F = \{\, f \in F_1 : \|Tf\| \leq 1 \,\}$$

for a linear surjection $T\colon F_1 \to X$ of linear spaces. As mentioned in Chapter 3, this is essentially equivalent to requiring that F be a convex balanced set.

(ii) The solution operator $S\colon F \to G$ is (the restriction of) a *linear* operator $S\colon F_1 \to G$.

(iii) The class Λ of permissible information operations is a subset of the class Λ' of *linear* functionals on F_1, i.e. $\lambda \in \Lambda$ if $\lambda\colon F_1 \to \mathbb{R}$ is linear.

For example, the problem studied in Chapter 2 is linear.

Notes and Remarks

NR 4.4.1:1 Note that the problem is linear iff all three conditions of this subsection hold. In particular, a problem involving a linear solution operator with linear information for which the set of problem elements is *not* convex and balanced is *not* a linear problem!

4.4.2   Adaptive vs. nonadaptive information

In this section, we mention a celebrated (and somewhat controversial) result, namely that adaptive information is not much more powerful than nonadaptive information for linear problems. This may seem surprising, since the class of nonadaptive information is a small subset of the class of adaptive information. Moreover, this result appears to fly in the face of much practical experience. We discuss this issue at greater length in §5.7, where we look at the question of adaption vs. nonadaption for linear elliptic partial differential equations.

For future reference, we first recall

LEMMA 4.4.2.1. *For any linear problem and for any nonadaptive linear information N,*

$$d(N) = 2 \sup_{h \in \ker N \cap F} \|Sh\|,$$

where

$$\ker N = \{\, h \in F_1 : Nh = 0 \,\}$$

is the kernel of N. Moreover, if $r(N) = \frac{1}{2}d(N)$, then

$$r(N) = \sup_{h \in \ker N \cap F} \|Sh\|. \qquad \square$$

Thus the diameter of nonadaptive information is determined by what happens when the computed information consists of zeros. We use this idea to explain how adaptive information is no stronger than nonadaptive information. Let N^{a} be adaptive linear information, i.e.

$$N^{\mathrm{a}}f = \begin{bmatrix} \lambda_1(f) \\ \lambda_2(f; y_1) \\ \vdots \\ \lambda_{n(f)}(f; y_1, \dots, y_{n(f)-1}) \end{bmatrix} \qquad \forall f \in F$$

where $y_1 = \lambda_1(f)$ and $y_i = \lambda_i(f; y_1, \dots, y_{i-1})$ for $2 \le i \le n(f)$. We choose a special nonadaptive linear information operator N^{non} of cardinality $n(0)$ by

$$N^{\mathrm{non}} f = \begin{bmatrix} \lambda_1(f) \\ \lambda_2(f; 0) \\ \vdots \\ \lambda_{n(0)}(f; 0, \dots, 0) \end{bmatrix} \qquad \forall f \in F.$$

Thus the ith information functional in N^{non} is $\lambda_i(\cdot; 0, \dots, 0)$ (where there are $i - 1$ zeros). Of course, $\mathrm{card}\, N^{\mathrm{non}} = n(0) \le \mathrm{card}\, N^{\mathrm{a}}$.

We may now state

THEOREM 4.4.2.1. *For any linear problem and any adaptive linear information N^{a}, let N^{non} be the nonadaptive information defined above. Then*

$$d(N^{\mathrm{non}}) \le d(N^{\mathrm{a}}),$$

and thus

$$r(N^{\mathrm{non}}) \le 2\, r(N^{\mathrm{a}}).$$

Moreover, if $r(N^{\mathrm{non}}) = \frac{1}{2} d(N^{\mathrm{non}})$, then

$$r(N^{\mathrm{non}}) \le r(N^{\mathrm{a}}).$$ □

In particular, suppose we have a linear problem in which the diameter of linear nonadaptive information is always twice the radius. We then see that given any adaptive information operator N^{a}, we can always find a nonadaptive information operator N^{non} such that $\mathrm{card}\, N^{\mathrm{non}} \le \mathrm{card}\, N^{\mathrm{a}}$, yet the optimal error among all algorithms using N^{non} is no more than the optimal error among all algorithms using N^{a}. This means that as far as optimal error is concerned, there is no need to consider algorithms using adaptive information, since we can always find an algorithm using the simpler nonadaptive information, without any degradation in the error.

Notes and Remarks

NR 4.4.2:1 For further discussion of adaption vs. nonadaption for linear problems, see Kon and Novak (1989).

4.4.3 Optimal information

We now wish to find the best possible information of given cardinality. That is, given $n \in \mathbb{N}$, we wish to find linear information of cardinality at most n with minimal radius. From the results of the previous section, we see that we can confine our search to nonadaptive information.

Recall that Λ is the set of permissible information operations. Let Λ_n denote the class of nonadaptive linear information of cardinality at most n, using functionals from Λ. Then the nth *minimal radius of information* in the class Λ is defined to be

$$r(n,\Lambda) = \inf_{N\in\Lambda_n} r(N).$$

Information $N_n \in \Lambda_n$ is said to be nth *optimal information* in the class Λ if $r(N_n^*) = r(n,\Lambda)$.

Since it is usually easier to compute the diameter of information than the radius, we define the nth *minimal diameter of information* in the class Λ to be

$$d(n,\Lambda) = \inf_{N\in\Lambda_n} d(N).$$

Clearly, $r(n,\Lambda) \leq d(n,\Lambda) \leq 2\,r(n,\Lambda)$, so that we can use the minimal diameter to determine the minimal radius to within a factor of two. Furthermore, for problems in which the diameter equals twice the radius, the nth minimal radius will be half of the nth minimal diameter.

In what follows, we will write $r(n)$ and $d(n)$ instead of $r(n,\Lambda')$ and $d(n,\Lambda')$ when the class of permissible information operations is the class Λ' of all linear functionals on F_1. Keep in mind that $r(n,\Lambda)$ and $d(n,\Lambda)$ refer to the nth minimal radius and diameter of *nonadaptive* information.

We first ask how much information is necessary to ensure that an algorithm with finite error exists. Let us define the *problem index* n^* as the dimension of an algebraic complement of $\ker T \cap \ker S$ in the space $\ker T$. Note that if $A(T,S)$ is such a complement and n^* is finite, then there exist linear functionals $\lambda_1^*,\dots,\lambda_{n^*}^*$ on F_1 and linearly independent elements $e_1^*,\dots,e_{n^*}^* \in F_1$ such that $f \in A(T,S)$ iff $f = \sum_{j=1}^{n^*} \lambda_j^*(f)e_j^*$. Of course, if F_1 is a Hilbert space, we will generally choose $A(T,S)$ to be the *orthogonal* complement of $\ker T \cap \ker S$ in $\ker T$.

Also, we need to recall the definition of a pseudoinverse $T^\dagger$ of T. Let $(\ker T)^\perp$ denote an algebraic complement of $\ker T$ in F_1. In particular, if F_1 is a Hilbert space, we let $(\ker T)^\perp$ denote the usual orthogonal complement of $\ker T$ in F_1. Since $T\colon F_1 \to X$ is a surjection, its restriction $T|_{(\ker T)^\perp}$ is a bijection of $(\ker T)^\perp$ onto X. Hence, we define the (Moore-Penrose) *pseudoinverse* $T^\dagger\colon X \to F_1$ of T by

$$T^\dagger = \left(T\big|_{(\ker T)^\perp}\right)^{-1}.$$

We then have

THEOREM 4.4.3.1.

   (i) If $\operatorname{card} N < n^*$, then $r(N) = \infty$.

(ii) *For finite n^*, define linear information N^* of cardinality n^* by*

$$N^*f = \begin{bmatrix} \lambda_1^*(f) \\ \vdots \\ \lambda_{n^*}^*(f) \end{bmatrix} \qquad \forall f \in F.$$

If $ST^\dagger \colon X \to G$ is continuous, then

$$r(N^*) \le 2\,\|ST^\dagger\|. \qquad\qquad \square$$

Hence for any Λ, we see that $r(n, \Lambda) = \infty$ if $n < n^*$. This means that for problems with infinite index, there is no information with finite radius. On the other hand, suppose that the index n^* is finite. If $ST^\dagger$ is continuous and if $\lambda_1^*, \dots, \lambda_{n^*}^*$ are permissible information operations (i.e. they belong to Λ), then N^* is permissible information of cardinality n^* with finite radius. Thus the index is the minimal cardinality for which finite-radius information exists.

With this in mind, we now restrict our search for optimal information of cardinality n to the case where $n \ge n^*$. We first consider the *Hilbert case*. That is, we let X and G be separable Hilbert spaces, whose inner products we denote by $\langle \cdot, \cdot \rangle$.

Let $K = ST^\dagger \colon X \to G$. Suppose that K is bounded. Let $K_1 = K^*K$, so that $K_1 \colon X \to X$ is a bounded self-adjoint operator that is compact iff K is compact. Let $\mathrm{sp}(K_1)$ denote the *spectrum* of K_1, i.e. $\mathrm{sp}(K_1) = \mathrm{p}(K_1) \cup \mathrm{c}(K_1)$, where $\mathrm{p}(K_1)$ and $\mathrm{c}(K_1)$ respectively denote the *point spectrum* and the *continuous spectrum* of K_1. Thus, $\mathrm{p}(K_1)$ is the set of all eigenvalues of K_1 (in which an eigenvalue of multiplicity k is counted k times) and $\mathrm{c}(K_1)$ is the set of all nonnegative numbers κ such that $(K_1 - \kappa I)^{-1}$ is densely defined but unbounded. Of course, if K is compact, then $\mathrm{c}(K_1) \subseteq \{0\}$, while if K is not compact, then $\mathrm{c}(K_1)$ contains an interval. Let

$$\kappa_i^2 = \inf_{|B| \le i} \sup\, \mathrm{sp}(S^*S) - B,$$

for $i \in \mathbb{N}$, with $\sup \emptyset = 0$. (Here, the supremum is over all $B \subseteq \mathrm{sp}(S^*S)$ with at most i points.) Thus κ_i is the ith largest singular value of K if it exists, and is the largest point of attraction in $\mathrm{sp}(K)$ otherwise.

We then have

THEOREM 4.4.3.2. *In the Hilbert case, the following hold:*

(i) *If K is bounded but not compact, then*

$$r(n) = \kappa_{n-n^*+1}.$$

Hence, there exist $\alpha_1, \alpha_2 \in \mathbb{R}$ such that

$$0 < \alpha_1 \le \alpha_2 < \infty,$$
$$r(n^*) = \alpha_2,$$

the sequence $\{r(n)\}_{n=n^*}^{\infty}$ is nonincreasing, and $\lim_{n \to \infty} r(n) = \alpha_1$.

(ii) If K is compact, let $z_1, z_2, \ldots$ denote a sequence of orthonormal eigenvectors corresponding to the eigenvalues $\kappa_1^2 \ge \kappa_2^2 \ge \ldots > 0$ of K_1. Let the linear functionals $\lambda_1^*, \ldots, \lambda_{n^*}^*$ be a dual basis to a basis for $A(T, S) = \ker T / (\ker T \cap \ker S)$. Define information N_n of cardinality n by

$$N_n f = \begin{bmatrix} \lambda_1^*(f) \\ \vdots \\ \lambda_{n^*}^*(f) \\ \langle Tf, z_1 \rangle \\ \vdots \\ \langle Tf, z_{n-n^*} \rangle \end{bmatrix} \qquad \forall f \in F.$$

Then N_n is nth optimal information, i.e.

$$r(n) = r(N_n) = \kappa_{n-n^*+1},$$

and

$$\lim_{n \to \infty} r(n) = 0. \qquad \square$$

Let us say that our problem is *convergent* if for any $\varepsilon > 0$, there exists information N of finite cardinality and an algorithm ϕ using N for which $e(\phi, N) \le \varepsilon$. This theorem tells us that the problem is convergent iff K is compact.

The previous result only holds in the Hilbert case. We now turn to the general case, showing a connection between minimal diameters of information and Gelfand widths for the case where S is an injection. Recall that the Gelfand n-width of a balanced subset A of a normed linear space G is defined to be

$$d^n(A, G) = \inf_{G^n \in \mathcal{G}^n} \sup_{a \in A \cap G^n} \|a\|,$$

where $G^n \in \mathcal{G}^n$ iff G^n is a subspace of G whose codimension is at most n. A subspace $G^n \in \mathcal{G}^n$ is said to be an nth *extremal subspace* of the set A (in the Gelfand sense) if

$$d^n(A, G) = \sup_{a \in A \cap G^n} \|a\|.$$

THEOREM 4.4.3.3. *Let S be an injection.*

(i) *For any $n \in \mathbb{N}$,*

$$d^n\big(S(F),G\big) \le r(n) \le 2\,d^n\big(S(F),G\big).$$

Hence, if $r(n) = \frac{1}{2}d(n)$, then

$$r(n) = d^n\big(S(F),G\big).$$

(ii) *Let G^n be an nth extremal subspace of $S(F)$, so that*

$$G^n = \{\, g \in G : \hat{\lambda}_1(g) = \ldots = \hat{\lambda}_n(g) = 0 \,\}$$

for linear functionals $\hat{\lambda}_1, \ldots, \hat{\lambda}_n$ on G. Define information N_n of cardinality n by

$$N_n f = \begin{bmatrix} \hat{\lambda}_1(Sf) \\ \vdots \\ \hat{\lambda}_n(Sf) \end{bmatrix} \qquad \forall f \in F.$$

Then N_n is nth optimal information in Λ', to within a factor of at most 2:

$$r(N_n) \le d(N_n) = d(n) \le 2\,r(n).$$

Hence, if $r(N_n) = \frac{1}{2}d(N_n)$, then N_n is nth optimal information:

$$r(N_n) = r(n) = d^n\big(S(F),G\big). \qquad \square$$

Notes and Remarks

NR 4.4.3:1 Part (ii) of Theorem 4.4.3.1 is based on a private communication of G. W. Wasilkowski.

NR 4.4.3:2 For more information about n-widths, see Pinkus (1985).

NR 4.4.3:3 Since for the linear problems that we will study in this book, the solution operator S is always injective, we need only the injective version of Theorem 4.4.3.3. For a non-injective version, see Theorem 6.1 of Chapter 2 in GTOA, as well as Corollary 5.4.1 of Chapter 4 in IBC.

4.4.4 Linear algorithms for linear problems

Let N be information of cardinality n; what can we say about the explicit form of an optimal error algorithm using N? In particular, when is there an optimal error algorithm that is easy to evaluate? The class of linear algorithms is an especially simple class; in this section, we give conditions under which a linear optimal error algorithm exists. Of particular importance will be the existence of linear algorithms which are nth *minimal error algorithms*, i.e. optimal error algorithms using nth optimal information.

From the results in §4.4.2, we can restrict our attention to nonadaptive information

$$Nf = \begin{bmatrix} \lambda_1(f) \\ \vdots \\ \lambda_n(f) \end{bmatrix} \qquad \forall f \in F.$$

We say that a mapping $\phi^L \colon N(F) \to G$ is a *linear algorithm* using N if there exist elements $q_1, \ldots, q_n \in G$ such that

$$\phi^L(Nf) = \sum_{j=1}^{n} \lambda_j(f) q_j \qquad \forall f \in F.$$

Why do we say that a linear algorithm is easy to evaluate? Since $q_1, \ldots, q_n$ are independent of any problem element $f \in F$, they may be *precomputed*, i.e. computed in advance. This means that for any $f \in F$, once we have computed $y = Nf$, we can compute $\phi^L(y)$ using at most n multiplications and $n-1$ additions in G. So, if ϕ^L is a linear algorithm using information N of cardinality n, then the combinatory cost of ϕ^L satisfies

$$\mathrm{cost}(\phi^L, Nf) \leq 2n - 1 \qquad \forall f \in F.$$

Now the informational cost satisfies

$$\mathrm{cost}(N, f) = c\,n \qquad \forall f \in F.$$

Assuming that $c \gg 1$, i.e. that the cost of evaluating a linear functional is much greater than that of an arithmetic operation in G, we see that

$$\mathrm{cost}(\phi^L, Nf) \ll \mathrm{cost}(N, f) \qquad \forall f \in F,$$

i.e. the informational cost dominates the combinatory cost. Moreover, since

$$\mathrm{cost}(\phi, N) \geq \sup_{f \in F} \mathrm{cost}(N, f) = c\,n \qquad \forall \phi \text{ using } N,$$

we see that

$$1 \leq \frac{\mathrm{cost}(\phi^L, N)}{\inf_{\phi \text{ using } N} \mathrm{cost}(\phi, N)} \leq \frac{c+2}{c} \qquad \forall \phi \text{ using } N.$$

Hence if $c \gg 1$, the cost of ϕ^L is nearly minimal among all algorithms using N.

Before discussing the existence of optimal error linear algorithms, we briefly consider the class of spline algorithms, which are homogeneous (but not always linear) and have almost optimal error. Let $y \in N(F)$. An element $\tilde{f} = \tilde{f}(y) \in F_1$ such that $N\tilde{f} = y$ and

$$\|T\tilde{f}\| = \min_{\substack{f \in F_1 \\ Nf = y}} \|Tf\|$$

is said to be a *spline interpolating* y. An algorithm ϕ^s of the form

$$\phi^s(y) = S\tilde{f}(y) \qquad \text{with } \tilde{f}(y) \text{ a spline interpolating } y \text{ for all } y \in N(F)$$

is said to be a *spline algorithm* using N.

Let us write

$$P(x) = \{\, h \in \ker N : \|x - Th\| = \inf_{z \in T(\ker N)} \|x - z\| \,\} \qquad \forall x \in X$$

for the set of elements whose T-image is a best $T(\ker N)$-approximation of x. The properties of splines and the spline algorithm are summed up in the following

LEMMA 4.4.4.1.

 (i) *There exists a spline interpolating* y *iff there exists* $f \in F_1$ *with* $Nf = y$ *such that* $P(Tf)$ *is nonempty.*
 (ii) *There exists a unique spline interpolating* y *iff* $P(Tf)$ *is a singleton for all* $f \in F_1$ *such that* $Nf = y$ *and* $\ker N \cap \ker T = \{0\}$.
 (iii) $\tilde{f}$ *is a spline interpolating* y *iff* $N\tilde{f} = y$ *and* $f - \tilde{f} \in P(Tf)$ *for all* $f \in F_1$ *such that* $Nf = y$.
 (iv) *If a spline algorithm* ϕ^s *exists and is unique, then it is homogeneous,* i.e.
$$\phi^s(\alpha y) = \alpha \phi^s(y) \qquad \forall \alpha \in \mathbb{R}, y \in N(F).$$
 (v) *Any homogeneous interpolatory algorithm is a spline algorithm.*
 (vi) *The spline algorithm is unique iff* $SP(Tf)$ *is a singleton for all* $f \in F_1$.
 (vii) *A spline algorithm* ϕ^s *is interpolatory, so that*
$$e(\phi^s, N, y) \le d(N, y) \le 2\, r(N, y) \qquad \forall y \in N(F)$$

and
$$e(\phi^s, N) \le d(N) \le 2\, r(N). \qquad \square$$

Hence a spline algorithm is almost strongly optimal.

As an illustration, suppose that X is strictly convex, i.e.

$$\|x_1\| = \|x_2\| \quad \text{and} \quad \|x_1+x_2\| = \|x_1\|+\|x_2\| \qquad \text{iff} \qquad \exists\, \alpha \in \mathbb{R} : x_2 = \alpha x_1.$$

For example, any Hilbert space is strictly convex, as are the Sobolev spaces $W^{m,p}(\Omega)$ for $m \geq 0$ and $1 \leq p < \infty$. Then the spline interpolating y is unique for all $y \in N(F)$ iff $\ker N \cap \ker T = \{0\}$, and the spline algorithm is unique iff $\ker N \cap \ker T \subseteq \ker S$. Note that this last condition is not restrictive in practice, since if it is violated, then card N is less than the problem index, so that there exists no algorithm using N whose error is finite.

We now show that linear optimal error algorithms exist for linear problems in the following cases:

(i) S is a linear functional,
(ii) G is a space of bounded real-valued functions under the sup norm,
(iii) X is a Hilbert space, and $T(\ker N)$ is closed in X.

Moreover, a linear algorithm that is an nth minimal error algorithm exists for all these cases, provided that nth optimal information exists.

We first consider case (i), the case of a real linear functional.

THEOREM 4.4.4.1. Let S be a linear functional and let

$$Nf = \begin{bmatrix} \lambda_1(f) \\ \vdots \\ \lambda_n(f) \end{bmatrix} \qquad \forall f \in F$$

be linear information.

(i)   We have
$$r(N) = \tfrac{1}{2}\, d(N) = \sup_{h \in \ker N \cap F} Sh.$$

(ii) There exist scalars $q_1, \ldots, q_n \in \mathbb{R}$ such that the algorithm ϕ^L given by
$$\phi^L(Nf) = \sum_{j=1}^{n} \lambda_j(f) q_j \qquad \forall f \in F$$

is an optimal error algorithm using N.

(iii) Let $\lambda_1, \ldots, \lambda_n$ be linearly independent. For $1 \leq j \leq n$, suppose that

$$r_j(x) = \sup\{\, Sf : f \in F \text{ such that } \lambda_i(x) = x\, \delta_{i,j} \text{ for } 1 \leq i \leq n \,\}$$

is a differentiable function of x at $x = 0$. Then the unique optimal error algorithm using N is given by setting $q_j = r_j'(0)$ for $1 \leq j \leq n$. □

We now deal with case (ii). Here G is a space of real bounded functions defined on a set D, whose norm is the sup norm, i.e.

$$\|g\| = \sup_{x \in D} |g(x)| \qquad \forall\, g \in G.$$

For any $f \in F$ and $x \in D$, we let $S_x f = (Sf)(x)$. Since S_x is a linear functional, there exist scalars $q_j = q_j(x)$ for $1 \leq j \leq n$ such that the algorithm $\sum_{j=1}^{n} \lambda_j(f) q_j(x)$ is an optimal error algorithm for S_x using N. We then have

THEOREM 4.4.4.2. *Let G be a space of bounded real-value functions on a set D, normed by the sup norm.*

(i) *We have*

$$r(N) = \tfrac{1}{2}\, d(N) = \sup_{\substack{h \in \ker N \cap F \\ x \in D}} (Sh)(x).$$

(ii) *For information N, define the algorithm ϕ^L using N by*

$$\phi^L(Nf)(x) = \sum_{j=1}^{n} \lambda_j(f) q_j(x) \qquad \forall\, x \in D, f \in F.$$

   Then ϕ^L is an optimal error algorithm for S using N. $\qquad\square$

Finally, we come to (iii) in the list above. That is, the range X of the restriction operator T is a Hilbert space, and $T(\ker N)$ is closed in X. Let us consider the spline algorithm more carefully for this case. First, note that for any $y \in N(F)$, a unique spline element interpolating y always exists, and is characterized by the conditions

$$N\tilde{f} = y \qquad \text{and} \qquad T\tilde{f} \perp T(\ker N).$$

From this, it may be seen that the spline element $\tilde{f}(y)$ is given by an orthogonal projection. Since orthogonal projections in a Hilbert space are linear, it follows that the spline algorithm is linear in the Hilbert case.

We develop an explicit representation formula for the spline algorithm in the Hilbert case. Without loss of generality, we assume that F_1 is infinite-dimensional and $\lambda_1, \ldots, \lambda_n$ are linearly independent. For each index j, let $f_j \in F_1$ be the spline interpolating the jth standard basis vector in $\mathbb{R}^n$, i.e.

$$\lambda_i(f_j) = \delta_{ij} \quad (1 \leq i \leq n) \qquad \text{and} \qquad Tf_j \perp T(\ker N).$$

Then

$$\tilde{f}(y) = \sum_{j=1}^{n} \lambda_j(f) f_j$$

is the spline interpolating $y = Nf$, and the spline algorithm has the form

$$\phi^s(Nf) = \sum_{j=1}^{n} \lambda_j(f) Sf_j$$

We then have the following

THEOREM 4.4.4.3. *Let the range X of the restriction operator be a Hilbert space. If N is linear information such that $T(\ker N)$ is a closed subspace of X, then the spline algorithm using N is a linear strongly optimal error algorithm, i.e.*

$$e(\phi^s, N, y) = r(N, y) = \tfrac{1}{2}\, d(N, y) \qquad \forall\, y \in \mathbb{R}^n,$$

and so

$$e(\phi^s, N) = r(N) = \tfrac{1}{2}\, d(N). \qquad\qquad \square$$

Notes and Remarks

NR 4.4.4:1 Packel (1986) has shown that a linear optimal error algorithm exists for *any* linear problem, provided that we are willing to extend its codomain. However, this extended codomain is the space of bounded functions over the unit ball in the conjugate space of G endowed with the weak-* topology. Since this extended codomain is so much bigger than the original codomain, Packel's beautiful result is only of theoretical interest.

4.4.5   Complexity

We now give tight bounds on the ε-complexity of the linear problem with linear solution operator S, a set F of problem elements that is given by restriction, and a class Λ of permissible information operations.

Recall that $m(\varepsilon)$ is the ε-cardinality number (i.e. the minimal number of possibly adaptive information operations needed to compute an ε-approximation) and that $r(n, \Lambda)$ denotes the nth minimal radius of non-adaptive information using operations from Λ. From Theorem 4.4.2.1, we see that

$$\inf\{\, n \in \mathbb{N} : r(n, \Lambda) \leq 2\varepsilon \,\} \leq m(\varepsilon) \leq \inf\{\, n \in \mathbb{N} : r(n, \Lambda) \leq \varepsilon + \delta \,\},$$

where $\delta = 0$ if there exists optimal information of any given cardinality, and may be chosen arbitrarily small otherwise. Using the results in §4.4.3 and §4.4.4, we then have

THEOREM 4.4.5.1.

(i) *The ε-complexity is bounded from below by*

$$\operatorname{comp}(\varepsilon) \geq c\, \inf\{\, n \in \mathbb{N} : r(n, \Lambda) \leq 2\varepsilon \,\}.$$

(ii) *Suppose that for any $n \in \mathbb{N}$,*

$$d(n, \Lambda) = 2\, r(n, \Lambda),$$

N_n *is nth optimal information in the class Λ, and*

there exists a linear optimal error algorithm ϕ_n^L using N_n.

Then

$$m(\varepsilon) = \inf\{\, n \in \mathbb{N} : r(n, \Lambda) \leq \varepsilon \,\},$$

and

$$c\, m(\varepsilon) \leq \mathrm{comp}(\varepsilon) \leq \mathrm{cost}(\phi^L_{m(\varepsilon)}, N_{m(\varepsilon)}) \leq (c+2)m(\varepsilon) - 1.$$

Hence for $c \gg 1$, the algorithm $\phi_{m(\varepsilon)}$ using information $N_{m(\varepsilon)}$ is an almost optimal complexity algorithm. □

For example, consider the cases discussed above, i.e. S a linear functional, G a space of functions under the sup norm, or X a Hilbert space, with $T(B)$ closed for any subspace B of finite codimension. For each of these cases, $d(N) = 2\,r(N)$ and there always exists a linear optimal error algorithm using N, for any information N. Thus if we can find nth optimal information N_n for all $n \in \mathbb{N}$, then the conclusions of (ii) in Theorem 4.4.5.1 hold.

In particular, we consider the convergent Hilbert case with $\Lambda = \Lambda'$, i.e. any linear functional is permissible. Let X and G be separable Hilbert spaces, with $T(B)$ closed for any subspace B of finite codimension. Let $K = ST^\dagger$ be compact. Let $\{e_1^*, \dots, e_{n^*}^*\}$ be a basis for $A(T, S) = \ker T / \ker T \cap \ker S$, with the linear functionals $\{\lambda_1^*, \dots, \lambda_{n^*}^*\}$ a dual basis. Let $z_1, z_2, \dots$, denote a sequence of orthonormal eigenvectors corresponding to the eigenvalues $\kappa_1^2 \geq \kappa_2^2 \geq \dots \geq 0$ of $K_1 = K^*K$. For $n \in \mathbb{N}$, let information N_n be defined as in (iii) of Theorem 4.4.3.2, and define an algorithm ϕ_n using N_n by

$$\phi_n(N_n f) = \sum_{j=1}^{n^*} \lambda_j^*(f) S e_j^* + \sum_{j=1}^{n-n^*} \langle Tf, z_j \rangle K z_j \qquad \forall f \in F.$$

COROLLARY 4.4.5.1. *Under the conditions mentioned above,*

(i) *For any $n \geq n^*$, ϕ_n is a linear nth minimal error algorithm, with*

$$e(\phi_n, N_n) = r(N_n) = r(n) = \kappa_{n-n^*+1}.$$

(ii) *For any $\varepsilon > 0$, the ε-cardinality number is given by*

$$m(\varepsilon) = \min\{\, n \geq n^* : \kappa_{n-n^*+1} \leq \varepsilon \,\}.$$

(iii) *For any $\varepsilon > 0$,*

$$c\, m(\varepsilon) \leq \mathrm{comp}(\varepsilon) \leq \mathrm{cost}(\phi_{m(\varepsilon)}, N_{m(\varepsilon)}) \leq (c+2)m(\varepsilon) - 1.$$ □

4.5   The residual error criterion

Many important problems of scientific computation arise as the solution of linear operator equations $Lu = f$. For such problems, we can use the residual $\|Lx - f\|$, rather than the absolute error $\|x - L^{-1}f\|$, to measure how well $x = x(f)$ approximates $u = L^{-1}f$. Although the residual can be used as an error criterion for any such operator equation, it is especially useful in the solution of ill-posed problems.

To illustrate this idea, consider the problem of solving a compact injective linear operator equation $Lu = f$ in the Hilbert case, the class F of problem elements being the intersection of the range of L with the unit ball of F_1. Since the restriction operator T is the identity on F_1, the operator $K = ST^\dagger$ is unbounded. In other words, the problem is *ill-posed*. In §6.4, we shall show that $r(n) = \infty$ for any $n \in \mathbb{N}$. so that there exists no algorithm with finite error for this problem. Hence, if we wish to solve such ill-posed problems in the worst case setting, we must find a different error criterion. It turns out the ill-posed problems are often solvable in the residual error criterion.

We outline the results of this section. First, we show how a linear problem under the residual error criterion can be reduced to the approximation problem under the usual error criterion. It then follows that the diameter of information for a linear operator equation under the residual error criterion equals the diameter for the approximation problem. Next, we discuss whether linear algorithms have optimal error. We find that there need not exist linear algorithms with finite error, even though the radius may be arbitrarily small; however, we find that when finite-error linear algorithms exist, linear algorithms are almost optimal in the Hilbert case. Finally, we determine tight complexity bounds, as well as nearly optimal complexity algorithms.

Notes and Remarks

NR 4.5:1 The results in this section are taken from Werschulz (1990).

4.5.1   Basic definitions

Let F_1 and G be Banach spaces, and let $L: G \to F_1$ be a bounded linear injection (which we call a *problem operator*) whose range D is dense in F_1. Let $T: F_1 \to X$ be a bounded surjective restriction operator, and let F be given by restriction, i.e.

$$F = \{\, f \in F_1 : \|Tf\| \leq 1 \,\}.$$

We then define our solution operator $S: F \cap D \to G$ by

$$u = Sf \quad \text{iff} \quad Lu = f \qquad \forall f \in F \cap D.$$

Since L is an injection, S is well-defined.

In addition, let W be a Banach space such that F_1 is *embedded* in W. That is, F_1 is a subspace of W, the inclusion mapping

$$E: F_1 \to W \qquad \text{defined by} \qquad Ef \equiv f \quad \forall f \in F_1$$

being continuous. For information N and an algorithm ϕ using N, we define the *residual* of ϕ to be

$$e^{\text{resid}}(\phi, N) = \sup_{f \in F \cap D} \left\| E\big(L\phi(Nf) - f\big) \right\|.$$

Note that under this residual error criterion, $\phi(Nf)$ is an ε-approximation to Sf if the W-norm of $L\big(\phi(Nf)\big) - f$ is at most ε. That is, the space W measures the quality of our approximation, whereas the spaces F_1 and X determine our problem elements.

As in §4.4.2, we let the *radius of information*

$$r^{\text{resid}}(N) = \inf_{\phi \text{ using } N} e^{\text{resid}}(\phi, N)$$

denote the minimal error among all algorithms using N. Defining the *diameter of information*

$$d^{\text{resid}}(N) = \sup_{f \in F \cap D} \sup_{\substack{\tilde{f} \in F \cap D \\ N\tilde{f} = Nf}} \left\| E\tilde{f} - Ef \right\|,$$

it is easy to see that we once again have

$$r^{\text{resid}}(N) \le d^{\text{resid}}(N) \le 2\, r^{\text{resid}}(N).$$

Furthermore, for a class Λ of permissible information operations, we define the *n*th *minimal radius* to be

$$r^{\text{resid}}(n, \Lambda) = \inf_{N \in \Lambda_n} r^{\text{resid}}(N),$$

while the *n*th *minimal diameter* is

$$d^{\text{resid}}(n, \Lambda) = \inf_{N \in \Lambda_n} d^{\text{resid}}(N).$$

Notes and Remarks

NR 4.5.1:1 Note that we have not assumed that D coincides with F_1 in our formulation of the residual error criterion. Recall that our main reason for introducing the residual error criterion was the study of ill-posed problems, i.e. linear operator equations $Lu = f$ with compact L. For many ill-posed problems, the range D of the compact operator L is hard to determine explicitly, although it is often easy to find a standard Banach space F_1 such that D is dense in F_1. Moreover for such problems, the assumption $D = F_1$ would imply that F_1 is not a Banach space, which would cause complications later on.

4.5.2 Reduction to the approximation problem

By the *approximation problem*, we mean the problem of approximating elements of the set F in the norm of the space W, i.e. we let the solution operator be the embedding $E: F_1 \to W$.

It is easy to see that our problem is equivalent to the approximation problem. Indeed for $\tilde{\phi} = L\phi$, we get

$$\left\| E\big(L\phi(Nf) - f\big) \right\| = \| f - \tilde{\phi}(Nf) \|_W \qquad \forall f \in F \cap D.$$

Therefore

$$e^{\text{resid}}(\phi, N) = e(\tilde{\phi}, N) = \sup_{f \in F \cap D} \| f - \tilde{\phi}(Nf) \|_W.$$

From the results of §4.4.2, we then find that the diameter of information is given by

$$d^{\text{resid}}(N) = 2 \sup_{h \in \ker N \cap F \cap D} \| Eh \|.$$

But since D is dense in F_1, it immediately follows that

$$d^{\text{resid}}(N) = 2 \sup_{h \in \ker N \cap F} \| Eh \|.$$

Thus the diameter for our problem is the same as that for the approximation problem.

Hence if a result for the approximation problem can be expressed and proved using only the diameter of information, it then applies immediately to our problem in the residual error criterion. We summarize these results in the following

THEOREM 4.5.2.1.

(i) *For any nonadaptive information N, the radius of information satisfies*

$$r^{\text{resid}}(N) = \alpha \sup_{h \in \ker N \cap F} \| Eh \|.$$

with $\alpha \in [1, 2]$.

(ii) *For any nonadaptive information N,*

$$r^{\text{resid}}(N) < \infty \qquad iff \qquad \ker T \cap \ker N = \{0\}.$$

Hence, there exists information N of finite cardinality with finite radius iff

$$n^* = \dim \ker T < \infty.$$

(iii) Let N^a be adaptive linear information, and let N^{non} be the non-adaptive linear information of cardinality at most $card(N^a)$ defined in §4.4.2. Then

$$d^{resid}(N^{non}) \leq d^{resid}(N^a),$$

so that

$$r^{resid}(N^{non}) \leq 2\, r^{resid}(N^a).$$

(iv) For $n \in \mathbb{N}$, the nth minimal diameter is given by

$$d^{resid}(n) = 2\, d^n\big(E(F), W\big),$$

with d^n denoting the Gelfand n-width. Moreover, nth optimal information for our problem is the same as that for the approximation problem.

(v) Our problem is convergent, i.e. $\lim_{n \to 0} r^{resid}(n) = 0$, iff E is compact. $\qquad\square$

However, not every result about the approximation problem holds for our problem. For instance, since $F \cap D$ is only dense in F, we cannot immediately infer that linear optimal error algorithms always exist in the Hilbert case. We shall pursue this matter more vigorously in the next subsection.

Since the remainder of this section deals only with the residual error criterion, we omit the superscript "resid" where this will cause no confusion. Hence all references in this section to $e(\phi, N)$, $r(N)$, $d(N)$, etc., are to be understood as referring to the residual error criterion.

4.5.3 Linear algorithms

We now turn to the question of the existence of linear (or nearly linear) optimal error algorithms using given information. Since we are only interested in cases for which the radius is finite, we will assume that $\dim \ker T < \infty$ in the remainder of this subsection.

THEOREM 4.5.3.1. *There exists a linear algorithm using continuous information of finite cardinality whose residual is finite iff $\ker T \subseteq D$.*

PROOF. Suppose that there exists information N and a linear algorithm ϕ using N with finite residual. Pick any $f \in \ker T$. For any $\alpha > 0$, we have

$\alpha f \in \ker T \subseteq F$. Since N is continuous and ϕ is linear, $\phi \circ N$ is continuous. Hence we have

$$
\begin{aligned}
\left\| E\big(L\phi(Nf) - f\big) \right\| &= \frac{1}{\alpha} \left\| E\Big(L\phi(N(\alpha f)) - \alpha f\Big) \right\| \\
&\leq \frac{1}{\alpha} \sup_{z \in F \cap D} \left\| E\big(L\phi(Nz) - z\big) \right\| \\
&= \frac{1}{\alpha} \sup_{z \in F} \left\| E\big(L\phi(Nz) - z\big) \right\| = \frac{1}{\alpha} e(\phi, N).
\end{aligned}
$$

Let $\alpha \to \infty$. Since $e(\phi, N)$ is finite, we find that

$$
\left\| E\big(L\phi(Nf) - f\big) \right\| = 0.
$$

Since $\| \cdot \|$ is a norm and E is an injection, $L\phi(Nf) = f$, and so $f \in D$. Thus $\ker T \subseteq D$.

Conversely, suppose that $\ker T \subseteq D$. Let $\{e_1^*, \dots, e_{n^*}^*\}$ be a basis for $\ker T$, and let $\{\lambda_1^*, \dots, \lambda_{n^*}^*\}$ be continuous linear functionals satisfying $\lambda_j^*(e_i^*) = \delta_{i,j}$ for $1 \leq i, j \leq n^*$. Define information N^* as in Theorem 4.4.3.1. Then we define an algorithm ϕ^* using N^* by

$$
\phi^*(N^* f) = \sum_{j=1}^{n^*} \lambda_j^*(f) L^{-1} e_j^* \qquad \forall f \in F_1,
$$

which is well-defined because $\ker T \subseteq D$. We claim that the residual of ϕ^* is finite. Indeed, for $f \in D \cap F$,

$$
\left\| E\big(L\phi^*(N^* f) - f\big) \right\| = \|Eh\|,
$$

where

$$
h = f - \sum_{j=1}^{n^*} \lambda_j^*(f) e_j^*.
$$

Note that $h \in \ker N^* = (\ker T)^{\perp}$, so that $h = T^{\dagger} T h$. Since T is bounded, the bounded inverse theorem implies that $T^{\dagger}$ is bounded. Since the set $\{e_1^*, \dots, e_{n^*}^*\}$ is a basis for $\ker T$, we see that $Th = Tf$, and so

$$
\|h\| \leq \|T^{\dagger}\| \, \|Th\| = \|T^{\dagger}\| \, \|Tf\| \leq \|T^{\dagger}\|.
$$

From this, it easily follows that

$$
e(\phi^*, N^*) \leq \|E\| \, \|T^{\dagger}\|. \qquad \square
$$

Using this result along with Proposition 3.3 in Chapter II of Pinkus (1985), we immediately have

COROLLARY 4.5.3.1. *If E is compact and $\ker T \not\subseteq D$, then:*
- *there exists no linear algorithm using continuous information of finite cardinality whose residual is finite, yet*
- *the problem is convergent.* □

We now consider the *Hilbert case*, in which the spaces F_1, G, X, and W are all separable Hilbert spaces. We now exhibit an almost linear algorithm using information N whose residual is almost minimal among all algorithms using N. Furthermore, when $\ker T \subseteq D$, this algorithm is linear.

Without serious loss of generality, we restrict our attention to information N of the form

$$
Nf = \begin{bmatrix} \langle f, e_1^* \rangle \\ \vdots \\ \langle f, e_{n^*}^* \rangle \\ \langle Tf, Ty_{n^*+1} \rangle \\ \vdots \\ \langle Tf, Ty_n \rangle \end{bmatrix} \qquad \forall f \in F_1.
$$

where $\{e_1^*, \ldots, e_{n^*}^*\}$ is an orthonormal basis for $\ker T$ and $y_{n^*+1}, \ldots, y_n \in (\ker T)^\perp$ satisfy

$$
\langle Ty_j, Ty_i \rangle = \delta_{ij} \qquad (n^* + 1 \le i, j \le n).
$$

Let $\delta > 0$. We now describe an algorithm ϕ_δ using N. Since D is dense in F, there exist $u_{n^*+1}, \ldots, u_n \in G$ such that

$$
\|E(Lu_j - y_j)\| \le \frac{\delta d(N)}{4(n - n^*)} \qquad (n^* + 1 \le j \le n).
$$

For any problem element $f \in D \cap F$, let

$$
f_0 = \sum_{j=1}^{n^*} \langle f, e_j^* \rangle e_j^*
$$

and

$$
f_1 = f - f_0.
$$

By denseness of D in F, there exists $u_\delta^* = u_\delta^*(Nf) \in G$ such that

$$
\|E(Lu_\delta^* - f_0)\| \le \tfrac{1}{4}\delta d(N).
$$

The algorithm ϕ_δ is defined to be

$$
\phi_\delta(Nf) = u_\delta^*(Nf) + \sum_{j=n^*+1}^{n} \langle Tf, Ty_j \rangle u_j \qquad \forall f \in D \cap F.
$$

THEOREM 4.5.3.2. *For any* $\delta > 0$,

$$r(N) \leq e(\phi_\delta, N) \leq (1 + \delta) \tfrac{1}{2} d(N) \leq (1 + \delta) r(N),$$

i.e. the residual of ϕ_δ is, to within a factor of at most $1 + \delta$, the minimal residual among all algorithms using N.

PROOF. Since ϕ_δ is an algorithm using N, the lower bound on the residual $e(\phi_\delta, N)$ is trivial. Furthermore, we have $\tfrac{1}{2} d(N) \leq r(N)$ from the results in §4.5.2. Hence, it only remains to show that the inequality $e(\phi_\delta, N) \leq (1 + \delta) \tfrac{1}{2} d(N)$ holds.

Let $f \in D \cap F$. Write

$$f = f_0 + f_1$$

as above, noting that

$$f_0 \in \ker T \quad \text{and} \quad f_1 \in (\ker T)^\perp.$$

We define

$$r_1 = L u_\delta^*(Nf) - f_0$$

$$r_2 = \sum_{j=n^*+1}^{n} \langle Tf, Ty_j \rangle (Lu_j - y_j)$$

$$r_3 = f_1 - \sum_{j=n^*+1}^{n} \langle Tf, Ty_j \rangle y_j,$$

so that

$$L\phi_\delta(Nf) - f = r_1 + r_2 - r_3.$$

By the way u_δ^* was chosen, we have

$$\|E r_1\| \leq \tfrac{1}{4} \delta d(N);$$

using $\|Ty_j\| = 1$ and $\|Tf\| \leq 1$, we find

$$\|E r_2\| \leq (n - n^*) \|Tf\| \max_{n^*+1 \leq j \leq n} \{\|E(Lu_j - y_j)\|\} \leq \tfrac{1}{4} \delta d(N).$$

We claim that

$$\|E r_3\| \leq \tfrac{1}{2} d(N).$$

Indeed, let $P \colon F_1 \to F_1$ be given by

$$Pz = \sum_{j=n^*+1}^{n} \langle z, Ty_j \rangle Ty_j \qquad \forall\, z \in F_1,$$

i.e. P is the orthogonal projector of F_1 onto $Y = \mathrm{span}\{Ty_{n^*+1}, \ldots, Ty_n\}$. From the definition of r_3, we have

$$\|Tr_3\| = \|Tf - PTf\| = \min_{x \in Y} \|Tf - x\| \le \|Tf\| \le 1,$$

and a short calculation shows that $r_3 \in \ker N$. Since $r_3 \in \ker N \cap F$, we have

$$\|Er_3\| \le \sup_{h \in \ker N \cap F} \|Eh\| = \tfrac{1}{2}d(N).$$

Combining our results, we find

$$\left\|E\big(L\phi_\delta(Nf) - f\big)\right\| \le \|Er_1\| + \|Er_2\| + \|Er_3\| \le (1+\delta)\tfrac{1}{2}d(N).$$

Since the problem element $f \in D \cap F$ is arbitrary, the desired bound for $e(\phi_\delta, N)$ follows immediately. □

Since $\delta > 0$ is arbitrary in Theorem 4.5.3.2, we have

COROLLARY 4.5.3.2. *In the Hilbert case,*

$$r(N) = \sup_{h \in \ker N \cap F} \|Eh\| \qquad \text{and} \qquad r(n) = d^n\big(E(F), W\big).$$

Furthermore, adaption is no more powerful than nonadaption, i.e. for any adaptive N^{a}, *there exists* N^{non} *with* $\mathrm{card}(N^{\mathrm{non}}) \le \mathrm{card}(N^{\mathrm{a}})$ *such that*

$$r(N^{\mathrm{non}}) \le r(N^{\mathrm{a}}).$$ □

Finally, a slight simplification of the proof of Theorem 4.5.3.2 allows us to see that when $\ker T \subseteq D$, then there exists a linear algorithm that is almost optimal:

THEOREM 4.5.3.3. *Suppose that* $\ker T \subseteq D$. *For any* $\delta > 0$, *define the algorithm* ϕ_δ *using* N *by*

$$\phi_\delta(Nf) = \sum_{j=1}^{n^*} \langle f, e_j^* \rangle L^{-1} e_j^* + \sum_{j=n^*+1}^{n} \langle Tf, Ty_j \rangle u_j \qquad \forall f \in D \cap F,$$

with $u_{n^*+1}, \ldots, u_n \in G$ *chosen so that*

$$\|E(Lu_j - y_j)\| \le \frac{\delta d(N)}{2(n - n^*)} \qquad (n^* + 1 \le j \le n).$$

Then

$$r(N) \le e(\phi_\delta, N) \le (1 + \delta)r(N),$$

i.e. ϕ_δ is a linear algorithm using N whose residual, to within a factor of at most $1 + \delta$, is minimal among all algorithms using N.

PROOF SKETCH. The proof follows that of the previous theorem, but is somewhat simpler because $L^{-1}e_j^*$ is well-defined. Since $r_1 = 0$, we can omit the factor "2" in the denominator of the criterion for choosing u_j for $n^* + 1 \le j \le n$. □

4.5.4   Complexity

We now give tight bounds on the ε-complexity of our problem. In what follows, we assume that $\Lambda \subseteq \Lambda^*$, i.e. that only continuous linear functionals are permissible. Following the reasoning that lead to Theorem 4.4.5.1, and using the results in §4.5.1–§4.5.3, we have

THEOREM 4.5.4.1.

(i) *The ε-complexity is bounded from below by*
$$\mathrm{comp}(\varepsilon) \ge c \inf\{\, n \in \mathbb{N} : r(n, \Lambda) \le 2\varepsilon \,\} \qquad \forall \varepsilon > 0.$$

(ii) *Suppose that for any $n \in N$,*
$$d(n, \Lambda) = 2\, r(n, \Lambda), \text{ and}$$
$$N_n \text{ is } n\text{th optimal information in the class } \Lambda.$$

In addition, suppose that for any $\delta > 0$, there exists a linear algorithm $\phi_{n,\delta}^L$ using N_n, with $e(\phi_{n,\delta}^L, N_n) \le (1 + \delta)r(N_n)$. Then
$$m(\varepsilon) = \inf\{\, n \in \mathbb{N} : r(n, \Lambda) \le \varepsilon \,\}. \qquad \forall \varepsilon > 0.$$

Moreover
$$c\, m(\varepsilon) \le \mathrm{comp}(\varepsilon) \le \mathrm{cost}(\phi_{m(\varepsilon/(1+\delta)),\delta}^L, N_{m(\varepsilon/(1+\delta))})$$
$$\le (c + 2)\, m\left(\frac{\varepsilon}{1+\delta}\right) - 1. \qquad \forall \varepsilon > 0.$$

(iii) *If*
$$m(\alpha\varepsilon) = \Theta\big(m(\varepsilon)\big) \qquad \text{as } \varepsilon \to 0, \qquad \forall \alpha \in [0, 1)$$

then for any $\delta > 0$,
$$\mathrm{cost}(\phi_{m(\varepsilon/(1+\delta)),\delta}^L, N_{m(\varepsilon/(1+\delta))}) = \Theta\big(m(\varepsilon)\big) \qquad \text{as } \varepsilon \to 0,$$

so that the algorithm $\phi_{m(\varepsilon/(1+\delta)),\delta}^L$ is an almost optimal complexity algorithm for any $\delta > 0$.

(iv) *If m is continuous at ε, then*
$$c\, m(\varepsilon) \le \mathrm{comp}(\varepsilon) \le (c + 2)\, m(\varepsilon) - 1. \qquad □$$

We now specialize this result to the Hilbert case. Let F_1, G, X, and W be separable Hilbert spaces. Let $K = ET^\dagger$ be compact. Suppose that $\Lambda = \Lambda'$. Let $\{e_1^*, \ldots, e_{n^*}^*\}$ be an orthonormal basis for $\ker T$. Let $z_1, z_2, \ldots$ denote a sequence of orthonormal eigenvectors corresponding to the eigenvalues $\kappa_1^2 \geq \kappa_2^2 \geq \ldots > 0$ of K^*K. For any $n \geq n^*$, define information N_n of cardinality n by

$$
N_n f =
\begin{bmatrix}
\langle f, e_1^* \rangle \\
\vdots \\
\langle f, e_{n^*}^* \rangle \\
\langle Tf, z_1 \rangle \\
\vdots \\
\langle Tf, z_{n-n^*} \rangle
\end{bmatrix}
\qquad \forall f \in F,
$$

choose $u_{n^*+1}, \ldots, u_n \in G$ such that

$$
\|E(Lu_j - T^\dagger z_j)\| < \frac{\delta\,\kappa_{n-n^*+1}}{n - n^*} \qquad (n^* + 1 \leq j \leq n),
$$

and define an algorithm $\phi_{n,\delta}$ using N_n by

$$
\phi_{n,\delta}(N_n f) = \sum_{j=1}^{n^*} \langle f, e_j^* \rangle L^{-1} e_j^* + \sum_{j=n^*+1}^{n} \langle Tf, z_j \rangle u_j \qquad \forall f \in D \cap F.
$$

THEOREM 4.5.4.2. *In the Hilbert case, the following hold:*

(i) *For any $\delta > 0$ and any $n \geq n^*$, $\phi_{n,\delta}$ is almost an nth minimal error algorithm:*

$$
r(n) = \kappa_{n-n^*+1} \leq e(\phi_{n,\delta}, N_n) \leq (1 + \delta)r(n).
$$

(ii) *For any $\varepsilon > 0$, the ε-cardinality number is given by*

$$
m(\varepsilon) = \min\{\, n \geq n^* : \kappa_{n-n^*+1} \leq \varepsilon \,\}.
$$

(iii) *If*

$$
m(\alpha\varepsilon) = \Theta\big(m(\varepsilon)\big) \qquad \text{as } \varepsilon \to 0, \qquad \forall\, \alpha \in [0, 1)
$$

then for any $\delta > 0$,

$$
\mathrm{cost}(\phi_{m(\varepsilon/(1+\delta)),\delta}, N_{m(\varepsilon/(1+\delta))}) = \Theta\big(m(\varepsilon)\big) \qquad \text{as } \varepsilon \to 0,
$$

so that the algorithm $\phi_{m(\varepsilon/(1+\delta)),\delta}$ is an almost optimal complexity algorithm for any $\delta > 0$.

(iv) *If m is continuous at ε, then*

$$
c\,m(\varepsilon) \leq \mathrm{comp}(\varepsilon) \leq (c + 2)\,m(\varepsilon) - 1. \qquad \qquad \square
$$

Note that these results are more complicated than the analogous results under the absolute error criterion. The complications arise because we do not generally have a minimal error linear algorithm; the best we can usually hope for (even in the Hilbert case) is an almost minimal error linear algorithm. This explains the presence of the $1 + \delta$ and the α appearing in the statements above.

Notes and Remarks

NR 4.5.4:1 We briefly comment on the conditions in Theorems 4.5.4.1 and 4.5.4.2.

First, we consider the condition $m(\alpha \varepsilon) = \Theta(m(\varepsilon))$ as $\varepsilon \to 0$ for $\alpha \in [0, 1)$. This essentially tells us that $r(n)$ cannot decrease too quickly as $n \to \infty$, i.e. the problem cannot be "too easy." For example, if $r(n) = \Theta(n^{-\beta})$ for some $\beta > 0$, then the condition is satisfied, whereas if $r(n) = \Theta(\gamma^{-n})$ for some $\gamma > 1$, then the condition is not satisfied.

Next, we consider the condition that m be continuous at ε. Since m is an integer-valued function, m will not generally be continuous on any interval of the form $(0, \varepsilon_0]$ (unless m is a constant function). Hence, the best we can expect is that m be continuous on such a half-open interval, except at a countable set of points.

5

Elliptic partial differential equations in the worst case setting

5.1 Introduction

In this chapter, we use the techniques of information-based complexity to find algorithms that are optimal in the worst case setting for elliptic partial differential equations (PDE). We will consider the approximate solution of linear elliptic $2m$th-order boundary value problems over a multidimensional domain $\Omega \subset \mathbb{R}^d$, error being measured in the Sobolev $H^l(\Omega)$-norm. Since we will be considering the variational formulation of such problems, we will restrict our attention to the case of $0 \le l \le m$.

The finite element method (FEM) has long been associated with such problems. We can gauge the popularity of the FEM by noting that Babuška (1988) mentioned that as of 1987, there were more than five hundred "user-oriented" finite element programs (see Frederikkson and Mackerle 1984) and more than two hundred monographs and conference proceedings (see Noor 1985) dealing with FEMs, with these numbers steadily increasing. Hence, we are particularly interested in the optimality of FEMs for these problems. A major task of this chapter will be to discover conditions that are necessary and sufficient for an FEM to be optimal.

We briefly outline the contents of this chapter.

First, we define $2m$th-order elliptic boundary value problems, including classical, weak, and variational solutions. Next, we describe how these problems are formulated in the framework of information-based complexity (i.e. we describe our solution operator S, problem elements F, and allowable information operations Λ).

We then analyze the normed case, in which F is the unit ball of the Sobolev space $H^r(\Omega)$, with $r \ge -m$. There are two subcases to examine. If $\Lambda = \Lambda'$ is the class of all linear functionals on $H^r(\Omega)$, then the inner products required by the FEM are permissible, and we find that the ε-complexity of the problem is $\Theta(\varepsilon^{-d/(r+2m-l)})$ as $\varepsilon \to 0$. Moreover, the FEM of degree k is optimal iff $k \ge 2m - 1 + r$. When $k < 2m - 1 + r$, we find that although the FEM is no longer optimal, the finite element information (FEI) used by the FEM is almost optimal information, i.e. the spline algorithm using FEI of degree k is almost optimal.

Let us momentarily consider these complexity results for the normed case. Since $r \geq -m$ and $l \leq m$, we know that $r + 2m \geq l$. Suppose that $r + 2m > l$. Then the problem complexity is exponential in its dependence on the dimension d of the domain, i.e. the problem is intractable (using the standard terminology of discrete complexity theory, see, for example, Garey and Johnson 1979). Things are even worse when $r + 2m = l$, since we find that the problem is not convergent in this case, i.e. the ε-complexity is infinite for ε sufficiently small. When can this happen? Suppose that we measure error in the natural norm for the problem (i.e. the energy norm, in which $l = m$) and that the problem elements have the least amount of smoothness to guarantee that solutions exist and are unique (i.e. $r = -m$). Then $r + 2m = l$ and the problem is not convergent. In short, the minimal conditions for existence and uniqueness of solutions to elliptic PDE are not strong enough to make the problem convergent. All of this means that if we wish to solve elliptic PDE in practice, we must avoid the intractability and non-convergence; to do this, we will have to abandon the worst case setting, a theme we will take up later in this book.

Of course, the assumption $\Lambda = \Lambda'$ is sometimes unrealistic, so we next analyze the case $\Lambda = \Lambda^{\text{std}}$, i.e. we assume that the only permissible information about a problem element is its values at a finite set of points. We find that the ε-complexity of the problem now becomes $\Theta(\varepsilon^{-d/r})$ as $\varepsilon \to 0$, i.e. there is a penalty associated with using standard information instead of arbitrary linear information. We also show that a finite element method with quadrature (FEMQ) of degree k is optimal iff $k \geq r + l - 1$.

Next, we turn to the seminormed case, in which F is the unit "semiball" of $H^r(\Omega)$, i.e. the set of elements whose whose Sobolev r-seminorm is bounded by one. Our motivation for doing this is twofold.

(i) We want to see how much the complexity and the optimal algorithms change when we change the definition of the problem.

(ii) The unit semiball is often used as a set of problem elements when studying the complexity of other problems such as the integration and approximation problems (see GTOA, IBC). If we wish to compare the complexity of our PDE with that of these other problems to see which is harder, we have to use the same class of problem elements.

We find that the ε-complexity of the problem is unchanged, whether arbitrary or standard information is used. However, finite element methods are usually not optimal; even worse, they have infinite error.

Finally, we look at the question of adaption for the approximate solution of elliptic PDE. Since the problem classes F studied so far have been convex and balanced, and the solution operator is linear, we know from §4.4.2

that nonadaptive information is as good as adaptive information. This assumption that F is convex and balanced is crucial, since if this does not hold, then the general theory does not tell us whether adaption helps in the solution of our problem. In the last section, we discuss this more fully. As a particular example, we study the sample problem of Chapter 2 for a nonconvex class of problem elements. Letting F be a class of piecewise constant functions with at most p breakpoints, we show that adaption helps exponentially if $p = 1$, but does not help at all if $p \geq 2$.

In the remainder of this chapter, we will follow a standard notational convention followed in most books on PDEs, such as Hormander (1983). The letter C will be used in inequalities to denote a generic finite, positive constant, independent of the functions and parameters appearing in that inequality.

5.2   Variational elliptic boundary value problems

In this section, we define precisely what is meant by a variational elliptic boundary value problem. Although the basic definitions are somewhat unwieldy, they are standard and taken from the literature. For further discussion, the reader should consult standard works on partial differential equations or the finite element method, such as Agmon (1965), Babuška and Aziz (1972), Ciarlet (1978), Hormander (1983), Lions and Magenes (1972), Nečas (1967), Oden and Carey (1983), and Oden and Reddy (1976). After we give these basic definitions, we illustrate them with several examples of variational boundary value problems. In what follows, we use the standard notation for Sobolev spaces, inner products, and norms, for multi-indices, etc., described in the Appendix.

Let $\Omega \subset \mathbb{R}^d$ be a bounded, simply connected, C^∞ region. For sufficiently smooth $v \colon \bar{\Omega} \to \mathbb{R}$, let

$$(Lv)(x) = \sum_{|\alpha|,|\beta| \leq m} (-1)^{|\alpha|} D^\alpha \big(a_{\alpha\beta}(x) D^\beta v(x) \big),$$

with real coefficients $a_{\alpha\beta} \in C^\infty(\overline{\Omega})$. We assume that L is a *uniformly strongly elliptic operator* in Ω. That is, letting

$$L_0(x, \xi) = \sum_{|\alpha|,|\beta| = m} (-1)^{|\alpha|} \xi^\alpha (a_{\alpha\beta} \xi^\beta v),$$

with $\xi^\alpha = \xi_1^{\alpha_1} \dots \xi_d^{\alpha_d}$, we assume that the following hold:

(i) For any point $x_0 \in \Omega$ and for any pair of linearly independent vectors $\xi, \eta \in \mathbb{C}^d$, the polynomial equation

$$L_0(x_0, \lambda\xi + \eta) = 0$$

has exactly m roots $\lambda_1, \ldots, \lambda_m$ with positive imaginary parts.

(ii) There exists a constant $\sigma > 0$ such that

$$L_0(x_0, \xi) \geq \sigma |\xi|^{2m} \qquad \forall \xi \in \mathbb{R}^d, x_0 \in \Omega,$$

where $|\cdot|$ is the Euclidean norm $|\xi| = (\xi_1^2 + \cdots + \xi_d^2)^{1/2}$ on $\mathbb{R}^d$.

To have appropriate boundary conditions, let

$$B_j v = \sum_{|\alpha| \leq q_j} b_{j\alpha} D^\alpha v \qquad (0 \leq j \leq m-1)$$

where $b_{j\alpha} \in C^\infty(\partial\Omega)$ are real-valued functions. We assume that $\{B_j\}_{j=0}^{m-1}$ is a *normal family of boundary operators* on $\partial\Omega$. That is, letting

$$B_{j0}(x, \xi) = \sum_{|\alpha|=q_j} b_{j\alpha}(x)\xi^\alpha v \qquad (0 \leq j \leq m-1),$$

we require that the following hold:

(i) For any $x \in \partial\Omega$ and for any vector ν normal to $\partial\Omega$ at x, we have $B_{j0}(x, \nu) \neq 0$.

(ii) $0 \leq q_0 < q_1 < \ldots < q_{m-1} \leq 2m - 1$.

Next, we assume that the boundary operators $\{B_j\}_{j=0}^{m-1}$ are *compatible* with the operator L. That is, for $x \in \partial\Omega$, let τ_x and ν_x respectively denote a unit tangent and an outward unit normal to $\partial\Omega$ at x. Then we require that the only function $u(s)$ for which

$$L_0\left(x, \tau_x - i\nu_x \frac{d}{ds}\right) u(s) = 0 \qquad \forall s > 0,$$

$$B_{j0}\left(x, \tau_x - i\nu_x \frac{d}{ds}\right) u(s)\bigg|_{s=0} = 0 \qquad (0 \leq j \leq m-1).$$

is the function $u(s) \equiv 0$. (Here, $i = \sqrt{-1}$.)

We divide our boundary operators into two classes. Let

$$m^* = \min\{j : q_j \geq m\}$$

(with $m^* = 0$ if no such j exists). Then the operators B_j with $j < m^*$ are said to be *essential boundary operators*, while those with $j \geq m^*$ are said to be *natural boundary operators*.

Finally, we want our family $(L, B_0, \ldots, B_{m-1})$ of operators to be *self-adjoint*. That is, we require that $a_{\alpha\beta} = a_{\beta\alpha}$ for all multi-indices α and β, and that

$$\{q_j\}_{j=0}^{m^*-1} \cup \{2m - 1 - q_j\}_{j=m^*}^{m-1} = \{0, \ldots, m-1\}.$$

We summarize our assumptions. We require that $(L, B_0, \ldots, B_{m-1})$ be a *self-adjoint elliptic boundary value problem*, i.e. that

(i) L is uniformly strongly elliptic,

(ii) $(B_0, \ldots, B_{m-1})$ is a normal family of boundary operators that is compatible with L, and

(iii) $(L, B_0, \ldots, B_{m-1})$ is a self-adjoint family of operators.

We can now give the *classical formulation* of our elliptic boundary value problem: For $f: \Omega \to \mathbb{R}$, find $u: \overline{\Omega} \to \mathbb{R}$ such that

$$
\begin{aligned}
Lu &= f \quad \text{in } \Omega \\
B_j u &= 0 \quad \text{on } \partial\Omega \quad (0 \le j \le m - 1).
\end{aligned}
\tag{1}
$$

The classical formulation of our elliptic problem is familiar to most readers. However, this formulation makes sense only for continuous functions f. Since we often need to solve (1) for non-continuous f, we present two alternative formulations of the problem. It turns out that all three formulations are equivalent for sufficiently smooth f, as mentioned in **NR 5.2:4**.

We first look at the weak formulation of (1). Let $f \in L_2(\Omega)$. We say that $u \in H^{2m}(\Omega)$ is a *weak solution* of (1) if

$$
\begin{aligned}
\langle Lu, v \rangle_{L_2(\Omega)} &= \langle f, v \rangle_{L_2(\Omega)} \qquad \forall\, v \in L_2(\Omega), \\
\langle B_j u, v \rangle_{L_2(\partial\Omega)} &= 0 \qquad \forall\, v \in L_2(\partial\Omega).
\end{aligned}
\tag{2}
$$

The standard result on the existence and uniqueness of weak solutions is given by the following "shift theorem:"

THEOREM 5.2.1. *Let $r \ge 0$. Then for any $f \in H^r(\Omega)$, there exists a unique $u \in H^{r+2m}(\Omega)$ satisfying (2). Furthermore, there exists a constant $\sigma \ge 1$, independent of f and u, such that*

$$
\sigma^{-1} \|u\|_{H^{r+2m}(\Omega)} \le \|f\|_{H^r(\Omega)} \le \sigma \|u\|_{H^{r+2m}(\Omega)}. \qquad \square
$$

Next, we consider the variational formulation of the problem. First, let $p \in [1, \infty]$ and $s \in \mathbb{R}$. The Sobolev embedding theorem (Lemma A.2.2.3) implies that if $v \in W^{s,p}(\Omega)$, then $B_j v$ is well-defined on $\partial\Omega$ iff $q_j < s - d/p$. Thus we write

$$
W_E^{s,p}(\Omega) = \{\, v \in H^m(\Omega) : B_j v = 0 \quad (q_j < s - d/p,\ 0 \le j \le m^* - 1) \,\}
$$

to denote the space of all functions in $W^{s,p}(\Omega)$ satisfying the essential boundary conditions. Adapting the usual notation, we also write $H_E^s(\Omega) = W_E^{s,2}(\Omega)$. In particular, $H_E^m(\Omega)$ is the *energy space*, i.e. the space of $H^m(\Omega)$-functions satisfying the essential boundary conditions.

We define a symmetric, continuous bilinear form B on $H_E^m(\Omega)$ by

$$B(v,w) = \sum_{|\alpha|,|\beta|\leq m} \int_\Omega a_{\alpha\beta} D^\alpha v D^\beta w \qquad \forall\, v,w \in H_E^m(\Omega).$$

An m-fold integration by parts establishes that

$$B(v,w) = \langle Lv,w \rangle_{L_2(\Omega)} \qquad \forall\, v \in H_E^{2m}(\Omega), w \in H_E^m(\Omega). \qquad (3)$$

We next assume that B is *weakly coercive* over $H_E^m(\Omega)$, see §A.3. Since B is symmetric, this means that there exists a positive constant γ such that for any nonzero $v \in H_E^m(\Omega)$, there exists nonzero $w \in H_E^m(\Omega)$ such that

$$|B(v,w)| \geq \gamma \|v\|_{H^m(\Omega)} \|w\|_{H^m(\Omega)}. \qquad (4)$$

When is B weakly coercive? The case $m = 0$ is easy, since it is clear that B is weakly coercive for $m = 0$ iff $a_{00} > 0$ almost everywhere. The following lemma gives a condition that is sufficient to establish weak coercivity for the case $m \geq 1$:

LEMMA 5.2.1. *Let $m \geq 1$. Suppose that*

(i) *the only solution of (1) with $f = 0$ is $u = 0$, and*

(ii) *B is $[H_E^m(\Omega), L_2(\Omega)]$-coercive; that is, there exist $\gamma_0 > 0$ and $\gamma_1 \geq 0$ such that*

$$|B(v,v)| \geq \gamma_0 \|v\|_{H^m(\Omega)}^2 - \gamma_1 \|v\|_{L_2(\Omega)}^2 \qquad \forall\, v \in H_E^m(\Omega). \qquad (5)$$

Then B is weakly coercive.

PROOF. Let $v \in H_E^m(\Omega)$. By (i) and Theorem 5.2.1, we see that there exists $z \in H^{3m}(\Omega)$ such that

$$Lz = \gamma_1 v \quad \text{in } \Omega,$$
$$B_j z = 0 \quad \text{on } \partial\Omega \qquad (0 \leq j \leq m-1),$$

with

$$\|z\|_{H^m(\Omega)} \leq \sigma\gamma_1 \|v\|_{H^{-m}(\Omega)} \leq \sigma\gamma_1 \|v\|_{H^m(\Omega)}.$$

Since L is self-adjoint, we may integrate by parts to find that $z \in H_E^m(\Omega)$ satisfies

$$B(v,z) = \langle v, Lz \rangle_{L_2(\Omega)} = \gamma_1 \langle v,v \rangle_{L_2(\Omega)} = \gamma_1 \|v\|_{L_2(\Omega)}^2.$$

Let $w = v + z$, so that

$$\|w\|_{H^m(\Omega)} \leq \|v\|_{H^m(\Omega)} + \|z\|_{H^m(\Omega)} \leq (1 + \sigma\gamma_1) \|v\|_{H^m(\Omega)}.$$

Since (5) holds, we have

$$B(v,w) = B(v,v) + \gamma_1 \|v\|_{L_2(\Omega)}^2 \geq \gamma_0 \|v\|_{H^m(\Omega)}^2$$
$$\geq \frac{\gamma_0}{1+\sigma\gamma_1} \|v\|_{H^m(\Omega)} \|w\|_{H^m(\Omega)},$$

as required. $\square$

REMARK. Suppose that our problem (1) is a *Dirichlet problem,* i.e. that B_j is the jth normal derivative operator ($0 \leq j \leq m - 1$). We claim that if (i) of the lemma holds, then B is weakly coercive. Indeed, note that $H_E^m(\Omega) = H_0^m(\Omega)$. Since (i) holds, it only remains to establish (5). But this is merely the well-known Gårding inequality for the Dirichlet problem, which may be found in Agmon (1965). Hence B is weakly coercive, as claimed. □

We are finally ready to define the *variational formulation* of our elliptic boundary value problem: For $f \in H^{-m}(\Omega)$, find $u = Sf \in H_E^m(\Omega)$ such that

$$B(u, v) = \langle f, v \rangle_{L_2(\Omega)} = \int_\Omega fv \qquad \forall v \in H_E^m(\Omega). \tag{6}$$

The standard result about existence and uniqueness to solutions of (6) is

THEOREM 5.2.2. *Let $r \geq -m$. Then for any $f \in H^{-r}(\Omega)$, there exists a unique $u = Sf \in H_E^{r+2m}(\Omega)$ satisfying (6). Furthermore, there exists $\sigma \geq 1$ such that*

$$\sigma^{-1}\|Sf\|_{H^{r+2m}(\Omega)} \leq \|f\|_{H^r(\Omega)} \leq \sigma\|Sf\|_{H^{r+2m}(\Omega)} \qquad \forall f \in H^r(\Omega). \tag{7}$$

PROOF. Since B is weakly coercive, symmetric, and bounded, the Generalized Lax-Milgram Lemma (Lemma A.3.1) gives existence and uniqueness. If $r \geq 0$, then (7) follows from Theorem 5.2.1. Suppose now that $-m \leq r \leq 0$. Let us momentarily write the restriction of S to $H^r(\Omega)$ as S_r. By Theorem 5.2.1 and the bounded inverse theorem, $S_0 \colon L_2(\Omega) \to H_E^{2m}(\Omega)$ is a *Hilbert space isomorphism,* i.e. S_0 is bounded with a bounded inverse. By the Generalized Lax-Milgram Lemma, $S_{-m} \colon H^{-m}(\Omega) \to H_E^m(\Omega)$ is a Hilbert space isomorphism. Using Lemma A.2.2.1, it then follows that $S_r \colon H^r(\Omega) \to H_E^{2m+r}(\Omega)$ is a Hilbert space isomorphism, which establishes (7). □

We now illustrate these definitions with several examples of elliptic boundary value problems, giving both the classical and variational formulations.

We first consider an elliptic problem with $m = 0$. This is the simplest possible elliptic problem.

EXAMPLE: *The approximation problem.* Define a bilinear form B over $L_2(\Omega)$ by

$$B(v, w) = \int_\Omega vw = \langle v, w \rangle_{L_2(\Omega)} \qquad \forall v, w \in L_2(\Omega).$$

For $f \in L_2(\Omega)$, we seek $u = Sf$ satisfying

$$B(u, v) = \int_\Omega fv \qquad \forall v \in L_2(\Omega).$$

Clearly, we have $u = Sf$ iff $u = f$ almost everywhere. Hence, S is the identity operator on $L_2(\Omega)$. We call this problem the *approximation problem*. □

We now consider three second-order elliptic problems ($m = 1$):

EXAMPLE: *The Poisson problem.*   Define a bilinear form B over $H_0^1(\Omega)$ by

$$B(v, w) = \int_\Omega \nabla v \cdot \nabla w \qquad \forall\, v, w \in H_0^1(\Omega).$$

For $f \in H^{-1}(\Omega)$, we seek $u = Sf \in H_0^1(\Omega)$ satisfying

$$B(u, v) = \int_\Omega fv \qquad \forall\, v \in H_0^1(\Omega). \tag{8}$$

This is the variational formulation of the *Poisson problem* with *Dirichlet boundary conditions*:

$$\begin{aligned} -\Delta u \equiv \operatorname{div}(\nabla u) &= f \qquad \text{in } \Omega, \\ u &= 0 \qquad \text{on } \partial\Omega. \end{aligned} \tag{9}$$

Problem (9) is the classical formulation of a famous elliptic boundary value problem arising in potential theory and in steady-state heat flow, and (8) is the corresponding variational elliptic boundary value problem. For the case $d = 2$, the solution to this problem gives the equilibrium position of an elastic membrane. □

EXAMPLE: *The generalized Poisson problem.*   By slightly generalizing the Poisson problem, we get a problem whose solution gives the steady-state heat flow for nonhomogeneous materials, the boundaries of which may be either held at a fixed temperature or insulated. First, we choose a boundary operator B as either the *Dirichlet boundary operator* $Bu = B_D u \equiv u$ or the *Neumann boundary operator* $Bu = B_N u \equiv \partial u/\partial n$. Let $a, b \colon \Omega \to \mathbb{R}$ be nonnegative functions. For $f \in H^{-1}(\Omega)$, we wish to solve the boundary value problem

$$\begin{aligned} -\operatorname{div}(a\nabla u) + bu &= f \qquad \text{in } \Omega, \\ Bu &= 0 \qquad \text{on } \partial\Omega. \end{aligned} \tag{10}$$

Since the Dirichlet boundary conditions are essential, while the Neumann conditions are natural, we have

$$H_E^1(\Omega) = \begin{cases} H_0^1(\Omega) & \text{if } B = B_D, \\ H^1(\Omega) & \text{if } B = B_N. \end{cases}$$

Our bilinear form is now

$$B(v, w) = \int_\Omega a\nabla v \cdot \nabla w + bvw \qquad \forall\, v, w \in H_E^1(\Omega).$$

The variational form of (10) is now to find, for $f \in H^{-1}(\Omega)$, an element $u = Sf \in H^1_E(\Omega)$ satisfying

$$B(u, v) = \int_\Omega fv \qquad \forall\, v \in H^1_E(\Omega). \qquad (11)$$

We now suppose that there exists $\alpha > 0$ such that $a(x) \geq \alpha$ for all $x \in \Omega$; furthermore, if $B = B_N$, we will require that there exists $\beta > 0$ such that $b(x) \geq \beta$ for all $x \in \Omega$. Then (10) and (11) are respectively the classical and variational formulation of an elliptic problem. □

EXAMPLE: *The Helmholtz problem.* Define a bilinear form B over $H^1_0(\Omega)$ by

$$B(v, w) = \int_\Omega \nabla v \cdot \nabla w - \lambda vw \qquad \forall\, v, w \in H^1_0(\Omega).$$

For $f \in H^{-1}(\Omega)$, we seek $u = Sf \in H^1_0(\Omega)$ satisfying

$$B(u, v) = \int_\Omega fv \qquad \forall\, v \in H^1_0(\Omega). \qquad (12)$$

This is the variational formulation of the *Helmholtz problem*

$$\Delta u + \lambda u = f \qquad \text{in } \Omega,$$
$$u = 0 \qquad \text{on } \partial\Omega. \qquad (13)$$

The Helmholtz problem, in both its classical formulation (13) and its variational formulation (12), is an elliptic boundary value problem iff λ is not an eigenvalue of $-\Delta$. It arises when using separation of variables to find solutions to the wave equation. □

We now consider two examples of fourth-order problems ($m = 2$):

EXAMPLE: *The clamped plate problem.* Let $d = 2$. Define a bilinear form B over $H^2_0(\Omega)$ by

$$B(v, w) = \int_\Omega [\Delta v \Delta w + (1 - \sigma)(2\,\partial_1\partial_2 v\,\partial_1\partial_2 w - \partial_1^2 v\,\partial_2^2 w - \partial_2^2 v\,\partial_1^2 w)]$$
$$= \int_\Omega [\sigma \Delta v \Delta w + (1 - \sigma)(\partial_1^2 v\,\partial_1^2 w + \partial_2^2 v\,\partial_2^2 w + 2\partial_1\partial_2 v\,\partial_1\partial_2 w)]$$
$$\forall\, v, w \in H^2_0(\Omega).$$

For $f \in H^{-2}(\Omega)$, we seek $u = Sf \in H^2_0(\Omega)$ satisfying

$$B(u, v) = \int_\Omega fv \qquad \forall\, v \in H^2_0(\Omega). \qquad (14)$$

Let $0 < \sigma < \frac{1}{2}$. Then there exists a unique solution u to (14). This solution is the equilibrium position of a plate of constant thickness e under the action of a transverse force whose density per unit area is $Ee^3/[12(1-\sigma)^2]f$. Here $E = \mu(3\lambda+2\mu)/(\lambda+\mu)$ is the Young's modulus, and $\sigma = \lambda/[2(\lambda+\mu)]$ is the Poisson coefficient, of the plate, with λ and μ being the material's Lamé coefficients. It may be shown that the (14) is the variational form of the *biharmonic problem*

$$\Delta^2 u = f \qquad \text{in } \Omega,$$
$$u = \frac{\partial u}{\partial n} = 0 \qquad \text{on } \partial\Omega. \tag{15}$$

Furthermore, problems (14) and (15) are respectively the classical and variational formulations of of an elliptic boundary value problem, see, for example, pp. 28–30 of Ciarlet (1978). □

EXAMPLE: *The biharmonic problem.* Define a bilinear form B over $H_0^2(\Omega)$ by

$$B(v, w) = \int_\Omega \Delta v \Delta w \qquad \forall v, w \in H_0^2(\Omega).$$

For $f \in H^{-2}(\Omega)$, we seek $u = Sf \in H_0^2(\Omega)$ satisfying

$$B(u, v) = \int_\Omega fv \qquad \forall v \in H_0^2(\Omega). \tag{16}$$

This is the canonical variational formulation of the biharmonic problem in (15), which (as we saw above) plays an important role in the deformation of thin plates. Problems (15) and (16) are respectively the classical and variational formulations of an elliptic boundary value problem. □

Note that our examples all have $m \in \{0, 1, 2\}$, i.e. m is usually small. Furthermore (possibly excepting the first example), all our examples require that Ω be a physical region in three-dimensional space, so that we have $d \in \{1, 2, 3\}$, i.e. d is also small in these examples. This is typical of problems arising in engineering. However, problems with large values of d do arise in other scientific disciplines. For example, the Poisson problem with d large arises in the study of Brownian motion of noninteracting many-particle systems, d being the number of particles in the system. Hence, the engineering-oriented reader should bear in mind that there are areas of scientific study in which we *do* have to solve problems over high-dimensional domains.

Notes and Remarks

NR 5.2:1 In our presentation, we are dealing only with self-adjoint problems with homogeneous boundary conditions. This has been done mainly for ease of exposition. Most of the results we develop hold also for non-self-adjoint or nonhomogeneous problems as well.

NR 5.2:2 We are only dealing with scalar-valued problems in our presentation (for ease of exposition). Vector-valued problems, such as those arising in linear elasticity, can be handled similarly. For further discussion, see pp. 23–28 of Ciarlet (1978).

NR 5.2:3 For a proof of Theorem 5.2.1, see Chapter 3 of Babuška and Aziz (1972) or Chapter 8 of Oden and Reddy (1976).

NR 5.2:4 Clearly a classical solution is a weak solution to our elliptic boundary value problem, and a weak solution is a variational solution. If $r > d/2$, then Theorem 5.2.1, the Sobolev embedding theorem, and (3) yield that the variational solution $u = Sf$ is the classical solution to (1). Hence all three formulations are equivalent when $r > d/2$.

NR 5.2:5 Many elliptic problems yield a bilinear form B such that there exists a positive constant γ for which

$$B(v,v) \geq \gamma \|v\|^2_{H^m(\Omega)} \qquad \forall\, v \in H^m_E(\Omega).$$

When this holds, the bilinear form is said to be *(strongly) coercive*, and the elliptic problem is said to be *definite*. On the other hand, if this condition does not hold, the elliptic problem is said to be *indefinite*. For example, the Poisson problem is a definite elliptic problem, whereas the Helmholtz problem is indefinite unless λ is smaller than the minimal eigenvalue of $-\Delta$.

NR 5.2:6 The result in Lemma 5.2.1 appears to be well-known. The proof given is based on that for the case $m = 1$ which is found in Chapter 5 of Babuška and Aziz (1972).

5.3 Problem formulation

We are now ready to give our problem formulation, which must consist of a solution operator, a description of permissible information, and a class of problem elements.

First, we define the solution operator. Recall that $(L, \{B_j\}_{j=0}^{m-1})$ is a self-adjoint elliptic boundary value problem, and B is the resulting weakly coercive bilinear form on $H^m_E(\Omega)$. Our class F of problem elements will be a subset of $H^{-m}(\Omega)$, to be specified shortly. Let $l \in \mathbb{R}$ satisfy $0 \leq l \leq m$. The space G of solution elements will be $H^l_E(\Omega)$ (as defined in the previous section). We define our solution operator $S \colon F \to H^l_E(\Omega)$ to be

$$u = Sf \quad \text{iff} \quad B(u,v) = \langle f, v \rangle_{L_2(\Omega)} \quad \forall\, v \in H^m_E(\Omega).$$

Since $F \subseteq H^{-m}(\Omega)$ and $G \subseteq H^m_E(\Omega)$, our solution operator S is well-defined. Note that we will be measuring the quality of our approximate solutions in the norm of G, i.e. the Sobolev l-norm.

Next, we choose our class Λ of permissible information operators. We will have one of two possibilities:

(i) $\Lambda = \Lambda'$, i.e. we will allow *arbitrary* linear information, or
(ii) $\Lambda = \Lambda^{\text{std}}$, i.e. we will allow only *standard* information.

In what follows, we will investigate complexity for several different problem element classes F. We are particularly interested in studying how the complexity changes as the smoothness of the problem elements change. With this in mind, we will do a complete investigation for two convex balanced classes of problem elements of "smoothness r":

(i) *The normed case:* Here we choose $r \geq -m$, and let F be the unit ball of $H^r(\Omega)$, i.e.

$$F = BH^r(\Omega) = \{\, f \in H^r(\Omega) : \|f\|_{H^r(\Omega)} \leq 1 \,\}.$$

(ii) *The seminormed case.* Here, we choose $r \in \mathbb{N}$, and let F be the unit "semiball" in $H^r(\Omega)$, i.e.

$$F = \mathcal{B}H^r(\Omega) = \{\, f \in H^r(\Omega) : |f|_{H^r(\Omega)} \leq 1 \,\},$$

where $|\cdot|_{H^r(\Omega)}$ is the Sobolev r-seminorm.

Although these classes seem to be similar (the only difference being between a norm and a seminorm), it turns out that there will be a few surprises in store for us when we ask to what extent standard algorithms are optimal in the seminormed case.

Of course, one might well argue that it is unrealistic to only solve problems in the unit ball or semiball of $H^r(\Omega)$, since we generally do not know an a priori bound on the Sobolev norm or seminorm of the problem elements. However, complexity of our problem over the unit ball is equivalent to complexity over the whole space $H^r(\Omega)$ using a normalized error criterion. Hence, we can remove the a priori bound that appears in the definition of F by going to a different error criterion. For further discussion of this, and other error criteria, see the Notes and Remarks.

Recall that for a given problem, we know the exact dimension d of Ω and the exact order $2m$ of the elliptic problem. However, it may be difficult to get a good value for r in practical situations. This explains why we are especially interested in how the complexity changes as a function of r. In particular, if we only know that r belongs to some set, we hope to find algorithms that are optimal for all r in that set.

Since these classes are convex and balanced, we know that adaption is no more powerful than nonadaption. Hence, we need only consider nonadaptive information for the normed and seminormed cases. The general theory makes no prediction whether adaption is better than nonadaption for problem classes that are not convex and balanced. Since many practitioners involved in scientific computation find it hard to believe that there is no advantage in considering adaptive information when solving elliptic problems (see Babuška 1987), we will also investigate a class that is not convex and balanced, to see whether adaption is helpful for nonconvex classes. In particular, we study a simple one-dimensional model problem, using a class of piecewise-constant functions. We show that adaption helps for this problem if and only if there is at most one discontinuity in the piecewise-constant, i.e. iff there are at most two "pieces."

Notes and Remarks

NR 5.3:1 We briefly consider what happens when we replace our absolute error criterion with other error criteria.

First, note that the results of this chapter have all been stated for the absolute error criterion. Using Theorems 5.2.1 and 5.2.2, it is easy to see that all our Θ-results in the $\|\cdot\|_{H^l(\Omega)}$-norm also hold for the residual error criterion in the $\|\cdot\|_{H^{l-2m}(\Omega)}$-norm.

One could consider a *normalized* error criterion, in which we let

$$e^{\mathrm{nor}}(\phi, N) = \sup_{f \in H^r(\Omega)} \frac{\|Sf - \phi(Nf)\|_{H^l(\Omega)}}{\|f\|_{H^r(\Omega)}}$$

for the normed case, and

$$e^{\mathrm{nor}}(\phi, N) = \sup_{f \in H^r(\Omega)} \frac{\|Sf - \phi(Nf)\|_{H^l(\Omega)}}{|f|_{H^r(\Omega)}}$$

for the seminormed case. This error criterion is often implicitly used in practice, when we have estimates of the form

$$\|Sf - \phi(Nf)\|_{H^l(\Omega)} \le \kappa(n)\nu(f)$$

with $\nu(f) = \|f\|_{H^r(\Omega)}$ or $\nu(f) = |f|_{H^r(\Omega)}$. For the problems of this chapter, homogeneous optimal error algorithms exist. It then easily follows that the complexity under the normalized criterion (over the whole space) and under the absolute criterion (over the ball) are the same. Moreover, an algorithm that is optimal under one criterion is optimal under the other.

Finally, we could consider complexity under a *relative* error criterion

$$e^{\mathrm{rel}}(\phi, N) = \frac{\|Sf - \phi(Nf)\|_{H^l(\Omega)}}{\|Sf\|_{H^l(\Omega)}}.$$

It is easy to see that for the problems of this chapter, the set $\ker N \cap F$ is balanced. From this, it is easy to see that $e^{\mathrm{rel}}(\phi, N) \ge 1$ for any algorithm ϕ using any information N. Hence $\mathrm{comp}^{\mathrm{rel}}(\varepsilon) = \infty$ for $\varepsilon < 1$.

For further discussion of different error criteria under the worst case setting, see Section 6 in Chapter 4 of IBC.

5.4 The normed case with arbitrary linear information

In this section, we analyze the normed case

$$F = BH^r(\Omega) = \{ f \in H^r(\Omega) : \|f\|_{H^r(\Omega)} \le 1 \},$$

under the assumption that arbitrary linear information is available. Here is a brief outline of what follows. First, we give tight bounds on the nth minimal radius. Next, we consider minimal error algorithms for our problem. Although the spline algorithm using nth optimal information is an nth minimal error algorithm, it is useful only when we can determine a priori spectral information about S. Since we can do this only for simple problems, we must find another algorithm that is easier to implement than the spline algorithm. One such algorithm is the finite element method (FEM), which approximates the solution by projection into a space of piecewise polynomials of degree k. We show that the FEM is an nth almost optimal error algorithm iff $k \geq 2m - 1 + r$; however, the finite element information (FEI) used by the FEM is always nth almost optimal information. Next, we determine ε-complexity of the problem, as well as the cost of using the FEM for ε-approximation. Finally, we compare the FEM and spline algorithms using FEI, showing that the latter is a generalized Galerkin method.

Since nonadaptive information is as powerful as adaptive information for our problem (see §4.4.2), we first give tight bounds on the nth minimal radius of (nonadaptive) information.

THEOREM 5.4.1.

$$r(n) = \Theta(n^{-(r+2m-l)/d}) \quad \text{as } n \to \infty.$$

PROOF. From Theorem 4.4.3.3, we know that $r(n)$ is the Gelfand n-width, i.e.

$$r(n) = d^n\big(SF, H_E^l(\Omega)\big). \tag{1}$$

For any $\theta > 0$, let

$$X(\theta) = \{\, u \in H_E^{r+2m}(\Omega) : \|u\|_{H^{r+2m}(\Omega)} \leq \theta \,\} = \theta\, BH_E^{r+2m}(\Omega). \tag{2}$$

Then Theorems 5.2.1 and 5.2.2 yield

$$X(\sigma^{-1}) \subseteq SF \subseteq X(\sigma), \tag{3}$$

and so (1) and (3) yield

$$d^n\big(X(\sigma^{-1}), H_E^l(\Omega)\big) \leq r(n) \leq d^n\big(X(\sigma), H_E^l(\Omega)\big). \tag{4}$$

But (2) implies that

$$d^n\big(X(\theta), H_E^l(\Omega)\big) = \theta\, d^n\big(BH_E^{r+2m}(\Omega), H_E^l(\Omega)\big) \qquad \forall\, \theta > 0, \tag{5}$$

so that (4) and (5) yield

$$\sigma^{-1}d^n\big(BH_E^{r+2m}(\Omega), H_E^l(\Omega)\big) \leq r(n) \leq \sigma d^n\big(BH_E^{r+2m}(\Omega), H_E^l(\Omega)\big). \tag{6}$$

From Theorem 2.5.1 of Babuška and Aziz (1972), we find

$$d^n \left(BH_E^{r+2m}(\Omega), H_E^l(\Omega) \right) = \Theta \left(d^n \left(BH_E^{1/2}(\Omega), L_2(\Omega) \right)^{2(r+2m-l)} \right). \quad (7)$$

But $H_E^{1/2}(\Omega) = H^{1/2}(\Omega) = H_0^{1/2}(\Omega)$, and so

$$\begin{aligned}
d^n \left(BH_E^{1/2}(\Omega), L_2(\Omega) \right) &= d^n \left(BH_0^{1/2}(\Omega), L_2(\Omega) \right) \\
&= \Theta \left(d^n \left(BH_0^1(\Omega), L_2(\Omega) \right)^{1/2} \right),
\end{aligned} \quad (8)$$

the last by another application of Theorem 2.5.1 of Babuška and Aziz (1972). Hence (6), (7), and (8) yield

$$r(n) = \Theta \left(d^n \left(BH_0^1(\Omega), L_2(\Omega) \right)^{r+2m-l} \right). \quad (9)$$

Now the results of Jerome (1968) yield that

$$d^n \left(BH_0^1(\Omega), L_2(\Omega) \right) = \Theta(n^{-1/d}). \quad (10)$$

The theorem follows immediately from (9) and (10). □

Using Theorem 4.4.3.2, we can find minimal error algorithms. Indeed, if $r = l - 2m$, then the theorem above states that our problem is not convergent. Hence, if we wish to solve our problem for small values of ε, we must assume that $r > l - 2m$. Noting that our restriction operator T is the identity operator on $H^r(\Omega)$, we let $z_1, z_2, \ldots$ denote a sequence of $H^r(\Omega)$-orthonormal eigenfunctions of S^*S corresponding to the eigenvalues $\kappa_1^2 \geq \kappa_2^2 \geq \ldots > 0$. For any $n \in \mathbb{N}$, we find that

$$N_n f = \begin{bmatrix} \langle f, z_1 \rangle_{H^r(\Omega)} \\ \vdots \\ \langle f, z_n \rangle_{H^r(\Omega)} \end{bmatrix} \qquad \forall f \in F$$

is nth optimal information, and that

$$\phi_n(N_n f) = \sum_{j=1}^n \langle f, z_j \rangle_{H^r(\Omega)} Sz_j \qquad \forall f \in F$$

is an nth minimal error algorithm.

Of course, ϕ_n is useful in practice only if we can find the eigenvectors of S^*S. Since we can do this only in the simplest cases, we must consider other algorithms, hoping to find an algorithm that is easier to implement than ϕ_n, yet whose error is $\Theta(r(n))$. With this in mind, we will consider finite element methods, our goal being to find conditions guaranteeing their optimality.

By a *triangulation* $\mathcal{T}$ of Ω, we mean a finite family of closed subsets of $\overline{\Omega}$ that satisfies the following conditions:

(i) For all $K \in \mathcal{T}$, the interior int K of K is a nonempty open polyhedron with Lipschitz-continuous boundary ∂K.

(ii) $\overline{\Omega} = \bigcup_{K \in \mathcal{T}} K$.

(iii) int $K \neq$ int K' for arbitrary distinct $K, K' \in \mathcal{T}$.

(iv) Any face of any $K \in \mathcal{T}$ is either a subset of $\partial \Omega$ or is a face of another element $K' \in \mathcal{T}$.

The elements are typically chosen as described in §A.2.3. That is, an element K of a triangulation $\mathcal{T}$ is the affine image of a fixed *reference element* $\hat{K}$ that is independent of $\mathcal{T}$ and K. This reference element is generally taken to be either a d-dimensional rectangle or a d-simplex in $\mathbb{R}^d$. So, we say that the triangulation $\mathcal{T}$ is *rectangular* or *simplicial*, according to whether the reference element is a rectangle or a simplex. Of course, it will generally not be possible to triangulate a smooth region Ω by simplices or rectangles; see **NR 5.4:3** for further discussion.

Suppose that $\mathcal{T}$ is a triangulation of Ω. Let $P_k(K)$ denote the space of all polynomials whose degree is most k, considered as functions over K. Then we say that $\mathcal{S}$ is a *finite element space* of *degree* k over the triangulation $\mathcal{T}$ if

$$\mathcal{S} = \left\{ s \in H_E^m(\Omega) : s\big|_K \in P_k(K) \ \forall K \in \mathcal{T} \right\}.$$

Note that the requirement that $\mathcal{S}$ be a piecewise polynomial space that is also a subspace of $H_E^m(\Omega)$ implies that functions in $\mathcal{S}$ have some amount of smoothness between adjacent elements:

LEMMA 5.4.1. *Let $m \geq 1$, and let $\mathcal{T}$ be a triangulation of Ω. Suppose that*

$$s\big|_K \in P_k(K) \qquad \forall K \in \mathcal{T}.$$

Then $s \in H_E^m(\Omega)$ iff

$$B_j s = 0 \qquad (0 \leq j \leq m^* - 1)$$

and

$$s \in C^{m-1}(\Omega). \qquad \square$$

PROOF. This lemma is proved for $m = 1$ in Sections 2.1 and 4.2 of Ciarlet (1978), the proof for $m \geq 2$ being a straightforward modification. $\square$

Using this lemma, we find a lower bound on the degree k of the finite element space.

LEMMA 5.4.2. *If* $\dim \mathcal{S} > \dim P_k(\Omega)$, *then* $k \geq m$.

PROOF. Suppose on the contrary that $k \leq m-1$. We claim that $\mathcal{S} \subseteq P_k(\Omega)$. To see this, let $s \in \mathcal{S}$, and let K_1 and K_2 be adjacent elements in the triangulation $\mathcal{T}$. Let s_1 and s_2 respectively denote the restriction of s to K_1 and K_2. By the previous lemma, $s \in C^{m-1}(K_1 \cup K_2)$. Since s_1 and s_2 are polynomials of degree at most $k \leq m-1$, it is easy to check that $s_1 = s_2$ on $K_1 \cup K_2$. From this, it follows that s is a polynomial, and not merely a piecewise polynomial, on $K_1 \cup K_2$. Repeating this argument to include all elements of the triangulation, we now find that $s \in P_k(\Omega)$. Thus, $\mathcal{S} \subseteq P_k(\Omega)$, as claimed. Since $\dim \mathcal{S} > \dim P_k(\Omega)$, this is impossible. $\square$

Although $k \geq m$ always holds by the previous lemma, we will assume that the somewhat stronger inequality

$$\text{if } l < m, \text{ then } k \geq 2m - 1 \tag{11}$$

holds. We will use this assumption to derive error bounds only when $l < m$, i.e. when we wish to find error bounds in norms that are lower than the natural norm on the energy space $H_E^m(\Omega)$. This assumption is not overly restrictive, since it often holds automatically:

(i) If $l = m$, then (11) holds vacuously because of the form of (11).

(ii) If $m = 0$ or $m = 1$, then (11) holds automatically by the previous lemma.

(iii) If $d \geq 2$ and simplicial elements are used, the results of Ženíšek (1972) and the inequality $l \geq 0$ imply that (11) holds.

Hence, the only possibility that (11) will fail is that we are measuring error in a lower norm, and that either $d = 1$ or that rectangular elements are used. However, it *is* possible for $k = m$ in either of these settings, and for (11) to not hold in those cases.

We next need to construct a basis $\{s_1, \ldots, s_n\}$ for $\mathcal{S}$ having "small" supports. That is, we require that the cardinality of the set of elements containing the support of any of the basis functions $s_1, \ldots, s_n$ be bounded, independent of n. Using the notation in §A.2.3, we see that this can be done by patching together the basis functions $\{s_{1,K}, \ldots, s_{J,K}\}$ of $P_k(K)$ for all $K \in \mathcal{T}$, and removing those basis functions that do not satisfy the essential boundary conditions.

Having constructed our basis $\{s_1, \ldots, s_n\}$, we let $\{\psi_1, \ldots, \psi_n\}$ denote the corresponding dual basis, i.e. a set of linear functionals on $H_E^m(\Omega)$ for which

$$\psi_i(s_j) = \delta_{i,j} \qquad (1 \leq i, j \leq n).$$

We call $\{\psi_1, \ldots, \psi_n\}$ the set of *degrees of freedom* for our finite element space $\mathcal{S}$. Since the basis for $\mathcal{S}$ is constructed by patching together the local basis functions, we see that the degrees of freedom for $\mathcal{S}$ consist of all the degrees of freedom of each of the individual elements $K \in \mathcal{T}$.

Let S be a finite element space. We define the S-*interpolant* of $v \in H_E^m(\Omega)$ to be

$$\Pi_S v = \sum_{j=1}^{n} \psi_j(v) s_j \qquad \forall v \in H_E^m(\Omega).$$

Since Π_K is the $P_k(K)$-interpolation operator (see §A.2.3), we have

$$(\Pi_S v)\big|_K = \Pi_K(v\big|_K) \qquad \forall K \in \mathcal{T}.$$

Since $\Pi_K\big|_{P_k(K)}$ is the identity operator, we see that

$$\Pi_S s = s \qquad \forall s \in S.$$

In what follows, we let $S_{n,k}$ denote a finite element subspaces of degree k over a triangulation $\mathcal{T}_n$, with $S_{n,k}$ being an n-dimensional subspace of $H_E^m(\Omega)$. As a slight abuse of notation, we will write Π_n for the interpolation operator into $S_{n,k}$, rather than $\Pi_{S_{n,k}}$.

We will require that the sequence $\{\mathcal{T}_n\}_{n=1}^{\infty}$ of triangulations not be too irregular. More precisely, for $K \in \mathcal{T}_n$, let

$$h_K = \operatorname{diam} K$$

and

$$\rho_K = \sup\{\operatorname{diam} B : \text{spheres } B \text{ containing } K\}.$$

Then we assume that $\{\mathcal{T}_n\}_{n=1}^{\infty}$ is a *quasi-uniform family* of triangulations of Ω, i.e. that

$$\varlimsup_{n \to \infty} \sup_{K \in \mathcal{T}_n} \frac{h_K}{\rho_K} < \infty.$$

It then easily follows that as $n \to \infty$,

$$h_K = \Theta(n^{-1/d}) \quad \forall K \in \mathcal{T}_n \qquad \text{and} \qquad \rho_K = \Theta(n^{-1/d}) \quad \forall K \in \mathcal{T}_n.$$

We describe local approximation properties over one element K of the triangulation $\mathcal{T}_n$ in Lemma A.2.3.3. Using these local results, we have

LEMMA 5.4.3.

(i) *Let* $p \in [1, \infty]$, $s \in \mathbb{N}$, *and* $j \in \{0, \dots, s\}$. *There exists a constant* $C > 0$ *such that for* $n \in \mathbb{P}$, $K \in \mathcal{T}_n$, *and for any* $v \in W_E^{s,p}(\Omega)$,

$$|v - \Pi_n v|_{W^{j,p}(\Omega)} \le C n^{-(\min\{k+1,s\}-j)/d} \|v\|_{W^{s,p}(\Omega)}.$$

(ii) *For any* $j \le m$, *there exists a constant* $C > 0$ *such that for* $n \in \mathbb{P}$ *and for any* $v \in H_E^s(\Omega)$, *there exists* $v_n \in S_{n,k}$ *such that*

$$\|v - v_n\|_{H^j(\Omega)} \le C n^{-(\min\{k+1,s\}-j)/d} \|v\|_{H^s(\Omega)}.$$

PROOF. To prove (i), let $j \in \{0, 1, \dots, s\}$. We assume without loss of generality that $p \in [1, \infty)$, the case $p = \infty$ being analogous. Suppose first that $k + 1 \leq s$. From Lemma A.2.3.3, we find that

$$
\begin{aligned}
|v - \Pi_n v|_{W^{j,p}(\Omega)} &= \left[\sum_{K \in \mathcal{T}_n} |v - \Pi_K v|^p_{W^{j,p}(K)} \right]^{1/p} \\
&\leq C n^{-(k+1-j)/d} \left[\sum_{K \in \mathcal{T}_n} |v|^p_{W^{s,p}(K)} \right]^{1/p} \\
&= C n^{-(k+1-j)/d} |v|_{W^{k+1,p}(\Omega)} \\
&\leq C n^{-(k+1-j)/d} \|v\|_{W^{k+1,p}(\Omega)} \\
&\leq C n^{-(k+1-j)/d} \|v\|_{W^{s,p}(\Omega)} .
\end{aligned}
$$

However, if $s \leq k$, then $P_s(K) \subseteq P_k(K)$ for $K \in \mathcal{T}_n$, and so

$$
\begin{aligned}
|v - \Pi_n v|_{W^{j,p}(\Omega)} &= \left[\sum_{K \in \mathcal{T}_n} |v - \Pi_K v|^p_{W^{j,p}(K)} \right]^{1/p} \\
&\leq C n^{-(s-j)/d} \left[\sum_{K \in \mathcal{T}_n} |v|^p_{W^{s,p}(K)} \right]^{1/p} \\
&= C n^{-(s-j)/d} |v|_{W^{s,p}(\Omega)} .
\end{aligned}
$$

These two inequalities yield (i).

We now turn to (ii). If $j = 0$ or $j = m$, then (ii) holds with $v_n = \Pi_n v$ by (i) of this Lemma. If $0 \leq j \leq m$, then we may use Hilbert space interpolation of the results for $j = 0$ and $j = m$ to see that

$$
\begin{aligned}
\|v - \Pi_n v\|_{H^j(\Omega)} &\leq \|v - \Pi_n v\|^{1-j/m}_{L_2(\Omega)} \|v - \Pi_n v\|^{j/m}_{H^m(\Omega)} \\
&\leq C n^{-(\min\{k+1,s\}-j)/d} \|v\|_{H^s(\Omega)}.
\end{aligned}
$$

Finally, we can use a duality argument to establish (ii) when $j < 0$; see Theorem 4.1.2 of Babuška and Aziz (1972) for details. $\qquad \square$

Recall that B is weakly coercive over $H^m_E(\Omega)$, see (4). We now additionally assume that B is *uniformly weakly coercive* over $\{S_{n,k}\}^\infty_{n=n_0}$ for some $n_0 \in \mathbb{P}$, i.e. there exists a positive constant γ', such that if $n \geq n_0$, then

$$
\forall v \in S_{n,k}, \exists w \in S_{n,k} : |B(v, w)| \geq \gamma \|v\|_{H^m(\Omega)} \|w\|_{H^m(\Omega)}.
$$

We often find that uniform weak coercivity on $\{S_{n,k}\}^\infty_{n=n_0}$ follows from weak coercivity over $H^m_E(\Omega)$, as in the following

LEMMA 5.4.4. *Let $m \geq 1$. Suppose that the only solution of (1) with $f = 0$ is $u = 0$, and that B is $[H_E^m(\Omega), L_2(\Omega)]$-coercive. Then there exists $n_0 \in \mathbb{P}$ such that B is uniformly weakly coercive over $\{S_{n,k}\}_{n=n_0}^\infty$.*

PROOF. By the previous lemma, we see that there is a positive constant C_1 such that for any $z \in H_E^m(\Omega)$ and $n \in \mathbb{P}$, we have

$$\|z - \Pi_n z\|_{H^m(\Omega)} \leq C_1 n^{-\mu/d} \|z\|_{H^{3m}(\Omega)}, \tag{12}$$

where

$$\mu = \min\{k + 1 - m, 2m\} > 0.$$

Since the coefficients of L are bounded, there exists $C_2 > 0$ such that

$$B(v, w) \leq C_2 \|v\|_{H^m(\Omega)} \|w\|_{H^m(\Omega)} \qquad \forall\, v, w \in H_E^m(\Omega).$$

We then let n_0 be an integer such that

$$n_0 > (C_1 C_2 \sigma \gamma_1 \gamma_0^{-1})^{d/\mu},$$

so that

$$\gamma_2 = \gamma_0 - C_1 C_2 \sigma \gamma_1 n_0^{-\mu/d}$$

is positive.

Let $n \geq n_0$ and $v \in S_{n,k}$. Choose $z \in H_E^m(\Omega)$ as in Lemma 5.2.1. Since $z \notin S_{n,k}$, we cannot choose $w = v + z$. However, setting $w = v + \Pi_n z$, we find that $w \in S_{n,k}$. We then have

$$
\begin{aligned}
B(v, w) &\geq B(v, v) + B(v, z) - |B(v, z - \Pi_n z)| \\
&\geq \gamma_0 \|v\|_{H^m(\Omega)}^2 - C_2 \|v\|_{H^m(\Omega)} \|z - \Pi_n z\|_{H^m(\Omega)}.
\end{aligned}
$$

Since (12), $n \geq n_0$, and Theorems 5.2.1 and 5.2.2 yield that

$$\|z - \Pi_n z\|_{H^m(\Omega)} \leq C_1 \sigma \gamma_1 n_0^{-\mu/d} \|v\|_{H^m(\Omega)},$$

we have

$$B(v, w) \geq \gamma_2 \|v\|_{H^m(\Omega)}^2.$$

But

$$
\begin{aligned}
\|w\|_{H^m(\Omega)} &\leq \|v\|_{H^m(\Omega)} + \|z\|_{H^m(\Omega)} + \|z - \Pi_n z\|_{H^m(\Omega)} \\
&\leq (1 + \sigma\gamma_1 + C_1 \sigma \gamma_1 n_0^{-\mu/d}) \|v\|_{H^m(\Omega)},
\end{aligned}
$$

so that

$$B(v, w) \geq \frac{\gamma_2}{1 + \sigma\gamma_1 + C_1 \sigma \gamma_1 n_0^{-\mu/d}} \|v\|_{H^m(\Omega)} \|w\|_{H^m(\Omega)},$$

as required. □

We are now ready to define the finite element method using $\{S_{n,k}\}_{n=n_0}^{\infty}$. Let $\{s_1, \ldots, s_n\}$ be a basis for $S_{n,k}$. For $f \in F$, we evaluate the inner products

$$N_{n,k}^{\text{FE}} f = \begin{bmatrix} \langle f, s_1 \rangle_{L_2(\Omega)} \\ \vdots \\ \langle f, s_n \rangle_{L_2(\Omega)} \end{bmatrix}. \tag{13}$$

Since B is uniformly weakly coercive, the Generalized Lax-Milgram Lemma (Lemma A.3.1) states that for any $n \geq n_0$ and any $f \in F$, there exists a unique solution $u_{n,k} \in S_{n,k}$ such that

$$B(u_{n,k}, s_i) = \langle f, s_i \rangle_{L_2(\Omega)} \qquad (1 \leq i \leq n). \tag{14}$$

The FEM yields $u_{n,k}$ which depends only on $N_{n,k}^{\text{FE}} f$; we write $u_{n,k} = \phi_{n,k}^{\text{FE}}(N_{n,k}^{\text{FE}} f)$. We call $N_{n,k}^{\text{FE}}$ *finite element information* (FEI) of *cardinality* n, and we will generally call the algorithm $\phi_{n,k}^{\text{FE}}$ the *finite element method* (FEM) using $N_{n,k}^{\text{FE}}$.

Let $f \in F$ and $n \geq n_0$. Note that the main difference between the exact solution $u = Sf$ and the finite element approximation $u_{n,k}$ is that $u \in H_E^m(\Omega)$ satisfies

$$B(u, v) = \langle f, v \rangle_{L_2(\Omega)} \qquad \forall v \in H_E^m(\Omega),$$

whereas $u_{n,k} \in S_{n,k}$ satisfies

$$B(u_{n,k}, s) = \langle f, s \rangle_{L_2(\Omega)} \qquad \forall s \in S_{n,k}.$$

In principle, we can compute the FEM $\phi_{n,k}^{\text{FE}}(N_{n,k}^{\text{FE}} f)$ using n information operations and $2n - 1$ arithmetic operations. Indeed, for $1 \leq j \leq n$, we first compute $u_j \in S_{n,k}$ satisfying

$$B(u_j, s_i) = \delta_{i,j} \qquad (1 \leq i \leq n).$$

That is, we precompute $u_1, \ldots, u_n$, independently of the problem element f. Then for $f \in F$, it is easy to see that

$$\phi_{n,k}^{\text{FE}}(N_{n,k}^{\text{FE}}) = u_{n,k} = \sum_{j=1}^{n} \langle f, s_j \rangle_{L_2(\Omega)} u_j.$$

Hence, the cost of computing $\phi_{n,k}^{\text{FE}}(N_{n,k}^{\text{FE}} f)$ is bounded by

$$\text{cost}(\phi_{n,k}^{\text{FE}}, N_{n,k}^{\text{FE}}) \leq cn + 2n - 1,$$

as claimed.

This approach works if the u_j are easy to obtain or if we wish to compute the FEM $\phi_{n,k}^{\text{FE}}$ at a fixed value of n for a large set of right-hand sides f. However, if the u_j are hard to obtain or we want to compute the FEM at only one f, we can compute $u_{n,k}$ by solving a system of linear equations. Indeed, define the Gram matrix

$$G = [g_{i,j}]_{1 \le i,j \le n}$$

by

$$g_{i,j} = B(s_j, s_i) \qquad (1 \le i,j \le n).$$

Then G is nonsingular for $n \ge n_0$. For $f \in F$, let

$$b = N_{n,k}^{\text{FE}} f.$$

Let

$$a = \begin{bmatrix} \alpha_1 \\ \vdots \\ \alpha_n \end{bmatrix}$$

be the unique solution to

$$Ga = b.$$

Then

$$\phi_{n,k}^{\text{FE}}(N_{n,k}^{\text{FE}} f) = u_{n,k} = \sum_{j=1}^{n} \alpha_j s_j.$$

If $d = 1$, the linear system $Ga = b$ can be solved in $\Theta(n)$ operations, since G is then a banded matrix with bandwidth independent of n. The efficient solution of this linear system has been widely investigated for $d \ge 2$. We can often compute a in $\Theta(n)$ or $\Theta(n \log n)$ arithmetic operations. Well-known techniques for the efficient solution of $Ga = b$ include the FFT methods of Swartztrauber (1977), the matrix capacitance methods in Dryja (1984) and Borgers and Widlund (1986), and the multigrid methods found in Hackbush (1985).

We are now ready to give tight bounds on the error of the FEM, as measured in the Sobolev l-norm.

THEOREM 5.4.2. *Let*

$$\mu = \min\{k + 1 - m, r + m\}.$$

Then

$$e(\phi_{n,k}^{\text{FE}}, N_{n,k}^{\text{FE}}) = \Theta(n^{-(\mu+m-l)/d}) \quad \text{as } n \to \infty.$$

PROOF. We first show that

$$e(\phi_{n,k}^{\text{FE}}, N_{n,k}^{\text{FE}}) = O(n^{-(\mu+m-l)/d}) \quad \text{as } n \to \infty. \tag{15}$$

First, we suppose that $l = m$. Let $f \in F$. Using Strang's Lemma (Lemma A.3.2) with $u = Sf$, $f_n = f$, $B_n = B$, and $V = W = H_E^m(\Omega)$, the approximation results in Lemma 5.3.3 (noting that $\mu = \min\{k+1, 2m+r\} - m$), and Theorem 5.2.2, there exist positive constants C, independent of n and f, such that

$$\begin{aligned}
\|Sf - \phi_{n,k}^{\text{FE}}(N_{n,k}^{\text{FE}}f)\|_{H^m(\Omega)} &\leq C\|Sf - \Pi_n Sf\|_{H^m(\Omega)} \\
&\leq Cn^{-\mu/d}\|Sf\|_{H^{2m+r}(\Omega)} \tag{16} \\
&\leq Cn^{-\mu/d}\|f\|_{H^r(\Omega)}.
\end{aligned}$$

Taking the supremum over $f \in F$, we get the desired upper bound (15) for the case $l = m$.

Suppose now that $l < m$. Let $f \in F$. Using the Aubin-Ciarlet-Nitsche Duality Argument (Lemma A.3.3), with $\tilde{S} = S$ (since B is symmetric), $u = Sf$, $f_n = f$, $B_n = B$, $V = H_E^m(\Omega)$, and $H = H^* = L_2(\Omega)$, we have

$$\|Sf - \phi_{n,k}^{\text{FE}}(N_{n,k}^{\text{FE}}f)\|_{L_2(\Omega)} \leq$$
$$M \sup_{g \in L_2(\Omega)} \frac{\|Sf - \phi_{n,k}^{\text{FE}}(N_{n,k}^{\text{FE}}f)\|_{H^m(\Omega)}\|Sg - \Pi_n Sg\|_{H^m(\Omega)}}{\|g\|_{L_2(\Omega)}}. \tag{17}$$

But (11) implies that $\min\{k+1, 2m\} - m = m$. Again using Theorems 5.2.1 and 5.2.2, along with the approximation results in Lemma 5.3.3, we have

$$\|Sg - \Pi_n Sg\|_{H^m(\Omega)} \leq Cn^{-m/d}\|Sg\|_{H^{2m}(\Omega)} \leq Cn^{-m/d}\|g\|_{L_2(\Omega)}.$$

Substituting this result and (16) into (17), we find

$$\|Sf - \phi_{n,k}^{\text{FE}}(N_{n,k}^{\text{FE}}f)\|_{L_2(\Omega)} \leq Cn^{-(\mu+m)/d}\|f\|_{H^r(\Omega)}. \tag{18}$$

Since $0 \leq l \leq m$, we may use interpolation of Hilbert spaces, (16), and (18) to find that for $n \geq n_0$ and for $f \in F$,

$$\begin{aligned}
\|Sf - \phi_{n,k}^{\text{FE}}(N_{n,k}^{\text{FE}}f)\|_{H^l(\Omega)} &\leq C\|Sf - \phi_{n,k}^{\text{FE}}(N_{n,k}^{\text{FE}}f)\|_{L_2(\Omega)}^{1-l/m} \\
&\quad \times \|Sf - \phi_{n,k}^{\text{FE}}(N_{n,k}^{\text{FE}}f)\|_{H^m(\Omega)}^{l/m} \\
&\leq Cn^{-(\mu+m-l)/d}\|f\|_{H^r(\Omega)}.
\end{aligned}$$

Taking the supremum over all $f \in F$, we finally get our upper bound (15) for the case $0 \leq l < m$.

We now turn to the lower bound

$$e(\phi_n, N_{n,k}^{\text{FE}}) = \Omega(n^{-(\mu+m-l)/d}) \quad \text{as } n \to \infty. \tag{19}$$

By Theorem 5.4.1, we have

$$e(\phi_{n,k}^{\text{FE}}, N_{n,k}^{\text{FE}}) \geq r(n) = \Theta(n^{-(r+2m-l)/d}). \tag{20}$$

It remains to show that

$$e(\phi_{n,k}^{\text{FE}}, N_{n,k}^{\text{FE}}) = \Omega(n^{-(k+1-l)/d}), \tag{21}$$

since the desired lower bound (19) follows immediately from (20) and (21).

Let Ω^0 be the interior of a hypercube whose closure is contained in Ω, and let

$$\mathcal{T}_n^0 = \{ K \in \mathcal{T}_n : K \subset \overline{\Omega^0} \}.$$

Fix $n_0 \in \mathbb{P}$, and let Ω^1 be an open region such that

$$\overline{\Omega^1} \subseteq \bigcup_{K \in \mathcal{T}_n^0} K \qquad \forall n \geq n_0.$$

Choose $u \in H_E^{r+2m}(\Omega)$ such that

(i) $u|_K \in P_{k+1}(K)$ for any $K \in \mathcal{T}_n^0$,

(ii) $|u|_{H^{k+1}(\Omega^1)}$ is bounded from below by a positive constant, independent of n, and

(iii) $\|Lu\|_{H^r(\Omega)}$ is bounded from above by a positive constant, independent of n.

For example, we may choose $u \in C_0^\infty(\Omega)$ such that $u|_{\Omega^1}$ is a polynomial of *exact* degree $k+1$. Then u is independent of n and conditions (i)–(iii) hold automatically.

We claim that there is a positive constant C, such that for any $n \in \mathbb{P}$ and any $K \in \mathcal{T}_n^0$,

$$\inf_{s \in P_k(K)} |u - s|_{H^l(K)} \geq C n^{-(k+1-l)/d} |u|_{H^{k+1}(K)} \tag{22}$$

Indeed, from Lemma A.2.3.3, we have

$$\inf_{s \in P_k(K)} |u - s|_{H^l(K)} \geq C h_K^{-l} (\text{vol } K)^{1/2} \inf_{\hat{s} \in P_k(\hat{K})} |\hat{u} - \hat{s}|_{H^l(\hat{K})}. \tag{23}$$

Recall that K is the affine image of a reference element $\hat{K}$ that is independent of K. It is straightforward to check that the functionals

$$\hat{v} \mapsto |\hat{v}|_{H^{k+1}(\hat{K})} \quad \text{and} \quad \hat{v} \mapsto \inf_{\hat{s} \in P_k(\hat{K})} |\hat{v} - \hat{s}|_{H^l(\hat{K})}$$

are seminorms on $P_{k+1}(\hat{K})$. Since $l \le m \le k$ (the last by Lemma 5.4.2), they have the same kernel $P_k(\hat{K})$. Since $P_{k+1}(\hat{K})$ is finite-dimensional, there is a positive constant $\hat{C} = \hat{C}(k, m, \hat{K})$ such that

$$\inf_{\hat{s} \in P_k(\hat{K})} |\hat{v} - \hat{s}|_{H^l(\hat{K})} \ge \hat{C}|\hat{v}|_{H^{k+1}(\hat{K})} \qquad \forall \hat{v} \in P_{k+1}(\hat{K}). \qquad (24)$$

Once again using Lemma A.2.3.3, we have

$$|\hat{u}|_{H^{k+1}(\hat{K})} \ge C\rho_K^{k+1}(\text{vol } K)^{-1/2}|u|_{H^{k+1}(K)}. \qquad (25)$$

Combining (23), (24), (25), we find

$$\inf_{s \in P_k(K)} |u - s|_{H^l(K)} \ge C\rho_K^{k+1} h_K^{-l}|u|_{H^{k+1}(K)}.$$

Since $\{\mathcal{T}_n\}_{n=1}^{\infty}$ is quasi-uniform, the desired conclusion (22) follows, as claimed.

Since $\mathcal{T}_n^0 \subset \mathcal{T}_n$, we see that (22) implies

$$\inf_{s \in \mathcal{S}_{n,k}} |u - s|_{H^l(\Omega)} \ge \left[\sum_{K \in \mathcal{T}_n^0} \inf_{s \in P_k(K)} |u - s|_{H^l(K)}^2 \right]^{1/2}$$

$$\ge Cn^{-(k+1-l)/d} \left[\sum_{K \in \mathcal{T}_n^0} |u|_{H^{k+1}(K)}^2 \right]^{1/2} \qquad (26)$$

$$\ge Cn^{-(k+1-l)/d}|u|_{H^{k+1}(\Omega^1)}.$$

Since $\|Lu\|_{H^r(\Omega)}$ is bounded from above by a fixed positive constant, independent of n, we may let

$$f^* = \frac{1}{\|Lu\|_{H^r(\Omega)}} Lu,$$

which is an element of $F = BH^r(\Omega)$. Since $\|\cdot\|_{H^l(\Omega)} \ge |\cdot|_{H^l(\Omega)}$ always holds and $\phi_{n,k}^{\text{FE}}(N_{n,k}^{\text{FE}} f^*) \in \mathcal{S}_{n,k}$, we have

$$\|Sf^* - \phi_{n,k}^{\text{FE}}(N_{n,k}^{\text{FE}} f^*)\|_{H^l(\Omega)} \ge |Sf^* - \phi_{n,k}^{\text{FE}}(N_{n,k}^{\text{FE}} f^*)|_{H^l(\Omega)}$$

$$\ge \inf_{s \in \mathcal{S}_{n,k}} |Sf^* - s|_{H^l(\Omega)} \qquad (27)$$

$$\ge \frac{1}{\|Lu\|_{H^r(\Omega)}} \inf_{s \in \mathcal{S}_{n,k}} |u - s|_{H^l(\Omega)},$$

the latter since $\mathcal{S}_{n,k}$ is a subspace of $H_E^m(\Omega)$ and

$$Sf^* = \frac{1}{\|Lu\|_{H^r(\Omega)}} u.$$

Letting C be the constant of (26), we set

$$C' = C \inf_{n \in \mathbb{P}} \frac{|u|_{H^{k+1}(\Omega^1)}}{\|Lu\|_{H^r(\Omega)}},$$

which is a positive constant. Then (26) and (27) imply that

$$\|Sf^* - \phi_{n,k}^{\mathrm{FE}}(N_{n,k}^{\mathrm{FE}} f^*)\|_{H^l(\Omega)} \geq C' n^{-(k+1-l)/d} \qquad \forall n \geq n_0.$$

Since $f^* \in F$, we find

$$
\begin{aligned}
e(\phi_{n,k}^{\mathrm{FE}}, N_{n,k}^{\mathrm{FE}}) &= \sup_{f \in F} \|Sf - \phi_{n,k}^{\mathrm{FE}}(N_{n,k}^{\mathrm{FE}} f)\|_{H^l(\Omega)} \\
&\geq \|Sf^* - \phi_{n,k}^{\mathrm{FE}}(N_{n,k}^{\mathrm{FE}} f^*)\|_{H^l(\Omega)} \geq C' n^{-(k+1-l)/d}
\end{aligned}
\qquad \forall n \geq n_0,
$$

which establishes the desired lower bound (21).   □

Comparing the previous two theorems, we immediately find that

COROLLARY 5.4.1. *The FEM of degree k is an almost minimal error algorithm iff $k \geq 2m - 1 + r$.*   □

Hence, the FEM $\phi_{n,k}^{\mathrm{FE}}$ of degree k is not an nth almost minimal error algorithm unless $k \geq 2m - 1 + r$. There are two possible explanations of why the error of the FEM is not almost minimal when this inequality is not satisfied. Either

(i) the FEM $\phi_{n,k}^{\mathrm{FE}}$ does not make good use of its information $N_{n,k}^{\mathrm{FE}}$, and there is another algorithm using FEI whose error is almost minimal, regardless of whether $k \geq 2m - 1 + r$, or

(ii) the FEM is an almost optimal error algorithm using FEI, and FEI is not strong enough information to admit an nth almost minimal error algorithm.

It turns out that possibility (i) is the reason the FEM (of too low a degree) does not have almost minimal error.

THEOREM 5.4.3. *The spline algorithm $\phi_{n,k}^{\mathrm{s}}$ using FEI $N_{n,k}^{\mathrm{FE}}$ is an optimal error algorithm using FEI and is an almost minimal error algorithm. That is,*

$$e(\phi_{n,k}^{\mathrm{s}}, N_{n,k}^{\mathrm{FE}}) = r(N_{n,k}^{\mathrm{FE}}) = \Theta(r(n)) = \Theta(n^{-(r+2m-l)/d}) \qquad \text{as } n \to \infty.$$

PROOF. As mentioned in Chapter 4, the spline algorithm using any information N is *always* an optimal error algorithm using N. Hence,

$$e(\phi_{n,k}^{\mathrm{s}}, N_{n,k}^{\mathrm{FE}}) = r(N_{n,k}^{\mathrm{FE}}).$$

From Theorem 5.4.1, we have the lower bound

$$r(N_{n,k}^{\text{FE}}) \geq r(n) = \Theta(n^{-(r+2m-l)/d}) \qquad \text{as } n \to \infty.$$

Hence, it remains to prove the upper bound

$$r(N_{n,k}^{\text{FE}}) = O(n^{-(r+2m-l)/d}) \qquad \text{as } n \to \infty. \tag{28}$$

To establish (28), note that if $z \in F \cap \ker N_{n,k}^{\text{FE}}$, then

$$\langle z, s \rangle_{L_2(\Omega)} = 0 \quad \forall s \in \mathcal{S}_{n,k} \qquad \text{and} \qquad \|z\|_{H^r(\Omega)} \leq 1. \tag{29}$$

Suppose first that $l = m$. By weak coercivity, there exists nonzero $w \in H_E^m(\Omega)$ such that

$$\gamma \|Sz\|_{H^m(\Omega)} \|w\|_{H^m(\Omega)} \leq |B(Sz, w)|. \tag{30}$$

Since $k + 1 \geq m$, Lemma 5.4.3 implies that there exists $s \in \mathcal{S}_{n,k}$ such that

$$\|w - s\|_{H^{-r}(\Omega)} \leq Cn^{-(r+m)/d} \|w\|_{H^m(\Omega)},$$

the constant C being independent of w and n. Using this bound with (29), we have

$$|B(Sz, w)| = |\langle z, w \rangle_{L_2(\Omega)}| = |\langle z, w - s \rangle_{L_2(\Omega)}| \leq \|z\|_{H^r(\Omega)} \|w - s\|_{H^{-r}(\Omega)}$$
$$\leq Cn^{-(r+m)/d} \|w\|_{H^m(\Omega)} \,.$$

Combining this result with (30), we have

$$\|Sz\|_{H^m(\Omega)} \leq Cn^{-(r+m)/d}.$$

Since $z \in \ker N_{n,k}^{\text{FE}} \cap F$ is arbitrary, we have

$$r(N_{n,k}^{\text{FE}}) = \sup_{z \in \ker N_{n,k}^{\text{FE}} \cap F} \|Sz\|_{H^m(\Omega)} \leq Cn^{-(r+m)/d},$$

which establishes (28) when $l = m$.

We now suppose that $l < m$. Symmetry of B and (29) yield

$$\begin{aligned}
\|Sz\|_{L_2(\Omega)}^2 &= |\langle Sz, Sz \rangle_{L_2(\Omega)}| = |B(Sz, S^2 z)| \\
&= |\langle z, S^2 z \rangle_{L_2(\Omega)}| = |\langle z, S^2 z - s \rangle_{L_2(\Omega)}| \\
&\leq \|z\|_{H^r(\Omega)} \|S^2 z - s\|_{H^{-r}(\Omega)} \\
&\leq \|S^2 z - s\|_{H^{-r}(\Omega)} \quad \forall s \in \mathcal{S}_{n,k}.
\end{aligned} \tag{31}$$

Note that (11) implies that $2m + r \leq k + 1 + r$. Hence by Lemma 5.3.3, there exists a positive constant C, independent of n and g, as well as an $s \in \mathcal{S}_{n,k}$ which *does* depend on g, such that

$$
\begin{aligned}
\|S^2 z - s\|_{H^{-r}(\Omega)} &\leq Cn^{-(r+2m)/d}\|S^2 z\|_{H^{2m}(\Omega)} \\
&\leq C\sigma n^{-(r+2m)/d}\|Sz\|_{L_2(\Omega)},
\end{aligned}
\tag{32}
$$

the positive constant σ coming from Theorems 5.2.1 and 5.2.2. Hence, (31) and (33) yield

$$
\|Sz\|_{L_2(\Omega)} \leq C\sigma n^{-(r+2m)/d}.
$$

Since $0 \leq l \leq m$, we may use Hilbert space interpolation, (31), and (32) to find that

$$
\|Sz\|_{H^l(\Omega)} \leq C\|Sz\|_{L_2(\Omega)}^{1-l/m}\|Sz\|_{H^m(\Omega)}^{l/m} \leq Cn^{-(r+2m-l)/d},
$$

the constant C being independent of z and n. Since $z \in \ker N_{n,k}^{\mathrm{FE}} \cap F$ is arbitrary, we have

$$
r(N) = \sup_{z \in \ker N_{n,k}^{\mathrm{FE}} \cap F} \|Sz\|_{H^l(\Omega)} \leq Cn^{-(r+2m-l)/d},
$$

establishing (28) for the case $l < m$, as required.   □

So the information $N_{n,k}^{\mathrm{FE}}$ is always nth almost optimal information. However, the FEM of degree k is an almost optimal error algorithm using FEI iff $k \geq 2m - 1 + r$.

For $\varepsilon > 0$, we now wish to find $\mathrm{comp}(\varepsilon)$, the ε-complexity of our problem. We also wish to determine conditions under which the FEM is an almost optimal complexity algorithm for our problem. Let

$$
\mathrm{cost}_k^{\mathrm{FE}}(\varepsilon) = \inf\{\, \mathrm{cost}(\phi_{n,k}^{\mathrm{FE}}, N_{n,k}^{\mathrm{FE}}) : e(\phi_{n,k}^{\mathrm{FE}}, N_{n,k}^{\mathrm{FE}}) \leq \varepsilon \,\}
$$

denote the minimal cost of using the FEM of degree k to compute an ε-approximation. We also let

$$
\mathrm{cost}_k^{\mathrm{spline}}(\varepsilon) = \inf\{\, \mathrm{cost}(\phi_{n,k}^{\mathrm{s}}, N_{n,k}^{\mathrm{FE}}) : e(\phi_{n,k}^{\mathrm{FE}}, N_{n,k}^{\mathrm{FE}}) \leq \varepsilon \,\}
$$

be the minimal cost of computing an ε-approximation with the spline algorithm using FEI of degree k. Using the previous three theorems, along with the results in §4.4.5, we have

THEOREM 5.4.4.

(i) *The ε-complexity is given by*

$$
\mathrm{comp}(\varepsilon) = \Theta(\varepsilon^{-d/(r+2m-l)}) \qquad \text{as } \varepsilon \to 0.
$$

(ii) *Let $\mu = \min\{k + 1 - m, m + r\}$. Then*

$$\text{cost}_k^{\text{FE}}(\varepsilon) = \Theta(\varepsilon^{-d/(\mu+m-l)}) \qquad \text{as } \varepsilon \to 0.$$

Hence, if $k \geq 2m - 1 + r$, then the FEM of degree k is an almost optimal complexity algorithm, i.e.

$$\text{cost}_k^{\text{FE}}(\varepsilon) = \Theta\big(\text{comp}(\varepsilon)\big) = \Theta(\varepsilon^{-d/(r+2m-l)}) \qquad \text{as } \varepsilon \to 0.$$

(iii) *The spline algorithm using FEI of degree k is an almost optimal complexity algorithm, i.e.*

$$\text{cost}_k^{\text{spline}}(\varepsilon) = \Theta\big(\text{comp}(\varepsilon)\big) = \Theta(\varepsilon^{-d/(r+2m-l)}) \qquad \text{as } \varepsilon \to 0. \qquad \square$$

REMARK. Recall that $\text{comp}(\varepsilon)$ is the inherent complexity of solving our problem. No matter how clever we are, we cannot design an algorithm that gives us an ε-approximation with fewer than $\Theta(\varepsilon^{-d/(r+2m-l)})$ operations. Note the "curse of dimensionality," namely that the complexity increases exponentially with the dimension d of the problem. Recall that in traditional "discrete" computational complexity, such an exponential growth means that a problem is inherently intractable, see Garey and Johnson (1979). This means that if one increases the dimension (and the other parameters are held fixed), then *the elliptic boundary value problem is inherently intractable.*

Of course, one might argue that for the problems we study, the dimension is fixed. Suppose for instance that we are solving a three-dimensional, second-order elliptic problem, with problem elements in $L_2(\Omega)$ and using the energy norm $\| \cdot \|_{H^1(\Omega)}$. That is, we have $d = 3$, $m = 1$, $l = 1$, and $r = 0$. Then we find that $\text{comp}(\varepsilon) = \Theta(\varepsilon^{-3})$. Let us assume, for the sake of simplicity, that the Θ-constant is order unity. Suppose that we need four-place accuracy, i.e. we want to find an ε-approximation with $\varepsilon = 10^{-4}$. Then $\text{comp}(\varepsilon)$ is around 10^{12} operations. This takes about two weeks on a megaflop machine, which is clearly too long to wait for an answer.

So we see that in the worst case setting, a bad combination of the parameters can yield an intractable elliptic problem. Hence, if we really want to compute an approximate solution to such a problem, we must change the setting. We will discuss different settings in later chapters of this book. $\qquad \square$

REMARK. Suppose that $r + 2m = l$. Then the exponent of ε^{-1} in this theorem is infinite. What does this mean? Note that for this case, the nth minimal radius is $\Theta(1)$ as $n \to \infty$, i.e. there is a positive constant ε_0 such that $r(n) \geq \varepsilon_0$ for all n. Hence $\text{comp}(\varepsilon) = \infty$ for $\varepsilon < \varepsilon_0$ if $r + 2m = l$.

The case $r + 2m = l$ is important. Indeed, suppose we decide to measure error in the energy norm, i.e. $l = m$. Furthermore, suppose we only know that the problem elements are smooth enough to guarantee existence and uniqueness of solutions, i.e. $r = -m$. Then $r + 2m = l$, and the ε-complexity is infinite for $\varepsilon < \varepsilon_0$. $\square$

REMARK.   Suppose we don't know the exact value of the regularity r of the problem elements, but that we know only that r belongs to some set. Of course, the ε-complexity is still $\Theta(\varepsilon^{-d/(r+2m-l)})$ as $\varepsilon \to 0$. Our problem is that since we don't know r, we cannot determine which spline algorithm or FEM to use. However, suppose that we know a finite upper bound on r, say $r \le r^*$. If $k \ge 2m - 1 + r^*$, then $\min\{k + 1 - m, m + r\} = m + r$, and so the previous theorem yields that $\mathrm{cost}_k^{\mathrm{FE}}(\varepsilon) = \Theta(\varepsilon^{-d/(r+2m-l)})$ as $\varepsilon \to 0$. Thus if we know only an upper bound on r, we can find an FEM that is an almost optimal complexity algorithm for our problem. $\square$

Note that there is a large asymptotic penalty to be paid when we use a FEM of too low a degree. That is, for $k < 2m - 1 + r$,

$$\frac{\mathrm{cost}_k^{\mathrm{FE}}(\varepsilon)}{\mathrm{comp}(\varepsilon)} = \Theta(\varepsilon^{-\delta d}) \qquad \text{as } \varepsilon \to 0,$$

where

$$\delta = \frac{1}{k + 1 - l} - \frac{1}{r + 2m - l} > 0.$$

Thus,

$$\lim_{\varepsilon \to 0} \frac{\mathrm{cost}_k^{\mathrm{FE}}(\varepsilon)}{\mathrm{comp}(\varepsilon)} = \infty. \tag{34}$$

On the other hand, the spline algorithm using FEI is always almost optimal. This means that there is a constant $C \ge 1$ such that

$$1 \le \frac{\mathrm{cost}_k^{\mathrm{spline}}(\varepsilon)}{\mathrm{comp}(\varepsilon)} \le C \qquad \forall \varepsilon > 0. \tag{35}$$

In particular, (34) and (35) imply that if $k < 2m - 1 + r$, then

$$\lim_{\varepsilon \to 0} \frac{\mathrm{cost}_k^{\mathrm{FE}}(\varepsilon)}{\mathrm{cost}_k^{\mathrm{spline}}(\varepsilon)} = \infty.$$

Of course, this implies that there is a cutoff value $\varepsilon_0 > 0$ for which

$$\mathrm{cost}_k^{\mathrm{spline}}(\varepsilon) \le \mathrm{cost}_k^{\mathrm{FE}}(\varepsilon) \qquad \text{for } 0 < \varepsilon < \varepsilon_0. \tag{36}$$

How big is ε_0? If ε_0 is unreasonably small, then it may still be more reasonable to use the FEM than the spline algorithm for "practical" values of ε. Although a general expression for ε_0 is not available, we can determine its value in the following

EXAMPLE: *A model problem.* Let $d = 1$, $\Omega = I = (0, \pi)$, $m = 1$, $r = 1$, $H_E^1(\Omega) = H_0^1(I)$, and consider the bilinear form $B \colon H_0^1(I) \times H_0^1(I) \to \mathbb{R}$ defined by

$$B(v, w) = \int_0^\pi v' w' \qquad \forall\, v,\, w \in H_0^1(I).$$

Hence for $f \in H^1(I)$, $u = Sf$ is the variational solution to the problem

$$-u'' = f \qquad \text{in } I$$
$$u(0) = u(\pi) = 0.$$

We choose $S_{n,1}$ to be the n-dimensional subspace of $H_0^1(I)$ consisting of piecewise linear polynomials with nodes at $x_j = j\pi/(n+1)$ for $0 \le j \le n+1$, so that the degree k is 1. Moreover, since any function in $S_{n,1}$ must vanish at the endpoints of I, we see that $\dim S_{n,1} = n$.

We first give a lower bound on the error $e(\phi_{n,1}^{\mathrm{FE}}, N_{n,1}^{\mathrm{FE}})$ of the nth FEM. Define $f \colon I \to \mathbb{R}$ by

$$f(x) = \frac{1}{\sqrt{\pi}} \qquad \forall\, x \in I.$$

Then $\|f\|_{H^1(I)} = 1$ and $Sf = u$, where

$$u(x) = \tfrac{1}{2} x \left(\sqrt{\pi} - \frac{x}{\sqrt{\pi}} \right).$$

Let $\Pi_n u$ denote the $S_{n,1}$-interpolate of u, i.e. $\Pi_n u$ is the unique element of $S_{n,1}$ for which

$$(\Pi_n u)(x_j) = u(x_j) \qquad (0 \le j \le n+1).$$

Using standard techniques found in (for example) Ciarlet (1978), Schultz (1973), and Strang and Fix (1973a), it is easy to show that

$$e(\phi_{n,1}^{\mathrm{FE}}, N_{n,1}^{\mathrm{FE}}) \ge \inf_{s \in S_{n,1}} |Sf - s|_{H^1(I)} = |u - \Pi_n u|_{H^1(I)}$$
$$= \frac{\pi}{\sqrt{12}\,(n+1)}, \tag{37}$$

giving the desired lower bound on the error of the FEM.

Now we can find a lower bound on $\mathrm{cost}_1^{\mathrm{FE}}(\varepsilon)$. Let n be chosen so that $e(\phi_{n,1}^{\mathrm{FE}}, N_{n,1}^{\mathrm{FE}}) \le \varepsilon$. Then (37) yields

$$n \ge \frac{\pi}{\sqrt{12}} \varepsilon^{-1} - 1.$$

Since $\text{cost}(\phi_{n,1}^{\text{FE}}, N_{n,1}^{\text{FE}}) \geq cn$, we have

$$\text{cost}_1^{\text{FE}}(\varepsilon) \geq c\left(\frac{\pi}{\sqrt{12}}\varepsilon^{-1} - 1\right). \tag{38}$$

Next, we need to find an upper bound on the error $e(\phi_{n,1}^{\text{s}}, N_{n,1}^{\text{FE}})$ of the spline algorithm $\phi_{n,1}^{\text{s}}$ using finite element information $N_{n,1}^{\text{FE}}$. Since the spline algorithm $\phi_{n,1}^{\text{s}}$ is an optimal error algorithm using $N_{n,1}^{\text{FE}}$, it suffices to compute the radius of FEI. Let $z \in \ker N_{n,1}^{\text{FE}} \cap F$. Letting P_n denote the orthogonal projector of $L_2(I)$ onto $\mathcal{S}_{n,1}$, we find

$$\begin{aligned}
|Sz|_{H^1(I)}^2 = B(Sz, Sz) &= \langle z, Sz\rangle_{L_2(I)} = \langle z, Sz - P_n Sz\rangle_{L_2(I)}\\
&= \langle z - \Pi_n z, Sz - P_n Sz\rangle_{L_2(I)}\\
&\leq \|z - \Pi_n z\|_{L_2(I)}\|Sz - P_n Sz\|_{L_2(I)}\\
&\leq \|z - \Pi_n z\|_{L_2(I)}\|Sz - \Pi_n Sz\|_{L_2(I)}.
\end{aligned} \tag{39}$$

Now Theorem 2.4 of Schultz (1973) states that for any $v \in H_0^1(I)$, we have

$$\|v - \Pi_n v\|_{L_2(I)} \leq \frac{1}{n+1}|v|_{H^1(I)}. \tag{40}$$

Since $z \in F$ (so that $|z|_{H^1(I)} \leq 1$), we may use (39) and (40) to find that

$$|Sz|_{H^1(I)} \leq \left(\frac{1}{n+1}\right)^2 \qquad \forall z \in \ker N_{n,1}^{\text{FE}} \cap F.$$

Using the Poincaré inequality $|\cdot|_{L_2(I)}^2 \leq \frac{1}{2}|\cdot|_{H^1(I)}^2$ on $H_0^1(I)$, see Chapter 1 of Schultz (1973), we find that

$$\|Sz\|_{H^1(I)} \leq \sqrt{\frac{3}{2}}|Sz|_{H^1(I)} \leq \sqrt{\frac{3}{2}}\frac{1}{(n+1)^2} \qquad \forall z \in \ker N_{n,1}^{\text{FE}} \cap F,$$

and so

$$e(\phi_{n,1}^{\text{s}}, N_{n,1}^{\text{FE}}) = r(N_{n,1}^{\text{FE}}) = \sup_{z \in \ker N_{n,1}^{\text{FE}} \cap F}\|Sz\|_{H^1(I)} \leq \sqrt{\frac{3}{2}}\frac{1}{(n+1)^2}. \tag{41}$$

Now we can find an upper bound on $\text{cost}_1^{\text{spline}}(\varepsilon)$. Let

$$n = \left(\frac{3}{2}\right)^{1/4}\varepsilon^{-1/2} - 1,$$

so that (41) yields

$$e(\phi_{n,1}^s, N_{n,1}^{\mathrm{FE}}) \le \varepsilon.$$

From the results in §4.4, we have

$$\mathrm{cost}(\phi_{n,1}^s, N_{n,1}^{\mathrm{FE}}) \le (c+2)n - 1.$$

Combining the last two inequalities, we have

$$\mathrm{cost}_1^{\mathrm{spline}}(\varepsilon) \le (c+2)\left(\left(\frac{3}{2}\right)^{1/4}\varepsilon^{-1/2} - 1\right) - 1. \tag{42}$$

We now wish to find $\varepsilon_0 = \varepsilon_0(c)$ such that (36) holds. Using (38) and (42), we find that ε_0 may be chosen as the smallest positive solution of

$$c\left(\frac{\pi}{\sqrt{12}}\varepsilon_0^{-1} - 1\right) = (c+2)\left(\left(\frac{3}{2}\right)^{1/4}\varepsilon_0^{-1/2} - 1\right) - 1.$$

Some algebra yields

$$\varepsilon_0 = \varepsilon_0(c) = \left[\frac{1}{4}\left(\frac{3}{2}\right)^{1/4}(c+2)\left(1 - \sqrt{1 - \frac{4\pi\sqrt{2}}{3}\frac{c}{(c+2)^2}}\right)\right]^2.$$

We now examine the value of $\varepsilon_0(c)$, noting that $\varepsilon_0(c)$ increases with (nonnegative) c. Let us assume that $c \ge 1$, i.e. that evaluating a linear functional is at least as hard as an arithmetic operation (it would be hard to imagine otherwise). We find that

$$\varepsilon_0(c) \ge \varepsilon_0(1) = \frac{3}{4}\left(\frac{3}{2}\right)^{1/4}\left(1 - \sqrt{1 - \frac{4\pi\sqrt{2}}{27}}\right) \doteq 0.3448.$$

Moreover, it is reasonable to assume that $c \gg 1$, i.e. that evaluating a linear functional is *much* harder than an arithmetic operation; see p. 85 of GTOA for further discussion. One may check that

$$\lim_{c \to \infty} \varepsilon_0(c) = \frac{\pi^2}{6^{3/2}} \doteq 0.6715,$$

giving an estimate of $\varepsilon_0(c)$ for large values of c. Hence we see that for our model problem, the cutoff value ε_0 is *not* unreasonably small. ☐

We have seen that the spline algorithm using FEI is always almost optimal, whereas the FEM is almost optimal iff $k \ge 2m - 1 + r$. Furthermore, the spline algorithm is a strongly optimal error algorithm using FEI (see Theorem 4.4.4.3). Hence, it is worthwhile to will look more closely at the spline algorithm. We will show that the spline algorithm may be realized as a generalized Galerkin method using different spaces of test and trial functions. Moreover, we will determine conditions that are necessary and sufficient for the FEM to be a spline algorithm.

We will only consider the case where B is a *definite* bilinear form on $H_E^m(\Omega)$, i.e. there is a positive constant γ for which

$$B(v,v) \geq \gamma \|v\|_{H^m(\Omega)}^2 \qquad \forall v \in H_E^m(\Omega).$$

Defining

$$\|v\|_B = \sqrt{B(v,v)} \qquad \forall v \in H_E^m(\Omega),$$

we see that $\|\cdot\|_B$ is a norm on $H_E^m(\Omega)$, which is equivalent to the usual Sobolev norm $\|\cdot\|_{H^m(\Omega)}$. Since we are mainly interested in almost optimal algorithms (i.e. optimality to within a constant), there is no loss of generality in using $\|\cdot\|_B$ as our norm on $H_E^m(\Omega)$, instead of $\|\cdot\|_{H^m(\Omega)}$. In what follows, we let $S^*\colon H_E^m(\Omega) \to H^r(\Omega)$ denote the adjoint to S, remembering that $H_E^m(\Omega)$ is a Hilbert space under the inner product given by the bilinear form B. That is,

$$\langle g, v \rangle_{L_2(\Omega)} = B(Sg, v) = \langle g, S^*v \rangle_{H^r(\Omega)} \qquad \forall v \in H_E^m(\Omega),\, g \in H^r(\Omega). \quad (43)$$

We first describe generalized Galerkin methods, sometimes referred to in the literature as Bubnov-Galerkin methods. Let $\{s_i\}_{i=1}^n$ and $\{t_i\}_{i=1}^n$ each be sets of linearly independent functions in $H_E^m(\Omega)$. Let

$$\mathcal{S} = \mathrm{span}\{s_i\}_{i=1}^n \quad \text{and} \quad \mathcal{T} = \{t_i\}_{i=1}^n$$

respectively denote the subspaces of *test* and *trial* functions. Define *Galerkin information* $N_\mathcal{S}$ based on $\mathcal{S}$ by

$$N_\mathcal{S}f = \begin{bmatrix} \langle f, s_1 \rangle_{L_2(\Omega)} \\ \vdots \\ \langle f, s_n \rangle_{L_2(\Omega)} \end{bmatrix} \qquad \forall f \in F.$$

Then the *generalized Galerkin method* $\phi_{\mathcal{S},\mathcal{T}}$ based on $\mathcal{S}$ and $\mathcal{T}$ is given by

$$\phi_{\mathcal{S},\mathcal{T}}(N_\mathcal{S}f) = u_{\mathcal{S},\mathcal{T}},$$

where for $f \in F$, $u_{\mathcal{S},\mathcal{T}} \in \mathcal{T}$ satisfies

$$B(u_{\mathcal{S},\mathcal{T}}, s) = \langle f, s \rangle_{L_2(\Omega)} \qquad \forall s \in \mathcal{S}.$$

When $\mathcal{T} = \mathcal{S}$, we write $\phi_\mathcal{S}$ instead of $\phi_{\mathcal{S},\mathcal{T}}$; the algorithm $\phi_\mathcal{S}$ is the *Galerkin method* based on the subspace $\mathcal{S}$. Of course, the FEM $\phi_{n,k}^{\mathrm{FE}}$ is a Galerkin method with $\mathcal{S} = \mathcal{S}_{n,k}$.

We first give a representation formula for the spline algorithm using Galerkin information $N_\mathcal{S}$ and for the generalized Galerkin algorithm based on the spaces $\mathcal{S}$ and $\mathcal{T}$ of test and trial functions.

LEMMA 5.4.5. *Suppose that the respective bases $\{s_i\}_{i=1}^n$ and $\{t_i\}_{i=1}^n$ of the subspaces S and $\mathcal{T}$ are chosen so that*

$$\langle S^* s_j, S^* s_i \rangle_{L_2(\Omega)} = \delta_{i,j} \qquad (1 \leq i, j \leq n) \tag{44}$$

and

$$B(t_j, s_i) = \delta_{i,j} \qquad (1 \leq i, j \leq n) \tag{45}$$

hold. Then the spline algorithm ϕ_S^s using N_S has the form

$$\phi_S^s(N_S f) = \sum_{j=1}^n \langle f, s_j \rangle_{L_2(\Omega)} \, SS^* s_j \,,$$

and the generalized Galerkin algorithm $\phi_{S,\mathcal{T}}$ has the form

$$\phi_{S,\mathcal{T}}(N_S f) = \sum_{j=1}^n \langle f, s_j \rangle_{L_2(\Omega)} \, t_j \,.$$

PROOF. Using (43), it is straightforward to check that $S^* S = (\ker N_S)^\perp$. Using the orthonormality of $\{S^* s_i\}_{i=1}^n$, we find that the orthogonal projector $P\colon H^r(\Omega) \to H^r(\Omega)$ is given by

$$Pf = \sum_{j=1}^n \langle f, s_j \rangle_{L_2(\Omega)} S^* s_j \qquad \forall f \in H^r(\Omega).$$

Since

$$\phi_S^s(N_S f) = SPf \qquad \forall f \in F,$$

we now have the formula for the spline algorithm. The formula for the generalized Galerkin algorithm follows immediately from its definition and from the biorthogonality of the bases for S and $\mathcal{T}$. $\qquad\square$

Using this lemma, we find that for any space of test functions, there is a unique space of trial functions for which the generalized Galerkin method is the spline algorithm.

THEOREM 5.4.5. *Let S and $\mathcal{T}$ be n-dimensional subspaces of $H_E^m(\Omega)$. Then the following are equivalent:*

(i) *The generalized Galerkin algorithm is the spline algorithm, i.e.*
$\phi_{S,\mathcal{T}} = \phi_S^s$.

(ii) $\mathcal{T} = SS^* S$. $\qquad\square$

PROOF. Suppose first that (i) holds. Choose bases $\{s_i\}_{i=1}^n$ for S and $\{t_i\}_{i=1}^n$ for $\mathcal{T}$ such that (44) and (45) hold. From the previous lemma, we have

$$t_i = \sum_{j=1}^n \langle S^* s_i, s_j \rangle_{L_2(\Omega)} t_j = \phi_{S,\mathcal{T}}(N_S S^* s_i)$$

$$= \phi_S^s(N_S S^* s_i) = \sum_{j=1}^n \langle S^* s_i, s_j \rangle_{L_2(\Omega)} S S^* s_j = S S^* s_i$$

for $1 \le i \le n$, which implies (ii).

Now suppose that (ii) holds. Choose a basis $\{s_i\}_{i=1}^n$ for S such that (44) holds. Let

$$t_i = S S^* s_i \qquad (1 \le i \le n). \tag{46}$$

Then (ii) and injectivity of SS^* show that $\{t_i\}_{i=1}^n$ is a basis for $\mathcal{T}$. Using (43) and (44), we easily find that (45) holds. Using (44), (45), and (46), along with the representation formulas of the lemma, we find that $\phi_{S,\mathcal{T}} = \phi_S^s$, establishing (i).   □

Hence, given any finite-dimensional subspace S of $H_E^m(\Omega)$, we see how to choose the unique subspace $\mathcal{T}$ of $H_E^m(\Omega)$ with $\dim \mathcal{T} = \dim S$ such that $\phi_S^s = \phi_{S,\mathcal{T}}$. However, the most natural choice of subspaces to pick $\mathcal{T} = S$, so that we get the standard Galerkin method ϕ_S. When is the standard Galerkin method ϕ_S the spline algorithm ϕ_S^s?

THEOREM 5.4.6. *Let S be an n-dimensional subspace of $H_E^m(\Omega)$. Then the following are equivalent:*

(i) *$\phi_S = \phi_S^s$.*
(ii) *$S = SS^* S$.*
(iii) *S is an eigenspace of SS^*.*
(iv) *$S = S\mathcal{F}$, where $\mathcal{F}$ is an n-dimensional invariant subspace (or, equivalently, an n-dimensional eigenspace) of $S^* S$.*

PROOF. The equivalence of the first two conditions is immediate from the previous theorem. Using a little simple linear algebra, the remainder of the proof follows immediately.   □

We now illustrate this theorem by two examples. In the first example, the Galerkin method is always a spline algorithm.

EXAMPLE.   Let $r = -m$. The S is the Riesz mapping, which is an isometric isomorphism of $H^{-m}(\Omega)$ (under the norm $\|S \cdot \|_B$, which is equivalent to $\| \cdot \|_{H^{-m}(\Omega)}$) onto $H_E^m(\Omega)$. Hence $SS^* = I$, the identity map on $H_E^m(\Omega)$, and so $S = SS^* S$ for *any* subspace S of $H_E^m(\Omega)$. So when $r = -m$, the Galerkin method is always the spline algorithm, no matter what the choice of S. Of course since $r = -m$, Theorem 5.4.1 shows that $\lim_{n\to\infty} r(n) \neq 0$, and so the problem is not convergent. ☐

In our second example, we exhibit a finite element method that is not a spline algorithm. This example is of particular interest because it gives an instance of an FEM that has optimal worst case error (to within a constant) yet is not a spline algorithm.

EXAMPLE. *The approximation problem.* We consider the L_2-approximation problem for H^1-functions on the unit interval $I = (0, 1)$. Choose $N = 1$, $m = 0$, $r = 1$, and let $S \colon H^1(I) \to L_2(I)$ be the canonical injection

$$Su = u \qquad \forall u \in H^1(I).$$

The variational form of this problem is to find, for $f \in H^1(I)$, a function $u = Sf \in L_2(I)$ for which

$$\langle u, v \rangle_{L_2(I)} = \langle f, v \rangle_{L_2(I)} \qquad \forall v \in L_2(I).$$

That is, the bilinear form B is merely the L_2-inner product. Of course, $u = f$.

We let $\mathcal{S}_{n,0}$ be an n-dimensional subspace of $L_2(I)$ consisting of piecewise constants, so that $k = 0$. Let

$$0 = x_0 < x_1 < \ldots < x_{n-1} < x_n = 1$$

be a partition of (I). Then $\mathcal{S}_{n,0}$ is the span of the functions $s_1, \ldots, s_n$, where for $1 \leq i \leq n$,

$$s_i(x) = \delta_{ij} \quad \forall x \in [x_{j-1}, x_j] \qquad (1 \leq j \leq n).$$

We next determine the space $SS^* \mathcal{S}_{n,0}$. Integrating by parts, we find that for any $s \in L_2(I)$, the function $w = SS^* s$ is the (weak) solution to the problem

$$-w'' + w = s \qquad \in I$$
$$w'(0) = w'(1) = 0,$$

so that

$$w(x) = \frac{\int_0^1 s(\xi) \cosh(1 - \xi)\, d\xi}{\sinh 1} \cosh x - \int_0^x s(\xi) \sinh(x - \xi)\, d\xi.$$

Hence, $SS^* \mathcal{S}_{n,0}$ is the span of the functions $w_1, \ldots, w_n$, where

$$w_i(x) = \frac{\sinh(1 - x_{i-1}) - \sinh(1 - x_i)}{\sinh 1} \cosh x$$
$$- \cosh(x - x_{i-1}) + \cosh(x - x_i) \qquad (1 \leq i \leq n).$$

Since none of the w_i is piecewise constant on I, we have $w_i \notin \mathcal{S}_{n,0}$. Hence $SS^* \mathcal{S}_{n,0} \neq \mathcal{S}_{n,0}$.

Using the previous theorem, we see that the FEM is not the spline algorithm. This is especially striking, since this FEM is an almost minimal error algorithm. ☐

These examples suggest the following

CONJECTURE 5.4.1. *Let* $r > -m$, *and let* $S_{n,k}$ *be a finite-element subspace of* $H_E^m(\Omega)$. *Then the FEM using* $S_{n,k}$ *is not a spline algorithm.* $\boxed{?}$

Clearly Theorem 5.4.4 implies that Conjecture 5.4.1 holds when $k < 2m - 1 + r$. Hence, it remains unproven only for the case $k \geq 2m - 1 + r$.

Notes and Remarks

NR 5.4:1 The results in this section are based on Werschulz (1982, 1986, 1987b).

NR 5.4:2 The results of Babuška and Aziz (1972) and Jerome (1968) concerning n-widths are not given for the Gelfand n-width, but are given only for the *Kolmogorov n-width*, which is defined to be

$$d_n(A, G) = \inf_{G_n \in \mathcal{G}_n} \sup_{a \in A} \inf_{g \in G_n} \|a - g\|,$$

where $G_n \in \mathcal{G}_n$ iff G_n is an n-dimensional subspace of G. However, since F is the unit ball of a Hilbert space and G is a Hilbert space, $d^n(SF, G) = d_n(SF, G)$, see Pinkus (1985).

NR 5.4:3 As pointed out in the text, we cannot generally triangulate the C^∞ region Ω using simplices or rectangles. This means that the inclusion $S_{n,k} \subseteq H_E^m(\Omega)$ will not hold in general. As is often done, we will ignore this source of error. If necessary, this error may be removed by using isoparametric elements as in Ciarlet and Raviart (1972).

NR 5.4:4 As an example of a basis with small supports, consider the model problem of Chapter 2. The basis whose ith element is a piecewise linear function whose value at the jth nodal point is $\delta_{i,j}$ is a basis with small supports, since the support of any such basis element is at most two elements. Of course, the corresponding dual basis consists of evaluations of functions at the nodal points, i.e. the degrees of freedom are function evaluations.

NR 5.4:5 Strang and Fix (1973b) show that the error estimate (15) cannot be improved in general. That is, they exhibit a particular family of finite element spaces, proving a lower bound on $\inf_{s \in S_{n,k}} \|u - s\|_{H^l(\Omega)}$ which is of the same form as the upper bound in (15). However, their result does not rule out the possibility that there might exist a special family of finite element spaces for which the upper bound can be improved. This explains why it is necessary for us to prove Theorem 5.4.2.

NR 5.4:6 Based on the example given in this section, it is reasonable to conjecture that for any regularly elliptic boundary value problem, (36) will hold for a reasonably large value of ε_0. Such a result will be difficult to establish. One problem is that being "sufficiently large" is a subjective criterion. However, there is an even greater source of difficulty, in that to determine ε_0, we must be able to compute explicit values for the Θ-constants appearing in this section. These values are not generally available for the general problem. This explains why we were content to analyze a simple model problem.

5.5   The normed case with standard information

We now analyze the normed case, under the condition that only standard information is available. That is, we assume that the $\Lambda = \Lambda^{\mathrm{std}}$, and the only permissible information functionals are the evaluation of a function (or one of its derivatives) at a point. Even though we allow the evaluation of derivatives as well as of function values, there is no advantage in doing this, since we find that nth almost optimal standard information consists of n evaluations of the problem element itself. Furthermore, we show that there is a loss when Λ changes from Λ' to Λ^{std}. We find that if $0 \leq l \leq m$, then the nth minimal radius of standard information is $\Theta(n^{-r/d})$ as $n \to \infty$, whereas the nth minimal radius of arbitrary linear information was $\Theta(n^{-(r+2m-l)/d})$. Similarly, the ε-complexity changes from $\Theta(\varepsilon^{-d/(r+2m-l)})$ as $\varepsilon \to 0$ to $\Theta(\varepsilon^{-d/r})$ when going from arbitrary linear information to standard information.

Note that the nth minimal radius of standard information is independent of l, as is the ε-complexity. Hence using lower-order Sobolev norms to measure the error does not improve the minimal error or complexity when only standard information is permissible. This should be contrasted with what happens if we are allowed to use arbitrary linear information, where the minimal radius and complexity decrease if error is measured in a lower Sobolev norm.

Since our problem is linear, there is no loss of generality in considering only nonadaptive information in what follows. We first show a lower bound on the nth minimal radius of (nonadaptive) standard information, remembering that we are permitting the evaluation of a function or one of its derivatives at a point. Recall that we are measuring error in the Sobolev $\|\cdot\|_{H^l(\Omega)}$-norm, our problem is a $2m$th-order elliptic problem, and that the unit ball of $H^r(\Omega)$ is our set of problem elements.

THEOREM 5.5.1. *There exists a positive constant C, independent of n and l, such that*

$$r(n, \Lambda^{\mathrm{std}}) \geq C n^{-r/d} \qquad \forall\, n \in \mathbb{P}.$$

PROOF. Let N be standard information of cardinality n. Then there exist $x_1, \dots, x_n \in \overline{\Omega}$ and multi-indices $\alpha^1, \dots, \alpha^n \in \mathbb{N}^d$ such that

$$Nf = \begin{bmatrix} D^{\alpha^1} f(x_1) \\ \vdots \\ D^{\alpha^n} f(x_n) \end{bmatrix} \qquad \forall\, f \in F.$$

Let Ω^0 be a C^∞ region whose closure is a subset of int Ω. By Poincaré's inequality, there exists a constant $\kappa_1 \geq 1$ such that

$$\|z\|_{H^r(\Omega^0)} \leq \kappa_1 |z|_{H^r(\Omega^0)} \qquad \forall\, z \in C_0^\infty(\Omega^0).$$

Let $\tilde{n}$ denote the number of sample points among $x_1, \dots, x_n$ which lie in Ω^0. Without loss of generality, we may assume that $\tilde{n} \geq 1$ and $x_1, \dots, x_{\tilde{n}} \in \Omega^0$. From pp. 301–304 of Bakhvalov (1977), there exists a positive constant κ_2

(independent of n and $\tilde{n}$) and a nonnegative function $z \in C_0^\infty(\Omega^0)$ such that

$$|(D^\alpha z)(x)| \leq \kappa_1^{-2}\binom{r+d-1}{r}^{-1}(\text{vol}\,\Omega^0)^{-1} \qquad \forall\, x \in \Omega, |\alpha| = r,$$

$$(D^\alpha z)(x_1) = \cdots = (D^\alpha z)(x_l) = 0 \qquad \forall\, \alpha \in \mathbb{N}^d, \text{ and}$$

$$\int_{\Omega^0} z(x)\,dx \geq \kappa_2 \tilde{n}^{-r/d}.$$

We extend z from Ω^0 to Ω by letting z be zero on the complement of Ω^0. Since z and all its derivatives vanish at $x_1, \ldots, x_l$ and

$$\|z\|_{H^r(\Omega)}^2 = \|z\|_{H^r(\Omega^0)}^2 \leq \kappa_1^2 |z|_{r,\Omega^0}^2 = \kappa_1^2 \sum_{|\alpha|=r} \int_{\Omega^0} |(D^\alpha v)(x)|^2\,dx$$

$$\leq \kappa_1^2 \,\text{vol}(\Omega^0)\binom{r+d-1}{r} \sup_{x\in\Omega^0} |(D^\alpha v)(x)| \leq 1,$$

we have $z \in \ker N \cap F$.

Choose a nonnegative function $v \in C_0^\infty(\Omega)$ such that $v \equiv 1$ on Ω^0. Since $v, z \in C_0^\infty(\Omega)$, we find that $LSz = z$ in Ω. Since L is self-adjoint and of $2m$th order, we see that there is a positive constant κ_3, depending only on Ω and the coefficients of L, such that

$$\langle z, v\rangle_{L_2(\Omega)} = \langle LSz, v\rangle_{L_2(\Omega)} = \langle Sz, Lv\rangle_{L_2(\Omega)} \leq \|Sz\|_{L_2(\Omega)}\|Lv\|_{L_2(\Omega)}$$

$$\leq \kappa_3 \|Sz\|_{L_2(\Omega)}\|v\|_{H^{2m}(\Omega)}.$$

However, since $\tilde{n} \leq n$ and $v \equiv 0$ outside of Ω, we have

$$\langle z, v\rangle_{L_2(\Omega)} = \int_{\Omega^0} z(x)\,dx \geq \kappa_2 \tilde{n}^{-r/d} \geq \kappa_2 n^{-r/d}.$$

Since $l \geq 0$, the last two inequalities imply that

$$\|Sz\|_{H^l(\Omega)} \geq \|Sz\|_{L_2(\Omega)} \geq \frac{\kappa_2}{\kappa_3\|v\|_{H^{2m}(\Omega)}} n^{-r/d}.$$

From Lemma 4.4.2.1, we thus find that there is a positive constant C such that

$$r(N) = \sup_{h \in \ker N \cap F} \|Sh\|_{H^l(\Omega)} \geq C n^{-r/d}.$$

Taking the infimum over all standard information of cardinality at most n, the result follows. $\qquad\square$

We now describe an almost minimal error algorithm using standard information consisting of function evaluations only, with no derivatives. Before proceeding further, we will require that $r > \frac{1}{2}d$. By the Sobolev embedding theorem, this means that $H^r(\Omega)$ is embedded in $C(\overline{\Omega})$. Hence for any $x \in \overline{\Omega}$, the mapping $f \mapsto f(x)$ is a bounded linear functional on $H^r(\Omega)$. Thus, any standard information operator using only function values is a well-defined, bounded linear operator on our class of problem elements.

Since the FEM of sufficiently high degree is optimal when $\Lambda = \Lambda'$, it is natural to try to modify the FEM so that it uses standard information and is optimal when $\Lambda = \Lambda^{\text{std}}$. Our modified FEM will be a finite element method with quadrature, in which all the integrals appearing in (14) in the previous section are approximated by quadrature rules. Note that we will be replacing the integrals appearing in the Gram matrix G, even though these do not depend on the problem element and can be precomputed. For further discussion, see **NR 5.3:1**.

Recall that $\hat{K}$ is our reference element. Choosing weights $\hat{\omega}_1, \ldots, \hat{\omega}_J \in \mathbb{R}$ and nodes $\hat{b}_1 \ldots, \hat{b}_J \in \hat{K}$, we define a quadrature rule

$$\hat{I}\hat{v} = \sum_{j=1}^{J} \hat{\omega}_j \hat{v}(\hat{b}_j) \qquad \forall \hat{v} \in C\left(\overline{\hat{K}}\right).$$

Denote the *reference element quadrature error* by

$$\hat{E}(\hat{v}) = \int_{\hat{K}} \hat{v} - \hat{I}\hat{v} \qquad \forall \hat{v} \in C\left(\overline{\hat{K}}\right).$$

We say that $\hat{I}$ is *exact* of degree q if

$$\hat{E}(\hat{p}) = 0 \qquad \forall \hat{p} \in P_q(\hat{K}).$$

Next, we move from the reference element to an element appearing in a triangulation. For $\tilde{n} \in \mathbb{P}$, we define weights and nodes over the element $K \in \mathcal{T}_{\tilde{n}}$ by

$$\omega_{j,K} = \det B_K \cdot \hat{\omega}_j \quad \text{and} \quad b_{j,K} = F_K(\hat{b}_j) \qquad (1 \le j \le J),$$

where $F_K \hat{x} \equiv B_K \hat{x} + b_K$ is the affine bijection of $\hat{K}$ onto K. Then the "reference" quadrature rule $\hat{I}$ over $\hat{K}$ induces a "local" quadrature rule I_K over K, which is defined by

$$I_K v = \sum_{j=1}^{J} \omega_{j,K} v(b_{j,K}).$$

Denote the *local quadrature error* over K by

$$E_K(v) = \int_K v - I_K v \qquad \forall\, v \in C(\overline{K}).$$

We say that I_K is *exact* of degree q if

$$E_K(p) = 0 \qquad \forall\, p \in P_q(K).$$

Note that I_K is exact of degree q over K iff $\hat{I}$ is exact of degree q over the reference element.

It is useful to let

$$\mathcal{N}_{\tilde{n}} = \bigcup_{K \in \mathcal{T}_{\tilde{n}}} \bigcup_{j=1}^{J} \{b_{j,K}\}$$

denote the set of all nodes in all the elements belonging to $\mathcal{T}_{\tilde{n}}$. Note that this is not a disjoint union, since a node on the boundary of one element may be on the boundary of another element.

We define a new bilinear form $B_{\tilde{n}}$ on $\mathcal{S}_{\tilde{n},k}$ by

$$B_{\tilde{n}}(v,w) = \sum_{|\alpha|,|\beta| \le m} \sum_{K \in \mathcal{T}_{\tilde{n}}} \sum_{j=1}^{J} \omega_{j,K} \cdot (a_{\alpha\beta} D^{\alpha} v D^{\beta} w)(b_{j,K}) \qquad \forall\, v, w \in \mathcal{S}_{\tilde{n},k}.$$

For $f \in F$, we define a linear functional $f_{\tilde{n}}$ on $\mathcal{S}_{\tilde{n},k}$ by

$$f_{\tilde{n}}(v) = \sum_{K \in \mathcal{T}_n} \sum_{j=1}^{J} \omega_{j,K} \cdot (fv)(b_{j,K}) \qquad \forall\, v \in \mathcal{S}_{\tilde{n},k}.$$

We are now ready to define our finite element method with quadrature. Let $n \in \mathbb{P}$. Choose $\tilde{n}$ to be the largest integer such that

$$\operatorname{card} \mathcal{N}_{\tilde{n}} \le n.$$

Note that $\tilde{n} = \Theta(n)$ as $n \to \infty$. Let $x_1, \ldots, x_{\tilde{n}}$ denote the elements of $\mathcal{N}_{\tilde{n}}$, and choose points $x_{\tilde{n}+1}, \ldots, x_n$ not belonging to $\mathcal{N}_{\tilde{n}}$. For $f \in F$, define information $N_{n,k}^{Q}$ to be

$$N_{n,k}^{Q} f = \begin{bmatrix} f(x_1) \\ \vdots \\ f(x_n) \end{bmatrix} \qquad \forall\, f \in F.$$

Then $N_{n,k}^{Q}$ is standard information of cardinality n. Suppose we can find $\tilde{u}_{\tilde{n},k} \in \mathcal{S}_{\tilde{n},k}$ satisfying

$$B_{\tilde{n}}(\tilde{u}_{\tilde{n},k}, s) = f_{\tilde{n}}(s) \qquad \forall\, s \in \mathcal{S}_{\tilde{n},k}. \tag{1}$$

Then $\tilde{u}_{\tilde{n},k}$ depends on f only through $N_{n,k}^{Q}f$, and so we write $\tilde{u}_{\tilde{n}} = \phi_{n,k}^{Q}(N_{n,k}^{Q}f)$. We say that $\phi_{n,k}^{Q}$ is the *finite element method with quadrature* (FEMQ).

As with the pure FEM, we can compute the FEMQ $\phi_{n,k}^{Q}(N_{n,k}^{Q}f)$ using n information operations and $2n - 1$ arithmetic operations by using precomputation. However, it is more common to reduce the problem (1) to a system of linear equations. Define

$$\tilde{G} = [\tilde{g}_{i,j}]_{1 \le i,j \le \tilde{n}} \qquad \text{and} \qquad \tilde{b} = \begin{bmatrix} f_{\tilde{n}}(s_1) \\ \vdots \\ f_{\tilde{n}}(s_{\tilde{n}}) \end{bmatrix},$$

where

$$\tilde{g}_{i,j} = B_{\tilde{n}}(s_j, s_i) \qquad (1 \le i, j \le \tilde{n}).$$

Let

$$\tilde{a} = \begin{bmatrix} \tilde{\alpha}_1 \\ \vdots \\ \tilde{\alpha}_{\tilde{n}} \end{bmatrix}$$

be the solution to the problem

$$\tilde{G}\tilde{a} = \tilde{b}.$$

Then

$$\phi_{n,k}^{Q}(N_{n,k}^{Q}f) = \tilde{u}_{\tilde{n},k} = \sum_{j=1}^{\tilde{n}} \tilde{\alpha}_j s_j.$$

In the rest of this section, we define

$$\nu = \min\{k + 1, r\}. \tag{2}$$

We assume that

(i) The smoothness r of the problem elements F satisfies $r \ge 1$ (as well as our previous requirement $r > \frac{1}{2}d$).

(ii) The degree k of the finite element subspaces $S_{\tilde{n},k}$ satisfies $k > \frac{1}{2}d - 1$.

(iii) $\hat{I}$ is exact of degree $2k + \nu - 1$ over the reference element $\hat{K}$.

Note that if $d \le 2$, then condition (ii) is the trivial inequality $k \ge 0$, while for $d = 3$, condition (ii) holds provided $k \ge 1$, which *must* be true for $m \ge 1$. Hence condition (ii) will often hold be true anyway. Condition (iii) means that our quadrature scheme must be accurate enough to guarantee that if the coefficients of L are piecewise polynomials of degree at most $\nu - 1$, then

$$B_{\tilde{n}}(v, w) = B(v, w) \qquad \forall v, w \in S_{\tilde{n},k}.$$

Before analyzing the FEMQ, we need to prove a preliminary result.

LEMMA 5.5.1. *Let $K \in \mathcal{T}_{\tilde{n}}$. There exists a positive constant C, independent of n and K, such that*

$$|E_K(avw)| \leq Ch_K^\nu \|a\|_{W^{r,\infty}(K)} \|v\|_{L_2(K)} \|w\|_{L_2(K)}$$
$$\forall a \in W^{r,\infty}(K), \forall v, w \in P_k(K)$$

and

$$|E_K(fv)| \leq Ch_K^\nu \|f\|_{H^r(K)} \|v\|_{H^m(K)} \qquad \forall f \in H^r(K), v \in P_k(K).$$

PROOF. In what follows, we assume without essential loss of generality that r is a positive integer, the case for real $r \geq 1$ following by interpolation between $\lfloor r \rfloor$ and $\lceil r \rceil$.

To establish the first inequality, let $a \in W^{r,\infty}(K)$ and $v, w \in P_k(K)$. Then

$$|E_K(avw)| = |\det B_K| \, |\hat{E}(\hat{a}\hat{v}\hat{w})|,$$

where $\hat{a} \in W^{r,\infty}(\hat{K})$ and $\hat{v}, \hat{w} \in P_k(\hat{K})$. We now define a functional $\lambda_{\hat{v},\hat{w}} : W^{\nu,\infty}(\hat{K}) \to \mathbb{R}$ by

$$\lambda_{\hat{v},\hat{w}}(\hat{a}) = \hat{E}(\hat{a}\hat{v}\hat{w}) \qquad \forall \hat{a} \in W^{\nu,\infty}(\hat{K}).$$

Since $k \geq 0$ and $r > \frac{1}{2}d > 0$, we find that (2) implies that $\nu > 0$, and so $W^{\nu,\infty}(\hat{K})$ is embedded in $C\left(\overline{\hat{K}}\right)$. We claim that $\lambda_{\hat{v},\hat{w}}$ is a bounded linear functional on $W^{\nu,\infty}(\hat{K})$, with

$$\|\lambda_{\hat{v},\hat{w}}\|^* \leq C\|\hat{v}\|_{L_2(\hat{K})} \|\hat{w}\|_{L_2(\hat{K})}.$$

Indeed, since $\nu \geq 0$ and all norms are equivalent on the finite dimensional space $P_k(\hat{K})$, we have

$$|\lambda_{\hat{v},\hat{w}}(\hat{a})| \leq C\|\hat{a}\hat{v}\hat{w}\|_{L_\infty(\hat{K})} \leq C\|\hat{a}\|_{L_\infty(\hat{K})} \|\hat{v}\|_{L_\infty(\hat{K})} \|\hat{w}\|_{L_\infty(\hat{K})}$$
$$\leq C\|\hat{a}\|_{W^{\nu,\infty}(\hat{K})} \|\hat{v}\|_{L_2(\hat{K})} \|\hat{w}\|_{L_2(\hat{K})},$$

as claimed. Since $\hat{E} = 0$ on $P_{2k+\nu-1}(\hat{K})$, we see that $\lambda_{\hat{v},\hat{w}} = 0$ on $P_{\nu-1}(\hat{K})$. Hence the Bramble-Hilbert Lemma (Lemma A.2.3.2) yields that

$$|\lambda_{\hat{v},\hat{w}}(\hat{a})| \leq C|\hat{a}|_{W^{\nu,\infty}(\hat{K})} \|\hat{v}\|_{L_2(\hat{K})} \|\hat{w}\|_{L_2(\hat{K})}.$$

Hence

$$|E_K(avw)| = |\det B_K| \, |\hat{E}(\hat{a}\hat{v}\hat{w})|$$
$$\leq C|\det B_K| \, |\hat{a}|_{W^{\nu,\infty}(\hat{K})} \|\hat{v}\|_{L_2(\hat{K})} \|\hat{w}\|_{L_2(\hat{K})}.$$

Now Lemma A.2.3.3 implies that

$$|\hat{a}|_{W^{\nu,\infty}(\hat{K})} \le Ch_K^\nu |a|_{W_K^\nu,\infty(K)} \le Ch_K^\nu \|a\|_{W^{\nu,\infty}(K)} \le Ch_K^\nu \|a\|_{W^{r,\infty}(K)},$$

$$\|\hat{v}\|_{L_2(\hat{K})} \le C|\det B_K|^{-1/2} \|v\|_{L_2(K)},$$

$$\|\hat{w}\|_{L_2(\hat{K})} \le C|\det B_K|^{-1/2} \|w\|_{L_2(K)}.$$

Substituting these into the previous inequality, the first inequality follows immediately.

To prove the second inequality, let $f \in H^r(K)$ and $v \in P_k(K)$. Then there exist $\hat{f} \in H^r(\hat{K})$ and $\hat{v} \in P_k(\hat{K})$ such that

$$|E_K(fv)| = |\det B_K| |\hat{E}(\hat{f}\hat{v})|.$$

For $\hat{v} \in P_k(\hat{K})$, define $\lambda_{\hat{v}} \colon H^\nu(\hat{K}) \to \mathbb{R}$ by

$$\lambda_{\hat{v}}(\hat{f}) = \hat{E}(\hat{f}\hat{v}) \qquad \forall \hat{f} \in H^\nu(\hat{K}).$$

Since $r > \frac{1}{2}d$ and $k + 1 > \frac{1}{2}d$, we see that $\nu > \frac{1}{2}d$, and so $H^\nu(\hat{K})$ is embedded in $C\left(\overline{\hat{K}}\right)$. Thus

$$|\lambda_{\hat{v}}(\hat{f})| \le C\|\hat{f}\hat{v}\|_{L_\infty(\hat{K})} \le C\|\hat{f}\|_{L_\infty(\hat{K})} \|\hat{v}\|_{L_\infty(\hat{K})}$$
$$\le C\|\hat{f}\|_{H^\nu(\hat{K})} \|\hat{v}\|_{H^m(\hat{K})},$$

which implies that $\lambda_{\hat{v}}$ is a bounded linear functional on $H^\nu(\hat{K})$, with norm bounded by

$$\|\lambda_{\hat{v}}\|^* \le C\|\hat{v}\|_{H^m(\hat{K})}.$$

Since $\hat{E} = 0$ on $P_{k+\nu-1}$, we see that $\lambda_{\hat{v}} = 0$ on $P_{\nu-1}(\hat{K})$. Hence the Bramble-Hilbert Lemma (Lemma A.2.3.2) implies that

$$|\lambda_{\hat{v}}(\hat{f})| \le C|\hat{f}|_{H^\nu(\hat{K})} \|\hat{v}\|_{H^m(K)}.$$

Hence

$$|E_K(fv)| = |\det B_K| |\hat{E}(\hat{f}\hat{v})| \le C|\det B_K| |f|_{H^\nu(\hat{K})} \|\hat{v}\|_{H^m(K)}.$$

Since Lemma A.2.3.3 implies that

$$|\hat{f}|_{H^\nu(\hat{K})} \le Ch_K^\nu |\det B_K|^{-1/2} |f|_{H^\nu(K)} \le Ch_K^\nu |\det B_K|^{-1/2} \|f\|_{H^r(K)}$$

and

$$\|\hat{v}\|_{H^m(\hat{K})} \le C|\det B_K|^{-1/2} \|v\|_{H^m(K)},$$

the second inequality follows immediately. $\qquad\square$

We now show that the FEMQ is well-defined, and give tight bounds on its error.

THEOREM 5.5.2. *There exists $n_1 \in \mathbb{P}$ such that $\phi_{n,k}^Q$ is well-defined for $n \geq n_1$. Furthermore,*

$$e(\phi_{n,k}^Q, N_{n,k}^Q) = \Theta(n^{-\mu/d}) \qquad \text{as } n \to \infty,$$

where

$$\mu = \min\{k+1-l, r\}.$$

PROOF. First, we show that there exists $n_1 \in \mathbb{P}$ such that the FEMQ is well-defined for all $n \geq n_1$. In what follows, let us write

$$A(r) = \sum_{|\alpha|, |\beta| \leq m} \|a_{\alpha,\beta}\|_{W^{r,\infty}(\Omega)}.$$

For any $\tilde{n} \in \mathbb{P}$, we may sum the result of the previous lemma over all $K \in \mathcal{T}_{\tilde{n}}$, finding that

$$|B(v,w) - B_n(v,w)| \leq Cn^{-\nu/d}A(r)\|v\|_{H^m(\Omega)}\|w\|_{H^m(\Omega)} \qquad (3)$$
$$\forall\, v, w \in \mathcal{S}_{\tilde{n},k}$$

and

$$|\langle f, w\rangle_{L_2(\Omega)} - f_n(w)| \leq Cn^{-\nu/d}\|f\|_{H^r(\Omega)}\|w\|_{H^m(\Omega)} \qquad (4)$$
$$\forall f \in H^r(\Omega), w \in \mathcal{S}_{\tilde{n},k}.$$

(Here we have used the inequality $\|\cdot\|_{L_2(\Omega)} \leq \|\cdot\|_{H^m(\Omega)}.$)

By Strang's Lemma (Lemma A.3.2), we see that (3) implies that there exists $n_1 \in \mathbb{P}$ such that for any $f \in F$ and any $n \geq n_1$, there exists a unique $u_{\tilde{n},k} \in \mathcal{S}_{\tilde{n},k}$ such that (1) holds. Hence the FEMQ $\phi_{n,k}^Q$ is well-defined for $n \geq n_1$.

We next prove the lower bound

$$e(\phi_{n,k}^Q, N_{n,k}^Q) = \Omega(n^{-\mu/d}) \qquad \text{as } n \to \infty.$$

Since $\phi_{n,k}^Q$ uses information $N_{n,k}^Q$ of cardinality n, Theorem 5.5.1 implies that

$$e(\phi_{n,k}^Q, N_{n,k}^Q) \geq r(n, \Lambda^{\text{std}}) = \Omega(n^{-r/d}).$$

Define f^* as in the proof of Theorem 5.4.2. Since $\phi_{n,k}^Q(N_{n,k}^Q f^*) \in \mathcal{S}_{\tilde{n},k}$, we see that

$$e(\phi_{n,k}^Q, N_{n,k}^Q) \geq \|Sf^* - \phi_{n,k}^Q(N_{n,k}^Q f^*)\|_{H^l(\Omega)}$$
$$\geq \inf_{s \in \mathcal{S}_{\tilde{n},k}} |Sf - s|_{H^l(\Omega)} \geq Cn^{-(k+1-l)/d}.$$

The desired lower bound for the FEMQ follows immediately from these last two lower bounds.

It remains to establish the upper bound

$$e(\phi_{n,k}^Q, N_{n,k}^Q) \leq Cn^{-\mu/d} \qquad \forall n \geq n_1. \tag{5}$$

We first suppose that $l = m$. For $n \geq n_1$ and $f \in F$, Lemma 5.4.3 and Theorems 5.2.1 and 5.2.2 yield that

$$
\begin{aligned}
\|Sf - \Pi_{\tilde{n}} Sf\|_{H^m(\Omega)} &\leq Cn^{-\min\{k+1-m, m+r\}/d} \|Sf\|_{H^{r+2m}(\Omega)} \\
&\leq Cn^{-\min\{k+1-m, m+r\}/d} \|f\|_{H^r(\Omega)} \\
&\leq Cn^{-\min\{k+1-m, r\}/d} \|f\|_{H^r(\Omega)}.
\end{aligned} \tag{6}
$$

Using Lemma 5.4.3 and Theorem 5.2.2, we have

$$
\begin{aligned}
\|\Pi_{\tilde{n}} Sf\|_{H^m(\Omega)} &\leq \|Sf\|_{H^m(\Omega)} + \|Sf - \Pi_{\tilde{n}} Sf\|_{H^m(\Omega)} \\
&\leq C\|Sf\|_{H^m(\Omega)} \leq C\|f\|_{H^{-m}(\Omega)} \\
&\leq C\|f\|_{H^r(\Omega)},
\end{aligned}
$$

and so (3) implies that

$$
\begin{aligned}
|B(\Pi_{\tilde{n}} Sf, w) - B_n(\Pi_{\tilde{n}} Sf, w)| &\leq CA(r)n^{-\nu/d}\|f\|_{H^r(\Omega)}\|w\|_{H^m(\Omega)} \\
&\leq CA(r)n^{-\min\{k+1-m, r\}/d} \\
&\quad \times \|f\|_{H^r(\Omega)}\|w\|_{H^m(\Omega)}
\end{aligned} \tag{7}
$$

for any $w \in S_{\tilde{n},k}$. Using (4), (6), and (7) along with Strang's Lemma (Lemma A.3.2), we see that there is a positive constant C, such that for $n \geq n_1$ and $f \in F$,

$$\|Sf - \phi_{n,k}^Q(N_{n,k}^Q f)\|_{H^m(\Omega)} \leq CA(r)n^{-\min\{k+1-m, r\}/d}\|f\|_{H^r(\Omega)}. \tag{8}$$

Taking the supremum over all $f \in F$, we find that (5) holds for the case $l = m$.

We now suppose that $l < m$. First, we bound the $L_2(\Omega)$-error. Let $f \in F$ and let $n \geq n_1$. Write $\tilde{u}_{\tilde{n},k} = \phi_{n,k}^Q(N_{n,k}^Q f)$. For $g \in L_2(\Omega)$, let $\tilde{S}g \in H_E^{2m}(\Omega)$ satisfy

$$B(v, \tilde{S}g) = \langle g, v \rangle_{L_2(\Omega)} \qquad \forall v \in H_E^m(\Omega).$$

Since B is symmetric, $\tilde{S} = S$ is well defined. Moreover, Theorem 5.2.1 yields that

$$\|\tilde{S}g\|_{H^{2m}(\Omega)} \leq C\|g\|_{L_2(\Omega)} \tag{9}$$

for some positive constant C independent of g. Since $l < m$, we have $k + 1 - m \geq m$. Hence

$$\|\tilde{S}g - \Pi_{\tilde{n}}\tilde{S}g\|_{H^m(\Omega)} \leq Cn^{-m/d}\|\tilde{S}g\|_{H^{2m}(\Omega)} \leq Cn^{-m/d}\|g\|_{L_2(\Omega)}. \qquad (10)$$

Using (8) and Theorem 5.2.2 , we have

$$\begin{aligned}\|\tilde{u}_{\tilde{n},k}\|_{H^m(\Omega)} &\leq \|u - \tilde{u}_{\tilde{n},k}\|_{H^m(\Omega)} + \|u\|_{H^m(\Omega)}\\&\leq C\left(\|f\|_{H^r(\Omega)} + \|f\|_{H^{-m}(\Omega)}\right) \leq \|f\|_{H^r(\Omega)}.\end{aligned} \qquad (11)$$

But (9) and (10) imply that

$$\begin{aligned}\|\Pi_{\tilde{n}}\tilde{S}g\|_{H^m(\Omega)} &\leq \|\tilde{S}g\|_{H^m(\Omega)} + \|\tilde{S}g - \Pi_{\tilde{n}}\tilde{S}g\|_{H^m(\Omega)}\\&\leq C\|\tilde{S}g\|_{H^m(\Omega)} \leq C\|\tilde{S}g\|_{H^{2m}(\Omega)} \leq C\|g\|_{L_2(\Omega)}.\end{aligned} \qquad (12)$$

So (3), (11), and (12) imply that

$$|B(\tilde{u}_{\tilde{n},k}, \Pi_{\tilde{n}}\tilde{S}g) - B_n(\tilde{u}_{\tilde{n},k}, \Pi_{\tilde{n}}\tilde{S}g)| \leq CA(r)n^{-\nu/d}\|f\|_{H^r(\Omega)}\|g\|_{L_2(\Omega)}. \qquad (13)$$

Using (4) with $w = \Pi_{\tilde{n}}\tilde{S}g$, (10) and (13), along with the Aubin-Ciarlet-Nitsche Duality Argument (Lemma A.3.3), we find that

$$\|Sf - \phi_{n,k}^Q(N_{n,k}^Q f)\|_{L_2(\Omega)} \leq CA(r)n^{-\zeta/d}\|f\|_{H^r(\Omega)},$$

where

$$\begin{aligned}\zeta &= \min\{\min\{k + 1 - m, r\} + m, \min\{k + 1, r\}\}\\&= \min\{\min\{k + 1, r + m\}, \min\{k + 1, r\}\}\\&= \min\{k + 1, r\}.\end{aligned}$$

Thus

$$\|Sf - \phi_{n,k}^Q(N_{n,k}^Q f)\|_{L_2(\Omega)} \leq CA(r)n^{-\min\{k+1,r\}/d}\|f\|_{H^r(\Omega)}, \qquad (14)$$

our desired bound on the $L_2(\Omega)$-error.

Now since $0 \leq l \leq m$, we may use interpolation of Hilbert spaces, (8), and (14) to find that for $n \geq n_1$ and for $f \in F$,

$$\begin{aligned}\|Sf - \phi_{n,k}^Q(N_{n,k}^Q f)\|_{H^l(\Omega)} &\leq C\,\|Sf - \phi_{n,k}^Q(N_{n,k}^Q f)\|_{L_2(\Omega)}^{1-l/m}\\&\quad \times \|Sf - \phi_{n,k}^Q(N_{n,k}^Q f)\|_{H^m(\Omega)}^{l/m}\\&\leq CA(r)n^{-\min\{k+1-l,r\}/d}\|f\|_{H^r(\Omega)}.\end{aligned}$$

Taking the supremum over all $f \in F$, we finally get our desired result.   □

Comparing the previous two theorems, we immediately find

COROLLARY 5.5.1. *The FEMQ of degree k is an almost minimal error algorithm iff $k \geq r + l - 1$.* ☐

For $\varepsilon > 0$, we now wish to find comp(ε), the ε-complexity of our problem, which we will compare to

$$\text{cost}_k^Q(\varepsilon) = \inf\{\,\text{cost}(\phi_{n,k}^Q, N_{n,k}^Q) : e(\phi_{n,k}^Q, N_{n,k}^Q) \leq \varepsilon\,\},$$

the minimal cost of computing an ε-approximation with the FEMQ. Using the last two theorems, along with the results in Chapter 4, we have

THEOREM 5.5.3.

(i) *The ε-complexity is given by*

$$\text{comp}(\varepsilon) = \Theta(\varepsilon^{-d/r}) \qquad \text{as } \varepsilon \to 0.$$

(ii) *Let $\mu = \min\{k + 1 - l, r\}$. Then*

$$\text{cost}_k^Q(\varepsilon) = \Theta(\varepsilon^{-d/\mu}) \qquad \text{as } \varepsilon \to 0.$$

Hence, if $k \geq r + l - 1$, then the FEMQ of degree k is an almost optimal complexity algorithm, i.e.

$$\text{cost}_k^Q(\varepsilon) = \Theta\big(\text{comp}(\varepsilon)\big) = \Theta(\varepsilon^{-d/r}) \qquad \text{as } \varepsilon \to 0. \qquad ☐$$

Hence we have shown that standard information is not as powerful as arbitrary linear information when solving elliptic PDE with $m > 0$ in the normed case.

As with the pure finite element method, we pay a large asymptotic penalty for using an FEMQ of too low a degree. That is, for $k < r + l - 1$,

$$\frac{\text{cost}_k^Q(\varepsilon)}{\text{comp}(\varepsilon)} = \Theta(\varepsilon^{-\delta d}) \qquad \text{as } \varepsilon \to 0,$$

where

$$\delta = \frac{1}{k + 1 - l} - \frac{1}{r} > 0.$$

Thus,

$$\lim_{\varepsilon \to 0} \frac{\text{cost}_k^Q(\varepsilon)}{\text{comp}(\varepsilon)} = \infty.$$

Notes and Remarks

NR 5.5:1 The material in this section is based on ideas from Section 1 in Chapter 4 of Ciarlet (1978) and on Section 5 in Chapter 5 of IBC.

NR 5.5:2 Although we have only considered the dependence of the solution $u = Sf$ on f, the solution u depends as well on the coefficients of L. If these coefficients are at all complicated, then it will generally not be possible to compute the integrals of the original Gram matrix. We briefly sketch how to deal with this additional source of uncertainty.

For multi-indices α, β with $|\alpha|, |\beta| \leq m$, we let $A_{\alpha,\beta}$ denote a class of permissible coefficients $a_{\alpha,\beta}$. Since we will be interested in problem elements of smoothness r, we assume that

$$A_{\alpha,\beta} \subset BW^{r,\infty}(\Omega) \cap C^{\infty}(\overline{\Omega}) \qquad \forall \, |\alpha|, |\beta| \leq m$$

Let $\mathcal{A}(\gamma)$ be the set of all vectors of coefficients

$$\mathbf{a} = [a_{\alpha,\beta}]_{|\alpha|,|\beta| \leq m} \in \prod_{|\alpha|,|\beta| \leq m} A_{\alpha,\beta}$$

such that $(L, \{B_j\}_{j=0}^{m-1})$ is a self-adjoint elliptic boundary value problem, and B is weakly coercive over $H_E^m(\Omega)$, with constant $\gamma > 0$. As before, we let $F = BH^r(\Omega)$ denote our class of right-hand sides. Setting $\tilde{F} = \mathcal{A}(\gamma) \times F$, we then consider the solution operator as a mapping $S \colon \tilde{F} \to G$.

Standard information of cardinality n now consists of n evaluations of f and $\mathbf{a} = [a_{\alpha,\beta}]_{|\alpha|,|\beta| \leq m}$. That is, our n evaluations must be allocated among the coefficients of L and the right-hand side f. We now wish to find $\text{comp}(\varepsilon)$ for this new problem, and to determine optimal complexity algorithms.

THEOREM.

(i) $r(n, \Lambda^{\text{std}}) = \Theta(n^{-r/d})$ *as* $n \to \infty$.

(ii)   *There exists standard information* $\tilde{N}_{n,k}^Q$ *of cardinality* n *consisting of function evaluations only (i.e. no derivatives), as well as a finite element method* $\tilde{\phi}_{n,k}^Q$ *of degree* k *with quadrature using* $\tilde{N}_{n,k}^Q$ *such that*

$$e(\tilde{\phi}_{n,k}^Q, \tilde{N}_{n,k}^Q) = \Theta(n^{-\mu/d}) \qquad \text{as } n \to \infty,$$

where

$$\mu = \min\{k + 1 - l, r\}.$$

Thus $\tilde{\phi}_{n,k}^Q$ *is an almost minimal error algorithm iff* $k \geq r + l - 1$.

(iii)   *The* ε-*complexity is given by*

$$\text{comp}(\varepsilon) = \Theta(\varepsilon^{-d/r}) \qquad \text{as } \varepsilon \to 0.$$

(iv)   *Let*

$$\text{cost}_k^Q(\varepsilon) = \inf\{\, \text{cost}(\tilde{\phi}_{n,k}^Q, \tilde{N}_{n,k}^Q) : e(\phi_{n,k}^Q, N_{n,k}^Q) \leq \varepsilon \,\}.$$

Then

$$\text{cost}_k^Q(\varepsilon) = \Theta(\varepsilon^{-d/\mu}) \qquad \text{as } \varepsilon \to 0.$$

Hence the FEMQ of degree k *is an almost optimal complexity algorithm iff* $k \geq r + l - 1$.

PROOF. Note that (iii) and (iv) follow immediately from (i) and (ii). To prove (ii), let $\kappa_{m,d}$ denote the maximum number of coefficients that can appear in a $2m$th-order symmetric elliptic operator over a d-dimensional domain. For $(\mathbf{a}, f) \in \tilde{F}$, let $\tilde{N}^Q_{n,k}(\mathbf{a}, f)$ consist of $\lfloor \tilde{n}/(\kappa_{m.d} + 1) \rfloor$ evaluations of each $a_{\alpha,\beta}$ and of f at nodes that are the affine images of a quadrature method that is exact of degree $2k - \nu + 1$ over the reference element. Then Theorem 5.5.2 implies that (ii) holds.

We need only prove (i). Let N be (possibly adaptive) information of cardinality at most n. Choose $\mathbf{a} \in \mathcal{A}(\gamma)$. Then

$$\mathbf{f}_1 = (\mathbf{a}, 0) \in \tilde{F}.$$

Of course, $S\mathbf{f}_1 = 0$. Suppose without loss of generality that $N\mathbf{f}_1$ evaluates its right-hand side (i.e. the zero function) at $x_1, \ldots, x_t$, and evaluates the coefficients of L at $x_{t+1}, \ldots, x_{n(\mathbf{f}_1)}$. From the proof of Theorem 5.5.1, there exists $z \in BH^r(\Omega)$ such that

$$D^\alpha z(x_1) = \ldots = D^\alpha z(x_t) = 0 \qquad \forall \, \alpha \in \mathbb{N}^d,$$
$$\int_\Omega z(x) \, dx \geq \kappa t^{-r/d}.$$

Let

$$\mathbf{f}_2 = (\mathbf{a}, z) \in \tilde{F}.$$

By the proof of Theorem 5.5.1, there exists $C(\mathbf{a}) > 0$ such that

$$\|S\mathbf{f}_2\| \geq C(\mathbf{a}) t^{-r/d} \geq C(\mathbf{a}) n^{-r/d}.$$

Since $\mathbf{f}_1, \mathbf{f}_2 \in F$ and $N\mathbf{f}_1 = N\mathbf{f}_2$, the results of §4.2 imply that

$$r(N) \geq \tfrac{1}{2} \|S\mathbf{f}_1 - S\mathbf{f}_2\|_{H^l(\Omega)} \geq C(\mathbf{a}) n^{-r/d}.$$

Since N is arbitrary possibly-adaptive information of cardinality at most n, (i) follows immediately from this inequality and (ii). □

NR 5.5:3 We can also create a *modified finite element method* (MFEM) that replaces inner products $\langle f, s \rangle_{L_2(\Omega)}$ by $\langle \Pi_n f, s \rangle_{L_2(\Omega)}$, where Π_n is an interpolation operator into a finite element space, this idea going back at least as far as Fried (1973). The MFEM uses standard information, and is an almost minimal error algorithm for $k \geq m + r - 1$. See Section 5 in Chapter 5 of IBC for details.

5.6 The seminormed case

In this section, we analyze the seminormed case

$$F = \mathcal{B}H^r(\Omega) = \{ f \in H^r(\Omega) : |f|_{H^r(\Omega)} \leq 1 \}.$$

We assume that either arbitrary linear information or standard information is permissible. Of course, we must assume that r is a nonnegative integer for the usual definition of $|\cdot|_{H^r(\Omega)}$ to make sense. Since the case $r = 0$ has been covered in the previous section, we can assume without loss of generality that r is a strictly positive integer.

We briefly outline the results of this section. First, suppose that $\Lambda = \Lambda'$, i.e. arbitrary linear information is permissible. We find that the nth minimal radius and the ε-complexity are the same as in the normed case, and that FEI is always almost optimal information. However, the FEM based on $\mathcal{S}_{n,k}$ has infinite error, unless the (restrictive) condition $SP_{r-1}(\Omega) \subseteq \mathcal{S}_{n,k}$ is satisfied; when this holds, the FEM is optimal iff $k \geq 2m - 1 + r$ (as in the normed case). We then suppose that $\Lambda = \Lambda^{\mathrm{std}}$, i.e. we assume that only standard information is permissible. Then the nth minimal radius and ε-complexity are the same as in the normed case. We find that the FEMQ has infinite error, unless $SP_{r-1}(\Omega) \subseteq \mathcal{S}_{n,k}$; when this holds, the FEMQ is optimal iff $k \geq r + l - 1$ (as in the normed case).

Before proceeding further, we state a preliminary result that is useful in its own right. For Hilbert spaces F_1, G, and X, let $S \colon F_1 \to G$ be a bounded linear solution operator and let $T \colon F_1 \to X$ be a bounded linear surjective restriction operator. Suppose we wish to solve a problem with solution operator S, problem elements

$$F = \{\, f \in F_1 : \|Tf\| \leq 1 \,\},$$

and permissible information Λ. Let

$$\hat{F}_1 = F_1/\ker T$$

denote the orthogonal complement of $\ker T$ in F_1, which is a Hilbert space under the norm $\|T \cdot \|_X$. Define a second class of problem elements for S by

$$\hat{F} = B\hat{F}_1 = F \cap \hat{F}_1.$$

If we wish to compare the minimal radii $r(\cdot, \Lambda, F)$ and $r(\cdot, \Lambda, \hat{F})$ for the respective problem element classes F and $\hat{F}$, we may use the following result (whose proof we omit):

LEMMA 5.6.1. *Let*

$$n^* = \dim S(\ker T).$$

Then

$$r(n, \Lambda, F) = \begin{cases} \infty & \text{if } n < n^*, \\ r(n - n^*, \Lambda, \hat{F}) & \text{if } n \geq n^*. \end{cases} \qquad \square$$

We now return to our elliptic problem. Let us consider the case $\Lambda = \Lambda'$, i.e. arbitrary linear information is permissible. First, we give an estimate of the nth minimal radius. Let

$$n^* = \binom{d + r - 1}{r - 1}.$$

THEOREM 5.6.1. *Let* $\Lambda = \Lambda'$.

   (i) *If* $n < n^*$, *then* $r(n) = \infty$.
   (ii) $r(n) = \Theta(n^{-(r+2m-l)/d})$ *as* $n \to \infty$.

PROOF. Let $\alpha^1, \dots, \alpha^s$ be the multi-indices $\alpha \in \mathbb{N}^d$ such that $|\alpha| = r$, where

$$s = \binom{d + r - 1}{r}.$$

Note that $L_2(\Omega)^s$ is a Hilbert space under the norm

$$\|\mathbf{x}\| = \left(\sum_{j=1}^{s} \|x_j\|_{L_2(\Omega)}^2 \right)^{1/2} \qquad \forall\, \mathbf{x} = \begin{bmatrix} x_1 \\ \vdots \\ x_s \end{bmatrix} \in L_2(\Omega)^s.$$

Let

$$X = \{\, \mathbf{x} \in L_2(\Omega)^s : \exists\, f \in H^r(\Omega) \text{ such that } D^{\alpha^j} f = x_j \ (1 \le j \le s) \,\}. \tag{1}$$

Since Ω is simply connected, an element $\mathbf{x} \in L_2(\Omega)^s$ belongs to X iff $\mathbf{x}$ satisfies a set of q equations in $H^{-1}(\Omega)$ of the form

$$\partial_i x_j - \partial_k x_l = 0,$$

where $\partial_i = \partial/\partial x_i$ and $i, j, k, l \in \{1 \dots, d\}$ satisfy

$$\partial_i D^{\alpha^j} = \partial_k D^{\alpha^l}$$

(which expresses the equality of the mixed partial derivatives of the function f for which (1) holds). Thus X is the kernel of a bounded linear operator from $L_2(\Omega)^s$ into $H^{-1}(\Omega)^q$, which implies that X is a closed subspace of $L_2(\Omega)$. Hence X is a Hilbert space.

We are now ready to prove (i). Define $T \colon H^r(\Omega) \to X$ by

$$Tf = \begin{bmatrix} D^{\alpha^1} f \\ \vdots \\ D^{\alpha^s} f \end{bmatrix} \qquad \forall f \in H^r(\Omega).$$

Then T is a bounded linear surjection, and

$$F = \{\, f \in H^r(\Omega) : \|Tf\| \le 1 \,\},$$

i.e. T is the restriction operator for F. Since S is injective, we have

$$\dim S(\ker T) = \dim \ker T = \dim P_{r-1}(\Omega) = n^*.$$

Hence (i) follows from Lemma 5.6.1.

To establish (ii), let

$$\hat{H}^r(\Omega) = H^r(\Omega)/P_{r-1}(\Omega)$$

under the quotient norm. Using Lemma 5.6.1 and the inclusion $B\hat{H}^r(\Omega) \subseteq BH^r(\Omega)$, we have

$$r(n) = r\big(n, \mathcal{B}H^r(\Omega)\big) = r\big(n-n^*, B\hat{H}^r(\Omega)\big) = O\Big(r\big(n-n^*, BH^r(\Omega)\big)\Big). \quad (2)$$

But Theorem 5.4.1 yields

$$r\big(n - n^*, BH^r(\Omega)\big) = \Theta\big((n - n^*)^{-(r+2m-l)/d}\big) = \Theta(n^{-(r+2m-l)/d}). \quad (3)$$

Hence (2) and (3) yield

$$r(n) = O(n^{-(r+2m-l)/d}). \quad (4)$$

On the other hand, $BH^r(\Omega) \subseteq \mathcal{B}H^r(\Omega) = F$ yields

$$r(n) \geq r\,(n, BH^r(\Omega)) = \Theta(n^{-(r+2m-l)/d}). \quad (5)$$

Then part (ii) follows from (4) and (5). □

Hence we see that the nth minimal radius for the seminormed case is roughly the same as that for the normed case.

We next investigate the error of the FEM for the seminormed case. We show that either the FEM has infinite error, or the results from the normed case hold. Recall that $S_{n,k}$ is an n-dimensional finite element subspace of $H_E^m(\Omega)$ consisting of piecewise polynomials of degree up through k, $N_{n,k}^{\mathrm{FE}}$ is finite element information based on $S_{n,k}$, and $\phi_{n,k}^{\mathrm{FE}}$ is the finite element method using $N_{n,k}^{\mathrm{FE}}$.

THEOREM 5.6.2. *Let*

$$\mu = \min\{k + 1 - m, m + r\}.$$

(i) *If* $SP_{r-1}(\Omega) \not\subseteq S_{n,k}$, *then*

$$e(\phi_{n,k}^{\mathrm{FE}}, N_{n,k}^{\mathrm{FE}}) = \infty.$$

(ii) *If* $SP_{r-1}(\Omega) \subseteq S_{n,k}$ *for all sufficiently large* n, *then*

$$e(\phi_{n,k}^{\mathrm{FE}}, N_{n,k}^{\mathrm{FE}}) = \Theta(n^{-(\mu+m-l)/d}) \qquad \text{as } n \to \infty.$$

PROOF. To prove (i), suppose that $SP_{r-1}(\Omega) \not\subseteq S_{n,k}$ and $e(\phi_{n,k}^{\mathrm{FE}}, N_{n,k}^{\mathrm{FE}}) < \infty$. Since the range of $\phi_{n,k}^{\mathrm{FE}}$ is $S_{n,k}$, there exists $f \in P_{r-1}(\Omega)$ such that $\phi_{n,k}^{\mathrm{FE}}(N_{n,k}^{\mathrm{FE}} f) \neq Sf$. Let $\alpha > 0$. Since $\alpha f \in P_{r-1}(\Omega) \subset F$, we have

$$\|Sf - \phi_{n,k}^{\mathrm{FE}}(N_{n,k}^{\mathrm{FE}} f)\|_{H^l(\Omega)} = \alpha^{-1} \left\|S(\alpha f) - \phi_{n,k}^{\mathrm{FE}}(N_{n,k}^{\mathrm{FE}}(\alpha f))\right\|_{H^l(\Omega)}$$
$$\leq \alpha^{-1} e(\phi_{n,k}^{\mathrm{FE}}, N_{n,k}^{\mathrm{FE}}).$$

Since $e(\phi_{n,k}^{\mathrm{FE}}, N_{n,k}^{\mathrm{FE}})$ is finite and independent of α, we may let $\alpha \to \infty$ in the above, finding that $\|Sf - \phi_{n,k}^{\mathrm{FE}}(N_{n,k}^{\mathrm{FE}} f)\|_{H^l(\Omega)} = 0$. Thus $Sf = \phi_{n,k}^{\mathrm{FE}}(N_{n,k}^{\mathrm{FE}} f) \in S_{n,k}$. Since $f \in P_{r-1}(\Omega)$ is arbitrary, this shows that if $e(\phi_{n,k}^{\mathrm{FE}}, N_{n,k}^{\mathrm{FE}}) < \infty$, then $SP_{r-1}(\Omega) \subseteq S_{n,k}$, establishing (i).

To prove (ii), suppose that $SP_{r-1}(\Omega) \subseteq S_{n,k}$. Since $BH^r(\Omega) \subseteq F$, Theorem 5.4.2 establishes that

$$e(\phi_{n,k}^{\mathrm{FE}}, N_{n,k}^{\mathrm{FE}}) = \sup_{f \in F} \|Sf - \phi_{n,k}^{\mathrm{FE}}(N_{n,k}^{\mathrm{FE}} f)\|_{H^l(\Omega)}$$
$$\geq \sup_{f \in BH^r(\Omega)} \|Sf - \phi_{n,k}^{\mathrm{FE}}(N_{n,k}^{\mathrm{FE}} f)\|_{H^l(\Omega)}$$
$$= \Theta(n^{-(\mu+m-l)/d}) \qquad \text{as } n \to \infty.$$

To complete the proof, we need only show that

$$e(\phi_{n,k}^{\mathrm{FE}}, N_{n,k}^{\mathrm{FE}}) = O(n^{-(\mu+m-l)/d}) \qquad \text{as } n \to \infty. \qquad (6)$$

Let $f \in F$. Then there exists a unique choice of

$$f_1 \in P_{r-1}(\Omega) \qquad \text{and} \qquad f_2 \in \hat{H}^r(\Omega) = H^r(\Omega)/P_{r-1}(\Omega)$$

such that

$$f = f_1 + f_2. \qquad (7)$$

From Lemma A.2.2.4, we see that there is a positive constant C_1 such that

$$\|\cdot\|_{H^r(\Omega)} \leq C_1 |\cdot|_{H^r(\Omega)} \qquad \text{on } \hat{H}^r(\Omega).$$

Since $f_1 \in P_{r-1}(\Omega)$, $f_2 \in \hat{H}^r(\Omega)$ and $f \in F$, we have

$$\|f_2\|_{H^r(\Omega)} \leq C_1 |f_2|_{H^r(\Omega)} = C_1 |f|_{H^r(\Omega)} \leq C_1.$$

Hence from Theorem 5.4.2, we see that there is a positive constant C_2, independent of f and n, such that

$$\|Sf_2 - \phi_{n,k}^{\mathrm{FE}}(N_{n,k}^{\mathrm{FE}} f_2)\|_{H^l(\Omega)} \leq C n^{-(\mu+m-l)/d} \|f_2\|_{H^r(\Omega)}$$
$$\leq C_2 n^{-(\mu+m-l)/d}. \qquad (8)$$

Now $f_1 \in P_{r-1}(\Omega)$ and $SP_{r-1}(\Omega)$ imply that $Sf_1 \in S_{n,k}$. Since $l \leq m$, we may use Strang's Lemma (Lemma A.3.2) to see that there is a positive constant C, independent of f and n, such that

$$\|Sf_1 - \phi_{n,k}^{FE}(N_{n,k}^{FE}f_1)\|_{H^l(\Omega)} \leq \|Sf_1 - \phi_{n,k}^{FE}(N_{n,k}^{FE}f_1)\|_{H^m(\Omega)}$$
$$\leq C \inf_{s \in S_{n,k}} \|Sf_1 - s\|_{H^m(\Omega)} = 0. \qquad (9)$$

Since S, $\phi_{n,k}^{FE}$, and $N_{n,k}^{FE}$ are linear, we may use (7), (8), and (9) to see that

$$\|Sf - \phi_{n,k}^{FE}(N_{n,k}^{FE}f)\|_{H^l(\Omega)} \leq C_2 n^{-(\mu+m-l)/d}.$$

Since $f \in F$ is arbitrary, this yields (ii). $\qquad \square$

The condition $SP_{r-1}(\Omega) \subseteq S_{n,k}$ of Theorem 5.6.2 means that the exact solution u of the problem $Lu = f$ (with f polynomial) must be a piecewise polynomial of degree at most k satisfying the boundary conditions. This is restrictive, as we see from the following

EXAMPLE. Suppose that $I = (0,1)$ and let

$$F = \{\, f \in H^1(I) : |f|_{H^1(I)} \leq 1 \,\}.$$

Consider the bilinear form

$$B(v,w) = \int_0^1 v'w' + vw,$$

which is coercive on $H_0^1(I)$ and on $H^1(I)$. We now define two solution operators $S_1 \colon F \to H_0^1(I)$ and $S_2 \colon F \to H^1(I)$ by

$$B(S_1 f, v) = \langle f, v \rangle_{L_2(I)} \qquad \forall v \in H_0^1(I)$$

and

$$B(S_2 f, v) = \langle f, v \rangle_{L_2(I)} \qquad \forall v \in H^1(I),$$

so that

if $f \in H_0^1(I)$, then $u = S_1 f$ satisfies
$$\begin{aligned} -u'' + u &= f \ \text{ in } I, \\ u(0) = u(1) &= 0, \end{aligned}$$

and

if $f \in H^1(I)$, then $u = S_2 f$ satisfies
$$\begin{aligned} -u'' + u &= f \ \text{ in } I, \\ u'(0) = u'(1) &= 0. \end{aligned}$$

Note that S_1 and S_2 differ only in their boundary conditions. We claim that the FEM has infinite error for S_1, but has finite error for S_2. Again, keep in mind that these problems are being solved for all $f \in H^1(I)$ such that $|f|_{H^1(I)} \le 1$, i.e., $r = 1$.

To see that the FEM has infinite error for S_1, note that $S_1(P_1(I))$ is spanned by the solution of the problem

$$-z'' + z = 1 \quad \text{in } I,$$
$$z(0) = z(1) = 0,$$

the solution of which is

$$z(x) = 1 - \left(\frac{e-1}{e^2-1}\right)e^x - \left(\frac{e^2-e}{e^2-1}\right)e^{-x}.$$

Since z is not a piecewise polynomial, we find that $S_1(P_1(I)) \not\subseteq S_{n,k}$, no matter how big the degree k of the subspace $S_{n,k}$ or the dimension n of $S_{n,k}$ are. Hence, the FEM has infinite error for the problem S_1.

We now consider the problem S_2. We find that $S_2(P_0(I)) = P_0(I)$, since the only solution to

$$-z'' + z = 1 \quad \text{in } I,$$
$$z'(0) = z'(1) = 0$$

is

$$z(x) \equiv 1.$$

Since $k \ge 1$ by Lemma 5.4.2, and there are no essential boundary conditions for this problem, we have $S_2(P_0(I)) \subseteq S_{n,k}$ for all $n \ge 1$ and any choice of k. This shows that the FEM has finite error for the problem S_2. Furthermore, the error of the FEM for this problem using $S_{n,k}$ is $\Theta(n^{-(\chi-l)})$, where $\chi = \min\{k+1, 3\}$. Comparing Theorems 5.6.1 and 5.6.2, we see that the FEM is an almost minimal error algorithm for this problem iff $k \ge 2$.   □

Of course, if the condition $SP_{r-1}(\Omega) \subseteq S_{n,k}$ is so easily violated in a simple one-dimensional problem with polynomial coefficients in L, then this condition is unlikely to hold for multi-dimensional problems with L having non-polynomial coefficients. Thus the FEM will generally have infinite error, unless we have been given an unreasonably lucky choice of the operator L, the boundary operators $B_0, \dots, B_{m-1}$, and the boundary Ω. Why does the FEM behave so unpleasantly? As in the normed case, there are two possible explanations: either it uses its information poorly, or finite element information is bad for the seminormed case. We now show that the former is the case, and that FEI is almost optimal information.

To show this, we must first establish

LEMMA 5.6.2. *There exists a positive integer n_0 and a positive constant C such that for any $n \geq n_0$,*

$$\|z\|_{H^r(\Omega)} \leq C|z|_{H^r(\Omega)} \qquad \forall z \in \ker N_{n,k}^{\mathrm{FE}}.$$

PROOF. Recall that $r \geq 1$. If the conclusion is false, then there is a subsequence $\{z_{n_i} \in \ker N_{n_i,k}^{\mathrm{FE}}\}_{i=1}^{\infty}$ such that

$$\|z_{n_i}\|_{H^r(\Omega)} = 1 \qquad \text{and} \qquad \lim_{i \to \infty} |z_{n_i}|_{H^r(\Omega)} = 0.$$

Since $\|z_{n_i}\|_{H^r(\Omega)} \leq 1$, we may use the Rellich-Kondrasov compactness theorem (Lemma A.2.2.3) to see that there exists $z \in H^{r-1}(\Omega)$ and a subsequence, which we again denote $\{z_{n_i} \in \ker N_{n_i,k}^{\mathrm{FE}}\}_{i=1}^{\infty}$, such that

$$\lim_{i \to \infty} \|z - z_{n_i}\|_{H^{r-1}(\Omega)} = 0. \tag{10}$$

Since $\lim_{i \to \infty} |z_{n_i}|_{H^r(\Omega)} = 0$ and the space $H^r(\Omega)$ is complete, we conclude that

$$\lim_{i \to \infty} \|z - z_{n_i}\|_{H^r(\Omega)} = 0,$$

i.e. the convergence is also in $H^r(\Omega)$. Since $\|z_{n_i}\|_{H^r(\Omega)} = 1$, this implies that

$$\|z\|_{H^r(\Omega)} = 1. \tag{11}$$

We claim that $z = 0$, contradicting (11). Indeed, let $\varepsilon > 0$. Since $C_0^{\infty}(\Omega)$ is dense in $L_2(\Omega)$, there exists $w \in C_0^{\infty}(\Omega)$ such that

$$\|z - w\|_{L_2(\Omega)} < \tfrac{1}{3}\varepsilon. \tag{12}$$

Since $C_0^{\infty}(\Omega) \subseteq H_E^m(\Omega) \cap H^1(\Omega)$, Lemma 5.4.2 yields that there is a positive constant C_1 (independent of z and w) such that for any $j > 0$, there exists $w_j \in \mathcal{S}_{j,k}$ for which

$$\|w - w_j\|_{L_2(\Omega)} \leq C_1 j^{-1/N} |w|_{H^1(\Omega)}.$$

Hence, there is an index $i_0(\varepsilon)$ such that for any $i \geq i_0(\varepsilon)$, there exists $w_{n_i} \in \mathcal{S}_{n_i,k}$ satisfying

$$\|w - w_{n_i}\|_{L_2(\Omega)} \leq \tfrac{1}{3}\varepsilon. \tag{13}$$

From (10) and $\|\cdot\|_{L_2(\Omega)} \leq \|\cdot\|_{H^r(\Omega)}$, there is an index $i_1(\varepsilon)$ such that for any $i \geq i_1(\varepsilon)$, there exists $z_{n_i} \in \ker N_{n_i,k}^{\mathrm{FE}}$ for which

$$\|z - z_{n_i}\|_{L_2(\Omega)} \leq \tfrac{1}{3}\varepsilon. \tag{14}$$

Let $i_2(\varepsilon) = \max\{i_0(\varepsilon), i_1(\varepsilon)\}$. Then (12)–(14) and the triangle inequality yield

$$\|z_{n_i} - w_{n_i}\|_{L_2(\Omega)} \le \varepsilon \qquad \forall\, i \ge i_2(\varepsilon). \tag{15}$$

But $z_{n_i} \in \ker N_{n_i,k}^{\mathrm{FE}} = H^r(\Omega) \cap (L_2(\Omega)/S_{n_i})$ and $w_{n_i} \in S_{n_i}$. Hence

$$\langle z_{n_i}, w_{n_i} \rangle_{L_2(\Omega)} = 0,$$

which, when combined with (15), yields

$$\begin{aligned}
\|z_{n_i}\|_{L_2(\Omega)}^2 &\le \|z_{n_i}\|_{L_2(\Omega)}^2 - 2\langle z_{n_i}, w_{n_i} \rangle_{L_2(\Omega)} + \|w_{n_i}\|_{L_2(\Omega)}^2 \\
&= \|z_{n_i} - w_{n_i}\|_{L_2(\Omega)}^2 \le \varepsilon^2.
\end{aligned}$$

Thus for any $\varepsilon > 0$, there is an index $i_2(\varepsilon)$ for which

$$\|z_{n_i}\|_{L_2(\Omega)} \le \varepsilon \qquad \forall\, i \ge i_2(\varepsilon).$$

Hence

$$\lim_{i \to \infty} z_{n_i} = 0 \quad \text{in } L_2(\Omega). \tag{16}$$

From (10) and (16), we have $z = 0$, the desired contradiction to (11). $\qquad\square$

We are now ready to show that the FEI is always quasi-optimal information for the seminormed case.

THEOREM 5.6.3. *The spline algorithm $\phi_{n,k}^{\mathrm{s}}$ using FEI $N_{n,k}^{\mathrm{FE}}$ is an almost minimal error algorithm. That is,*

$$e(\phi_{n,k}^{\mathrm{s}}, N_{n,k}^{\mathrm{FE}}) = r(N_{n,k}^{\mathrm{FE}}) = \Theta\big(r(n)\big) = \Theta(n^{-(r+2m-l)/d}) \qquad \text{as } n \to \infty.$$

PROOF. Using Theorem 2.5.1, we have

$$r(N_{n,k}^{\mathrm{FE}}) \ge r(n) = \Theta(n^{-(r+2m-l)/d}),$$

establishing (i). To prove (ii), let $z \in F \cap \ker N_{n,k}^{\mathrm{FE}}$. Then from Lemma 5.6.2, we see that there is a positive constant C, independent of n and z, such that

$$\langle z, s \rangle_{L_2(\Omega)} = 0 \quad \forall\, s \in S_{n,k} \qquad \text{and} \qquad \|z\|_{H^r(\Omega)} \le C. \tag{17}$$

Since (17) holds, we can now repeat the proof of the upper bound in Theorem 5.4.3, which establishes (ii). $\qquad\square$

Summarizing, we have found that

(i) FEI $N_{n,k}^{\mathrm{FE}}$ is always nth almost optimal information, but

(ii) the FEM $\phi_{n,k}^{\mathrm{FE}}$ using $N_{n,k}^{\mathrm{FE}}$ is an nth almost minimal error algorithm iff $k \ge 2m - 1 + r$ and $SP_{r-1}(\Omega) \subseteq S_{n,k}$ for sufficiently large n.

For $\varepsilon > 0$, we now wish to find comp(ε), the ε-complexity of our problem, as well as determine when the FEM is an almost optimal complexity algorithm for our problem. As in the previous section, we let

$$\text{cost}_k^{\text{FE}}(\varepsilon) = \inf\{\, \text{cost}(\phi_{n,k}^{\text{FE}}, N_{n,k}^{\text{FE}}) : e(\phi_{n,k}^{\text{FE}}, N_{n,k}^{\text{FE}}) \le \varepsilon \,\}$$

denote the minimal cost of using the FEM of degree k to compute an ε-approximation. We also let

$$\text{cost}_k^{\text{spline}}(\varepsilon) = \inf\{\, \text{cost}(\phi_{n,k}^{\text{s}}, N_{n,k}^{\text{FE}}) : e(\phi_{n,k}^{\text{FE}}, N_{n,k}^{\text{FE}}) \le \varepsilon \,\}$$

be the minimal cost of computing an ε-approximation with the spline algorithm using FEI of degree k. Using the previous three theorems, along with the results in §4.4, we have

THEOREM 5.6.4.

(i) *The ε-complexity is given by*

$$\text{comp}(\varepsilon) = \Theta(\varepsilon^{-d/(r+2m-l)}) \qquad \text{as } \varepsilon \to 0.$$

(ii) *If there exists no integer $n \in \mathbb{N}$ for which $SP_{r-1}(\Omega) \subseteq \mathcal{S}_{n,k}$, then*

$$\text{cost}_k^{\text{FE}}(\varepsilon) = \infty \qquad \forall \varepsilon \ge 0.$$

(iii) *Suppose that there exists an integer $n_0 \in \mathbb{N}$ such that $SP_{r-1}(\Omega) \subseteq \mathcal{S}_{n,k}$ for $n \ge n_0$. Then*

$$\text{cost}_k^{\text{FE}}(\varepsilon) = \Theta(\varepsilon^{-d/(\mu+m-l)}) \qquad \text{as } \varepsilon \to 0,$$

where

$$\mu = \min\{k+1-m, m+r\}.$$

Hence, if $SP_{r-1}(\Omega) \subseteq \mathcal{S}_{n,k}$ for $n \ge n_0$ and $k \ge 2m - 1 + r$, then the FEM of degree k is an almost optimal complexity algorithm, i.e.

$$\text{cost}_k^{\text{FE}}(\varepsilon) = \Theta\big(\text{comp}(\varepsilon)\big) = \Theta(\varepsilon^{-d/(r+2m-l)}) \qquad \text{as } \varepsilon \to 0.$$

(iv) *The spline algorithm using FEI of degree k is an almost optimal complexity algorithm, i.e.*

$$\text{cost}_k^{\text{spline}}(\varepsilon) = \Theta\big(\text{comp}(\varepsilon)\big) = \Theta(\varepsilon^{-d/(r+2m-l)}) \qquad \text{as } \varepsilon \to 0. \qquad \square$$

Having completed the analysis of arbitrary linear information for the seminormed case, we now consider the case where only standard information is permissible, i.e. $\Lambda = \Lambda^{\text{std}}$. As in the previous section, we must assume $r > \frac{1}{2}d$ for standard information of cardinality n to be a well-defined, bounded linear operator on F.

We first give an estimate of the nth minimal radius. Recall that

$$n^* = \binom{d+r-1}{r-1}.$$

THEOREM 5.6.5.

(i) If $n < n^*$, then $r(n, \Lambda^{\text{std}}) = \infty$.

(ii) $r(n, \Lambda^{\text{std}}) = \Theta(n^{-r/d})$ as $n \to \infty$.

PROOF. To prove (i), suppose that $n < n^*$. Then $\Lambda^{\text{std}} \subset \Lambda'$ implies that

$$r(n, \Lambda^{\text{std}}) \geq r(n, \Lambda') = \infty,$$

the latter by (i) of Theorem 5.6.1.

To establish (ii), let

$$\hat{H}^r(\Omega) = H^r(\Omega)/P_{r-1}(\Omega)$$

under the quotient norm. Using Lemma 5.6.1 and the inclusion $B\hat{H}^r(\Omega) \subseteq BH^r(\Omega)$, we have

$$\begin{aligned} r(n, \Lambda^{\text{std}}) &= r\big(n, \mathcal{B}H^r(\Omega), \Lambda^{\text{std}}\big) = r\big(n - n^*, B\hat{H}^r(\Omega), \Lambda^{\text{std}}\big) \\ &= O\big(r\big(n - n^*, BH^r(\Omega)\big), \Lambda^{\text{std}}\big). \end{aligned} \tag{18}$$

Using Theorems 5.5.1 and 5.5.2, we have

$$r\big(n - n^*, BH^r(\Omega), \Lambda^{\text{std}}\big) = \Theta\big((n - n^*)^{-r/d}\big) = \Theta(n^{-r/d}). \tag{19}$$

Hence (18) and (19) yield

$$r(n, \Lambda^{\text{std}}) = O(n^{-r/d}). \tag{20}$$

On the other hand, $BH^r(\Omega) \subseteq \mathcal{B}H^r(\Omega) = F$ yields

$$r(n, \Lambda^{\text{std}}) \geq r\,(n, BH^r(\Omega)) = \Theta(n^{-r/d}), \tag{21}$$

the last by Theorem 5.4.1. Hence (ii) follows from (20) and (21). $\qquad\square$

Hence we see that the nth minimal error is roughly the same for the normed and seminormed cases when only standard information is permissible.

We now consider the finite element method with quadrature (FEMQ) for the seminormed case. Recall that $\phi_{n,k}^Q$ is the FEMQ based on the subspace $S_{\tilde{n},k}$, and that $\phi_{n,k}^Q$ is an algorithm using standard information $N_{n,k}^Q$ of cardinality at most n. Here, $\tilde{n}$ is the largest integer such that

$$\text{card}\, N_{\tilde{n}} \leq n$$

(see the discussion of the FEMQ in §5.5 for further details). Note that $\tilde{n} = \Theta(n)$ as $n \to \infty$. Our main result about the error of the FEMQ is analogous to Theorem 5.6.2 for the FEM; the proof of this new result closely follows that of the older one (making the obvious changes):

THEOREM 5.6.6. *Let*

$$\mu = \min\{k + 1 - l, r\}.$$

(i) *If* $SP_{r-1}(\Omega) \not\subseteq \mathbb{S}_{\tilde{n},k}$, *then*

$$e(\phi_{n,k}^Q, N_{n,k}^Q) = \infty.$$

(ii) *If* $SP_{r-1}(\Omega) \subseteq \mathbb{S}_{\tilde{n},k}$ *for all sufficiently large* $\tilde{n}$, *then*

$$e(\phi_{n,k}^Q, N_{n,k}^Q) = \Theta(n^{-\mu/d}) \qquad \text{as } n \to \infty. \qquad \square$$

Since the condition $SP_{r-1}(\Omega) \subseteq \mathbb{S}_{\tilde{n},k}$ is unlikely to hold for any $\tilde{n}$, we describe an algorithm for our problem whose error is always almost minimal, regardless of whether this condition holds. Let $\{\mathfrak{T}_n\}_{n=1}^\infty$ denote a quasi-uniform family of triangulations of Ω, and let $\tilde{\mathbb{S}}_{n,k}$ be a finite element subspace of $H^r(\Omega)$, with degree k and dimension n. Note that the only difference between $\mathbb{S}_{n,k}$ and $\tilde{\mathbb{S}}_{n,k}$ is that the former is a subspace of $H_E^m(\Omega)$ and the latter is a subspace of $H^r(\Omega)$. Thus, elements of these spaces have different inter-element continuity requirements, and different behavior at the boundary of Ω. Let $\{s_1, \ldots, s_n\}$ be a basis for $\tilde{\mathbb{S}}_{n,k}$ having small supports, and let $\{f(x_1), \ldots, f(x_n)\}$ be the corresponding dual basis. The $\tilde{\mathbb{S}}_{n,k}$-interpolant $\tilde{\Pi}_n f$ of $f \in H^r(\Omega)$ is given by

$$\tilde{\Pi}_n f = \sum_{j=1}^n f(x_j) s_j.$$

We then have the following

THEOREM 5.6.7. *For* $n \in \mathbb{N}$, *define standard information* N_n *of cardinality* n *by*

$$N_n f = \begin{bmatrix} f(x_1) \\ \vdots \\ f(x_n) \end{bmatrix} \qquad \forall f \in F$$

(where $x_1, \ldots, x_n$ *are as described above), and an algorithm* ϕ_n *using* N_n *by*

$$\phi_n(N_n f) = S\tilde{\Pi}_n f \qquad \forall f \in F.$$

If $k \geq r$, *then*

$$e(\phi_n, N_n) = \Theta\big(r(n, \Lambda^{\text{std}})\big) = \Theta(n^{-r/d}) \qquad \text{as } n \to \infty,$$

and so ϕ_n is an nth almost minimal error algorithm.

PROOF. Let $f \in F$. Since $0 \le l \le m$, Theorems 5.2.1 and 5.2.2 imply that

$$\|Sf - \phi_n(N_n f)\|_{H^l(\Omega)} \le \|Sf - \phi_n(N_n f)\|_{H^m(\Omega)}$$
$$\le C\|f - \tilde{\Pi}_n f\|_{H^{-m}(\Omega)} \le \|f - \tilde{\Pi}_n f\|_{L_2(\Omega)}.$$

Since $k \ge r$, the proof of Lemma 5.4.3 implies that

$$\|f - \Pi_n f\|_{L_2(\Omega)} \le C n^{-r/d} |f|_{H^r(\Omega)} \le C n^{-r/d},$$

the last since $|f|_{H^r(\Omega)} \le 1$ for $f \in F$. Combining these two estimates, and taking the supremum over all $f \in F$, the result follows.   □

For $\varepsilon > 0$, we now wish to find the ε-complexity comp$(\varepsilon, \Lambda^{\mathrm{std}})$ for our problem when only standard information is permissible. We also wish to determine when the FEMQ is an almost minimal complexity algorithm for our problem. Let

$$\mathrm{cost}_k^Q(\varepsilon) = \inf\{\,\mathrm{cost}(\phi_{n,k}^Q, N_{n,k}^Q) : e(\phi_{n,k}^Q, N_{n,k}^Q) \le \varepsilon\,\}$$

denote the minimal cost of using the FEMQ of degree k to compute an ε-approximation. From our previous results, we have

THEOREM 5.6.8.

(i) *The ε-complexity is given by*

$$\mathrm{comp}(\varepsilon, \Lambda^{\mathrm{std}}) = \Theta(\varepsilon^{-d/r}) \qquad \text{as } \varepsilon \to 0.$$

(ii) *If there exists no integer $n \in \mathbb{N}$ for which $SP_{r-1}(\Omega) \subseteq S_{n,k}$, then*

$$\mathrm{cost}_k^Q(\varepsilon) = \infty \qquad \forall \varepsilon \ge 0.$$

(iii) *Suppose that there exists an integer $n_0 \in \mathbb{N}$ such that $SP_{r-1}(\Omega) \subseteq S_{n,k}$ for $n \ge n_0$. Then*

$$\mathrm{cost}_k^Q(\varepsilon) = \Theta(\varepsilon^{-d/(\mu)}) \qquad \text{as } \varepsilon \to 0,$$

where
$$\mu = \min\{k + 1 - l, r\}.$$

Hence, if $SP_{r-1}(\Omega) \subseteq S_{n,k}$ for sufficiently large n and $k \ge r + l - 1$, then the FEMQ of degree k is an almost optimal complexity algorithm, i.e.

$$\mathrm{cost}_k^Q(\varepsilon) = \Theta\big(\mathrm{comp}(\varepsilon)\big) = \Theta(\varepsilon^{-d/r}) \qquad \text{as } \varepsilon \to 0.   \qquad \square$$

Hence we have shown that standard information is not as powerful as arbitrary linear information when solving elliptic PDE with $m > 0$ in the seminormed case.

Notes and Remarks

NR 5.6:1 Some of the material in this section is based on Werschulz (1986). For a proof of Lemma 5.6.1, see Lemma 5.3 of Werschulz (1984).

5.7   Can adaption ever help?

In the preceding sections, we have dealt with problems of the following form. Let $(L, B_0, \ldots, B_{m-1})$ be a self-adjoint $2m$th-order elliptic boundary value problem on a region $\Omega \subseteq \mathbb{R}^d$. Let F denote a class of real-valued functions defined on Ω. For $f \in F$, we wished to approximate the variational solution $u = Sf \in H_E^l(\Omega)$ of the problem

$$
\begin{aligned}
Lu &= f & \text{in } \Omega, \\
B_0 u &= \cdots = B_{m-1}u = 0 & \text{on } \partial\Omega.
\end{aligned}
\tag{1}
$$

One of the most powerful tools for helping us to find optimal algorithms was the result that adaption is no stronger than nonadaption for convex, balanced F. That is, since the F we dealt with was a unit ball or semiball of some Hilbert space (and hence convex and balanced), the results in §4.4.2 implied that the nth minimal radii over adaptive and nonadaptive information were equal. Hence, we could restrict our attention to algorithms using nonadaptive information. Since nonadaptive information is much simpler than adaptive information, this made our task much easier.

This result (adaption is no better than nonadaption) has not been enthusiastically accepted by many practitioners of scientific computation, since the conventional wisdom is that for "practical" problems, adaption is better than nonadaption. How can we reconcile the general theorem of information-based complexity with the results of many years of experience? The only possible answer (barring either the unlikely possibility of a mistake in the simple proof of this theorem or the possibility that the conventional wisdom is mistaken) is that in practice, the problem (1) is not a linear problem. This can happen iff at least one of the following holds:

(i) N is not linear.
(ii) S is not linear.
(iii) F is not balanced and convex.

We can generally rule out (i), since most algorithms use standard information consisting of evaluation of problem elements. Hence, we are left to deal with (ii) and (iii).

We first consider (ii). Although (1) involves the solution of a linear boundary value problem, the linearity is only in the dependence of u on f. However, the solution u also depends on the coefficients in the operators $L, B_0, \ldots, B_{m-1}$ and on the region Ω over which the problem is to be solved. Furthermore, this additional dependence is highly nonlinear, and not nearly as well-understood as the dependence on f. With this in mind, we see that our problem becomes nonlinear if we reformulate it more realistically. Namely, let $\mathcal{D}$ denote the class of domains $\Omega \subseteq \mathbb{R}^d$ over which we wish to solve our problems, let $\mathcal{A}$ denote the class of elliptic operators $(L, B_0, \ldots, B_{m-1})$ (including boundary conditions), and let F denote the class of right-hand sides. Then we should define our solution operator as $S \colon \mathcal{A} \times \mathcal{D} \times F \to H_E^l(\Omega)$, with $u = S\big((L, B_0, \ldots, B_{m-1}), \Omega, f\big)$ iff u satisfies (1).

Suppose first that $\mathcal{A}$ is a singleton, i.e. we only need to solve the problem for a fixed $(L, B_0, \ldots, B_{m-1})$. Hence our solution operator is $S \colon \mathcal{D} \times F \to H_E^l(\Omega)$. We claim that there are important classes $\mathcal{D}$ of domains such that dealing with $\mathcal{D}$ may not be intrinsically harder than dealing with a fixed domain Ω. Indeed, suppose that $\mathcal{D}$ consists of a class of polyhedral domains with a bounded number of faces (as is often the case). Then we can have complete information about any $\Omega \in \mathcal{D}$, using only information of finite cardinality. Hence for $(\Omega, f) \in \mathcal{D} \times F$, we compute $S(\Omega, f)$ from complete information about Ω and partial information about f. This means that from the viewpoint of information-based complexity, $S(\Omega, f)$ really only depends on f, and so if F is convex and balanced, our problem has been reduced to a linear problem. Thus if $\mathcal{A}$ is a singleton, $\mathcal{D}$ consists of polyhedral domains and F is convex and balanced, then adaption is no better than nonadaption.

The opposite extreme is dealt with in Leyk (1990). He supposes that $\mathcal{D}$ is a singleton, and that $\mathcal{A}$ is a class of elliptic boundary value problems. In particular, he assumes that there exist positive constants γ and M such that the bilinear form B appearing in the variational formulation of any $(L, B_0, \ldots, B_{m-1}) \in \mathcal{A}$ is γ-strongly coercive and M-bounded. He then shows that for any Galerkin information N, the Galerkin method is an almost optimal error algorithm using N (see §5.4 for discussion of Galerkin information and algorithms). However, Leyk does not address the question of adaption.

Another way of looking at the case where $\mathcal{D}$ is a singleton is to let $\mathcal{A}$ be a class of elliptic boundary value problems for which we only have partial information about the coefficients. This is perhaps more realistic than Leyk's approach. We have taken this viewpoint in our treatment of the FEMQ in §5.5 (see especially **NR 5.5:2** for further development). Our result from §5.5 showed that for fixed Ω, a class $\mathcal{A}$ of elliptic problems for which a fixed bound was given on the $W^{r,\infty}(\Omega)$-norm of the coefficients, and right-hand sides f belonging to the unit ball of $H^r(\Omega)$, the FEMQ of sufficiently high order was optimal, and adaption is no better than nonadaption.

Of course, the most important case occurs when $\mathcal{A}$, $\mathcal{D}$, and F are all (non-singleton) sets. To date, there has been no work on the complexity of such problems. More research is required, with special attention to the sensitivity of the solution to simultaneous changes in the region Ω, the coefficients $(L, B_0, \ldots, B_{m-1})$, and the right-hand side f.

In the remainder of this section, we consider possibility (iii) in the previous list. We consider a simple one-dimensional model problem. Let F be a (nonconvex) class of piecewise-constant functions with p breakpoints. We show that if $p = 1$, then adaption is exponentially better than nonadaption, while if $p \geq 2$, adaption is no better than nonadaption. Hence, it appears that for problem classes that are not convex and balanced, we can only determine whether adaption is better than nonadaption on a case-by-case basis.

We first define our class F of problem elements. Let $I = (0,1)$ be the unit interval. Let p be a nonnegative integer, and let $M > 0$. Then F is the set of all functions $f \in L_\infty(I)$ for which there exist breakpoints $x_i = x_i(f)$ (where $1 \leq i \leq p$) with $0 < x_0 < x_1 < \cdots < x_p < 1$ such that if we set $x_0 = 0$ and $x_{p+1} = 1$, then the following hold:

(i) For $0 \leq i \leq p$, the restriction $f|_{(x_i, x_{i+1})}$ of f to (x_i, x_{i+1}) is a constant.

(ii) Either $f(x_i^+) = f(x_i)$ or $f(x_i^-) = f(x_i)$ for $1 \leq i \leq p$.

(iii) $\|f\|_{L_\infty(I)} \leq M$.

Note that F is not convex.

Letting B be the bilinear form on $H^1(I)$ given by

$$B(v, w) = \langle v, w \rangle_{H^1(I)} = \int_0^1 \big(v(x)w(x) + v'(x)w'(x)\big)\, dx \qquad \forall\, v, w \in H^1(I),$$

we define our solution operator $S\colon F \to H^1(I)$ to be

$$B(Sf, v) = \langle f, v \rangle_{L_2(I)} \qquad \forall\, v \in H^1(I). \tag{2}$$

Note that if $f \in F$, then $u = Sf$ is the variational solution of the two-point boundary value problem

$$\begin{aligned} -u''(x) + u(x) &= f(x) \qquad (0 < x < 1), \\ u'(0) &= u'(1) = 0. \end{aligned} \tag{3}$$

Of course, the problems (2) and (3) are (respectively) the variational and classical formulations of the second-order elliptic problem studied in Chapter 2.

Before proceeding further, we will need two simple lemmas.

LEMMA 5.7.1. *Let*

$$\|f\|_{H^{-1}(I)} = \sup_{v \in H^1(I)} \frac{|\langle f, v \rangle_{L_2(I)}|}{\|v\|_{H^1(I)}} = \sup_{\substack{v \in H^1(I) \\ \|v\|_{H^1(I)} \leq 1}} |\langle f, v \rangle_{L_2(I)}| \qquad \forall f \in F.$$

Then

$$\|Sf\|_{H^1(I)} = \|f\|_{H^{-1}(I)} \qquad \forall f \in F.$$

PROOF. Let $f \in F$. The second equality being obvious, we prove the first. Since

$$\langle Sf, v \rangle_{H^1(I)} = \langle f, v \rangle_{L_2(I)} \qquad \forall v \in H^1(I),$$

we have

$$\|Sf\|_{H^1(I)} = \sup_{v \in H^1(I)} \frac{|\langle Sf, v \rangle_{H^1(I)}|}{\|v\|_{H^1(I)}} = \sup_{v \in H^1(I)} \frac{|\langle f, v \rangle_{L_2(I)}|}{\|v\|_{H^1(I)}}. \qquad \square$$

LEMMA 5.7.2. *Let*

$$\beta = \sqrt{\frac{4}{3}} \doteq 1.1547.$$

For any $a, b \in I$ with $a < b$,

$$b - a \leq \sup_{\substack{v \in H^1(I) \\ \|v\|_{H^1(I)} \leq 1}} \left| \int_a^b v(x)\, dx \right| \leq \beta(b - a).$$

PROOF. For the lower bound, choose $v \equiv 1$.

For the upper bound, let $v \in H^1(I)$ with $\|v\|_{H^1(I)} \leq 1$. Integrating by parts, we have

$$\int_a^b v(x)\, dx = (b - a) \left[v(b) - \int_a^b \left(\frac{x - a}{b - a} \right) v'(x)\, dx \right]. \tag{4}$$

Let

$$\sigma(x) = x - \chi_{[b,1]}(x),$$

where χ_A denotes the characteristic function of $A \subseteq [0, 1]$. Using another integration by parts, we have

$$\int_0^1 (v(b) - v(x))\, dx = v(b) - v(1) + \int_0^1 x v'(x)\, dx$$

$$= -\int_b^1 v'(x)\, dx + \int_0^1 x v'(x)\, dx$$

$$= \int_0^1 \sigma(x) v'(x)\, dx,$$

and so

$$v(b) = \int_0^1 v(x)\,dx + \int_0^1 \big(v(b) - v(x)\big)\,dx$$

$$= \int_0^1 v(x)\,dx + \int_0^1 \sigma(x)v'(x)\,dx. \tag{5}$$

Let

$$\tau(x) = x - \chi_{[b,1]}(x) - \frac{x-a}{b-a}\chi_{[a,b]}(x).$$

Substituting (5) into (4), we have

$$\frac{1}{b-a}\int_a^b v(x)\,dx = \left| \int_0^1 v(x)\,dx + \int_0^1 \tau(x)v'(x)\,dx \right|$$

$$\le \left| \int_0^1 v(x)\,dx \right| + \left| \int_0^1 \tau(x)v'(x)\,dx \right| \tag{6}$$

$$\le \|v\|_{L_2(I)} + \|\tau\|_{L_2(I)}\|v'\|_{L_2(I)}.$$

Now

$$\|\tau\|_{L_2(I)}^2 = \tfrac{1}{3}(b^2 + ab - 2b + a^2 - a + 1).$$

Since this last expression is maximized over $(a,b) \in [0,1] \times [0,1]$ by choosing (a,b) to be any corner point, we find that

$$\|\tau\|_{L_2(I)} \le \sqrt{\frac{1}{3}}.$$

Substituting this result into (6) and using the inequality

$$|x + \alpha y| \le \sqrt{1+\alpha^2}\sqrt{x^2 + y^2},$$

the desired upper bound follows immediately. $\qquad\square$

We now wish to compare $r^{\mathrm{a}}(n, \Lambda^{\mathrm{std}})$ and $r^{\mathrm{non}}(n, \Lambda^{\mathrm{std}})$, the nth minimal radii of adaptive and nonadaptive standard information. First, we consider the case $p = 1$, i.e. one breakpoint.

THEOREM 5.7.1. *Let $p = 1$.*

(i) *The nth minimal radius of adaptive standard information satisfies*

$$\frac{M}{2^{n-2}} \le r^{\mathrm{a}}(n, \Lambda^{\mathrm{std}}) \le \frac{M\beta}{2^{n-2}}.$$

(ii) *The nth minimal radius of nonadaptive standard information satisfies*

$$\frac{M}{n+1} \le r^{\mathrm{non}}(n, \Lambda^{\mathrm{std}}) \le \frac{M\beta}{n+1}.$$

PROOF. To show the upper bound for (i), let N_n be bisection information of cardinality n. That is, we do the following to evaluate $N_n f$:

begin

    $t_1 \leftarrow 0;\ y_1 \leftarrow f(t_1);$

    $t_2 \leftarrow 1;\ y_2 \leftarrow f(t_2);$

    $t_L \leftarrow t_1;\ t_R \leftarrow t_2;\ y_L \leftarrow y_1;\ y_R \leftarrow y_2;$

    for $i \leftarrow 3$ **to** n **do**

        begin

            $t_i \leftarrow \frac{1}{2}(t_L + t_R);$

            $y_i \leftarrow f(t_i);$

            if $y_i = y_L$ **then**

                begin $t_L \leftarrow t_i;\ y_L \leftarrow y_i$ **end**

            else

                begin $t_R \leftarrow t_i;\ y_R \leftarrow y_i$ **end**

        end

end

Define f_1 and f_2 by

$$f_1 = \begin{cases} y_L & \text{in } [0, t_L] \\ y_R & \text{in } (t_L, 1] \end{cases} \quad \text{and} \quad f_2 = \begin{cases} y_L & \text{in } [0, t_R) \\ y_R & \text{in } [t_R, 1] \end{cases}.$$

Then f_1 and f_2 form an envelope for $N_n^{-1}(y)$, the set of all $f \in F$ such that $N_n f = y$ (see §4.2). Moreover, let

$$M = \tfrac{1}{2}(L + R).$$

Then

$$f^* = \begin{cases} y_L & \text{in } [0, t_M) \\ y_R & \text{in } [t_M, 1] \end{cases}$$

is a center of $N_n^{-1}(y)$. Hence, the algorithm

$$\phi(y) = S f^*$$

is a strongly optimal error algorithm. Using Lemma 5.7.1, we find that the local error of ϕ at y is given by

$$e(\phi, N, y) = \sup_{f \in N_n^{-1}(y)} \| S f^* - S f \|_{H^1(I)} = \sup_{f \in N_n^{-1}(y)} \| f^* - f \|_{H^{-1}(I)}.$$

Since f_1 and f_2 form an envelope for $N_n^{-1}(y)$, we may use Lemma 5.7.2 to see that

$$e(\phi, N, y) = \tfrac{1}{2}\|f_1 - f_2\|_{H^{-1}(I)}$$

$$= \sup_{\substack{v \in H^1(I) \\ \|v\|_{H^1(I)} \leq 1}} \left| \int_0^1 (f_1(x) - f_2(x))v(x)\,dx \right|$$

$$= \tfrac{1}{2}|y_R - y_L| \sup_{\substack{v \in H^1(I) \\ \|v\|_{H^1(I)} \leq 1}} \left| \int_{t_L}^{t_R} v(x)\,dx \right|$$

$$\leq \tfrac{1}{2}(|y_R| + |y_L|)\,\beta|t_R - t_L|.$$

Since $|y_L|$ and $|y_R|$ are at most M and $|t_R - t_L| \leq 2^{-(n-2)}$, this implies that

$$e(\phi, N, y) \leq \frac{M\beta}{2^{n-2}}.$$

Since $f \in F$ is arbitrary, we may take the supremum over $y \in N_n(F)$ to find that

$$e(\phi, N_n) \leq \frac{M\beta}{2^{n-2}}.$$

Since ϕ is an algorithm using adaptive standard information of cardinality n, we thus have

$$r^{\mathrm{a}}(n, \Lambda^{\mathrm{std}}) \leq e(\phi, N_n) \leq \frac{M\beta}{2^{n-2}}.$$

We now have to prove the lower bound for (i). Let N be adaptive standard information of cardinality n. Let $y \in N(F)$. From §4.4 and Lemma 5.7.1, we see that

$$r(N, y) \geq \tfrac{1}{2} \sup_{f, \tilde{f} \in N^{-1}(y)} \|Sf - S\tilde{f}\|_{H^1(I)} = \tfrac{1}{2} \sup_{f, \tilde{f} \in N^{-1}(y)} \|f - \tilde{f}\|_{H^{-1}(I)}.$$

Let $t_1, \ldots, t_n$ be the evaluation points for $N^{-1}(y)$. Reordering if necessary, we may assume that $0 \leq t_1 < \cdots < t_n \leq 1$. Choose $i = i(y)$ by

$$i = \max\{j \geq 1 : y_j = y_1\},$$

and define functions $f_1, f_2 \in N^{-1}(y)$ by

$$f_1 = \begin{cases} y_1 & \text{in } [0, t_i], \\ y_n & \text{in } (t_i, 1] \end{cases} \quad \text{and} \quad f_2 = \begin{cases} y_1 & \text{in } [0, t_{i+1}), \\ y_n & \text{in } [t_{i+1}, 1]. \end{cases}$$

Since f_1 and f_2 form an envelope for $N^{-1}(y)$, we may use Lemmas 5.7.1 and 5.7.2 to see that

$$d(N, y) = \|Sf_1 - Sf_2\|_{H^1(I)} = \|f_1 - f_2\|_{H^{-1}(I)}$$

$$= \sup_{\substack{v \in H^1(I) \\ \|v\|_{H^1(I)} \leq 1}} \left| \int_0^1 (f_1(x) - f_2(x)) v(x)\, dx \right|$$

$$= |y_1 - y_n| \sup_{\substack{v \in H^1(I) \\ \|v\|_{H^1(I)} \leq 1}} \left| \int_{t_i}^{t_{i+1}} v(x)\, dx \right|$$

$$= |y_1 - y_n| |t_{i+1} - t_i|.$$

Hence

$$r(N^a, y) \geq \tfrac{1}{2} d(N, y) \geq \tfrac{1}{2} |y_1 - y_n| |t_{i+1} - t_i|.$$

Since $y \in N(F)$ is arbitrary, we thus have

$$r(N) = \sup_{y \in N(F)} r(N, y) \geq \tfrac{1}{2} \sup_{y \in N(F)} |y_1 - y_n| s(N),$$

where

$$s(N) = \sup_{\{t_j\}_{j=1}^n} |t_{i+1} - t_i|,$$

the supremum being over all sets of evaluation points for N. However, $s(N)$ is bounded from below by the maximum *diameter* of uncertainty in determining the discontinuity point after n evaluations. This problem is completely analogous to the problem of locating a zero of a function $f: I \to \mathbb{R}$ for which $f(0)f(1) < 0$. Kung (1976) shows that the diameter of this zero-finding problem is $2 \cdot 2^{-n}$. Following his arguments (and realizing that we have to use two initial evaluations at the endpoints), we find that

$$s(N) \geq \frac{2}{2^{n-2}} = \frac{1}{2^{n-1}}.$$

Since

$$\tfrac{1}{2} \sup_{y \in N(F)} |y_1 - y_n| = M,$$

we find that

$$r(N) \geq \frac{M}{2^{n-2}}.$$

Since N is arbitrary adaptive information of cardinality n, the desired lower bound follows, completing the proof of (i).

Next, we turn to the proof of (ii). Let N be nonadaptive information of cardinality n. Let $y = Nf$. Then

$$
y = \begin{bmatrix} y_1 \\ \vdots \\ y_n \end{bmatrix}
$$

with $y_i = f(t_i)$ for some $f \in F$, where we may assume that the t_i are increasing without loss of generality. We also define $t_0 = 0$ and $t_{n+1} = 1$. Define an index $i = i(y)$, as well as functions $f_1, f_2 \in N^{-1}(y)$ as was done in the proof of the lower bound for the adaptive case. As before, since f_1 and f_2 form an envelope for $N^{-1}(y)$, we find that

$$
r(N, y) \geq \tfrac{1}{2} d(N, y) = \tfrac{1}{2} |y_1 - y_n|(t_{i+1} - t_i).
$$

Moreover, $f^* = \tfrac{1}{2}(f_1 + f_2)$ is a center of the set $N^{-1}(y)$. Hence the algorithm

$$
\phi(y) = Sf^*
$$

is an optimal error algorithm. Following the proof in the case (i), we find that the local error satisfies the inequality

$$
e(\phi, N, y) = \tfrac{1}{2} d(N, y) \leq \tfrac{1}{2} |y_1 - y_n| \, \beta(t_{i+1} - t_i).
$$

Taking the supremum over all $y \in N(F)$, we find

$$
M \sup_{0 \leq i \leq n} (t_{i+1} - t_i) \leq r(N) \leq M\beta \sup_{0 \leq i \leq n} (t_{i+1} - t_i).
$$

Now take the infimum over all sample points $t_1, \ldots, t_n$ to find

$$
M\kappa(n) \leq r^{\mathrm{non}}(n, \Lambda^{\mathrm{std}}) \leq M\beta\kappa(n),
$$

where

$$
\kappa(n) = \inf_{0 \leq t_1 \leq \cdots \leq t_n \leq 1} \sup_{0 \leq i \leq n} (t_{i+1} - t_i).
$$

Since

$$
\kappa(n) = \frac{1}{n+1},
$$

the desired conclusion (ii) follows immediately. □

Note that the algorithm ϕ in the proof of (i) of Theorem 5.7.1 is easy to implement because the problem $Lu = f$ involves a simple operator L. For more complicated L, implementing the corresponding ϕ will be more difficult, since ϕ involves the computation of $S = L^{-1}$ at an interpolating problem element. Since the optimality of finite element methods has already been shown for other problem classes and since FEMs do not require exact solution at an interpolant, it is only natural to look for an FEM that is optimal for our problem.

Let $\alpha_1, \alpha_2 \in (0, 1)$ with $\alpha_1 < \alpha_2$ and let $k \in \mathbb{N}$. Set $m = 2k + 2$. We define the finite element space $S_m(\alpha_1, \alpha_2, k)$ to be the m-dimensional space of continuous functions $s \colon I \to \mathbb{R}$ such that

$$s\big|_{[0,\alpha_1]} \in P_k([0, \alpha_1]),$$
$$s\big|_{[\alpha_1,\alpha_2]} \in P_1([\alpha_1, \alpha_2]), \text{ and}$$
$$s\big|_{[\alpha_2,1]} \in P_k([\alpha_2, 1]).$$

Choosing basis functions $s_1, \ldots, s_m \in S_m(\alpha_1, \alpha_2, k)$ satisfying

$$s_i(x_j) = \delta_{i,j} \qquad (1 \le i, j \le m),$$

the $S_m(\alpha_1, \alpha_2, k)$ interpolation operator Π_m is defined by

$$\Pi_m f = \sum_{j=1}^{m} f(x_j) s_j \qquad \forall f \in C(I).$$

where the nodes $x_1, \ldots, x_m$ are given by

$$x_i = \begin{cases} \dfrac{i-1}{k}\alpha_1 & (1 \le i \le k+1), \\[2ex] t_R + \dfrac{i-(k+2)}{k}(1 - \alpha_2) & (k+2 \le i \le m). \end{cases}$$

We are now ready to define adaptive information of cardinality at most n. Let

$$l = \left\lfloor \frac{\ln \pi(n+2)}{\ln \pi + 2\ln 2} \right\rfloor$$

and

$$k = \left\lfloor \frac{l \ln 2}{\ln \pi} \right\rfloor.$$

As in Theorem 5.7.1, we let N_l be bisection information of cardinality l and let $[t_L, t_R]$ be the final interval of uncertainty containing the breakpoint. Let $S_m = S_m(t_L, t_R, k)$, and define

$$\tilde{N}_m f = \begin{bmatrix} f(x_1) \\ \vdots \\ f(x_m) \end{bmatrix} \qquad \forall f \in F.$$

Then we define our information $\hat{N}_n$ to be

$$\hat{N}_n f = \begin{bmatrix} N_l f \\ \tilde{N}_m f \end{bmatrix} \qquad \forall f \in F.$$

Since x_1, x_{k+1}, x_{k+2}, and x_m are evaluation points for N_l and for $\tilde{N}_m$, we find that

$$\text{card } \hat{N}_n \leq l + m - 4 = n.$$

Hence $\hat{N}_n$ is adaptive information of cardinality at most n.

Next, we define our FEM. For $f \in F$ and $n \in \mathbb{N}$, there exists a unique $u_m \in S_m$ such that

$$B(u_m, s_i) = \langle \Pi_m f, s_i \rangle_{L_2(I)} = \sum_{j=1}^{m} f(x_j) \langle s_j, s_i \rangle_{L_2(I)} \qquad (1 \leq i \leq n). \quad (7)$$

Since u_m depends on f only through $\hat{N}_n f$, we may write

$$u_m = \hat{\phi}_n(\hat{N}_n f),$$

with $\hat{\phi}_n$ an algorithm using $\hat{N}_n$.

Note that since we are using standard information, we use an interpolant of f, rather than f itself, in the right-hand side of (7). Hence, $\hat{\phi}_n$ is not a "pure" finite element method, but a "modified" FEM.

We are now ready to state

THEOREM 5.7.2. *Let*

$$\gamma = 2^{\ln \pi / (\ln \pi + 2 \ln 2)} \doteq 1.3682\,.$$

Then

$$e(\hat{\phi}_n, \hat{N}_n) = O(\gamma^{-n}) \qquad \text{as } n \to \infty.$$

Recall that $r^a(n) = \Theta(2^{-n})$ as $n \to \infty$, so that the FEM will not be an almost minimal error algorithm if the upper bound given by this theorem is sharp. So why is this a good result? Recall that the ε-cardinality number $m(\varepsilon)$ is the minimal cardinality of information whose radius is at most ε. From Theorem 5.6.1, we know that

$$m(\varepsilon) = \Theta\big(\log(\varepsilon^{-1})\big) \qquad \text{as } \varepsilon \to 0.$$

Let $m^{\text{FE}}(\varepsilon)$ denote the ε-cardinality number of our FEM, i.e. the minimal number of evaluations it needs to guarantee an error of at most ε. Then this result states that

$$m^{\text{FE}}(\varepsilon) = \Theta\big(\log(\varepsilon^{-1})\big) \qquad \text{as } \varepsilon \to 0.$$

Thus our FEM uses an almost minimal number of information evaluations.

Before proving this theorem, we state the following

LEMMA 5.7.3.

(i) Let $g \in H^1(a,b)$, with $g(a) = 0$ or $g(b) = 0$. Then

$$\|g\|_{L_2(a,b)} \leq \frac{2}{\pi}(b-a)\|g'\|_{H^1(a,b)}.$$

(ii) Let $g \in H^1(a,b)$, with $g\left(\frac{1}{2}(a+b)\right) = 0$. Then

$$\|g\|_{L_2(a,b)} \leq \frac{1}{\pi}(b-a)\|g'\|_{H^1(a,b)}.$$

(iii) Let $g \in H^2(a,b)$, with $g(a) = g(b) = 0$. Then

$$\|g^{(i)}\|_{L_2(a,b)} \leq \frac{1}{\pi}(b-a)\|g^{(i+1)}\|_{H^1(a,b)} \qquad (i = 0,1).$$

PROOF. To prove (i), let $h = b - a$. Since $g(0) = 0$, we may expand g in a sine series

$$g(x) = \sum_{j=1}^{\infty} \alpha_j \sin\frac{j\pi(x-a)}{2h} \qquad (a \leq x \leq b).$$

The result now follows by a straightforward calculation, as in Chapter 1 of Strang and Fix (1973a). The proofs of the remaining items are similar. □

We are now ready for the

PROOF OF THEOREM 5.7.2. Let $f \in F$, $u = Sf$, and $u_m = \hat{\phi}_n(\hat{N}_n f)$. Define a (new) interpolant $\hat{\Pi}_m u \in S_m$ of u satisfying

$$(\hat{\Pi}_m u)^{(i)}(\tfrac{1}{2}t_L) = u^{(i)}(\tfrac{1}{2}t_L) \qquad (0 \leq i \leq k-1),$$
$$(\hat{\Pi}_m u)(t_L) = u(t_L),$$
$$(\hat{\Pi}_m u)(t_R) = u(t_R),$$
$$(\hat{\Pi}_m u)^{(i)}\left(\tfrac{1}{2}(1-t_R)\right) = u^{(i)}\left(\tfrac{1}{2}(1-t_R)\right) \qquad (0 \leq i \leq k-1).$$

Since

$$\|Sf - \hat{\phi}_n(\hat{N}_n f)\|_{H^1(I)} = \inf_{s \in S_m}\|u - s\|_{H^1(I)} \leq \|u - \hat{\Pi}_m u\|_{H^1(I)},$$

it suffices to show that

$$\|u - \hat{\Pi}_m u\|_{H^1(I)} = O(\gamma^{-n}) \qquad \text{as } n \to \infty.$$

Let $e = u - \hat{\Pi}_m u$. We have

$$\|e\|^2_{H^1(0,t_L)} = \|e\|^2_{H^1(0,t_L)} + \|e\|^2_{H^1(t_L,t_R)} + \|e\|^2_{H^1(t_R,1)}.$$

We first look at the left-hand subinterval $(0, t_L)$. Since $e^{(k)}(t_L) = 0$ and $(\hat{\Pi}_m u)^{(k+1)} \equiv 0$ on this subinterval, the previous lemma implies that

$$\|e^{(k)}\|_{L_2(0,t_L)} \leq \frac{2t_L}{\pi} \|e^{(k+1)}\|_{L_2(0,t_L)} = \frac{2t_L}{\pi} \|u^{(k+1)}\|_{L_2(0,t_L)}.$$

Since $e^{(i)}(\frac{1}{2}t_L) = 0$ for $0 \leq i \leq k - 1$, we have

$$\|e^{(i)}\|_{L_2(0,t_L)} \leq \frac{t_L}{\pi} \|e^{(i+1)}\|_{L_2(0,t_L)} \qquad (0 \leq i \leq k - 1).$$

Using these results, we find that

$$\|e'\|_{L_2(0,t_L)} \leq \left(\frac{t_L}{\pi}\right)^{k-1} \|e^{(k)}\|_{L_2(0,t_L)} \leq 2\left(\frac{t_L}{\pi}\right)^k \|u^{(k+1)}\|_{L_2(0,t_L)}$$

and

$$\|e\|_{L_2(0,t_L)} \leq \frac{t_L}{\pi} \|e'\|_{L_2(0,t_L)} \leq 2\left(\frac{t_L}{\pi}\right)^{k+1} \|u^{(k+1)}\|_{L_2(0,t_L)}.$$

Combining these results, we thus have

$$\|e\|^2_{H^1(0,t_L)} = \|e\|^2_{L_2(0,t_L)} + \|e'\|^2_{L_2(0,t_L)}$$

$$\leq \left[2\left(\frac{t_L}{\pi}\right)^k \|u^{(k+1)}\|_{L_2(0,t_L)}\right]^2 \left[1 + \left(\frac{t_L}{\pi}\right)^2\right]$$

$$\leq (1 + \pi^{-2}) \left[2\left(\frac{t_L}{\pi}\right)^k \|u^{(k+1)}\|_{L_2(0,t_L)}\right]^2.$$

Now on $(0, t_L)$, we have $f \equiv f(0)$. Differentiating the equation

$$-u'' + u = f(0),$$

we find that

$$u^{(k+1)} = \begin{cases} u' & k \text{ even}, \\ u - f & k \text{ odd}. \end{cases}$$

Hence for even k, we have

$$\|u^{(k+1)}\|_{L_2(0,t_L)} = \|u'\|_{L_2(0,t_L)} \leq \|u\|_{H^1(I)} \leq \|f\|_{L_2(I)} \leq M,$$

while for odd k,

$$\|u^{(k+1)}\|_{L_2(0,t_L)} \leq \|u\|_{L_2(0,t_L)} + \|f\|_{L_2(0,t_L)}$$
$$\leq \|u\|_{H^1(I)} + \|f\|_{L_2(I)} \leq 2\|f\|_{L_2(I)} \leq 2M.$$

(See **NR 2.3:2** for a proof that $\|u\|_{H^1(I)} \leq \|f\|_{L_2(I)}$.) Thus we have

$$\|e\|^2_{H^1(0,t_L)} \leq (1+\pi^{-2})\left[2M\left(\frac{t_L}{\pi}\right)^k\right]^2 \leq (1+\pi^{-2})4M^2\left(\frac{1}{\pi}\right)^{2k}.$$

By the way k and l were chosen, we have $\pi^k = \Theta(2^l) = \Theta(\gamma^n)$. Hence

$$\|e\|^2_{H^1(0,t_L)} = O(\gamma^{-2n}). \tag{8}$$

Similarly, we have

$$\|e\|^2_{H^1(t_R,1)} = O(\gamma^{-2n}). \tag{9}$$

It remains to show that

$$\|e\|^2_{H^1(t_L,t_R)} = O(\gamma^{-2n}). \tag{10}$$

Since $e(t_L) = e(t_R) = 0$, we find that

$$\|e'\|_{L_2(t_L,t_R)} \leq \frac{t_R - t_L}{\pi}\|e''\|_{L_2(t_L,t_R)} \leq \frac{t_R - t_L}{\pi}\|u''\|_{0,(t_L,t_R)}$$

and

$$\|e\|_{L_2(t_L,t_R)} \leq \frac{t_R - t_L}{\pi}\|e'\|_{L_2(t_L,t_R)} \leq \pi^{-1}\|e'\|_{L_2(t_L,t_R)}.$$

So

$$\|e\|^2_{H^1(t_L,t_R)} \leq (1+\pi^{-2})\left[\left(\frac{t_R - t_L}{\pi}\right)\|u''\|_{L_2(t_L,t_R)}\right]^2.$$

As in **NR 2.3:2**, we find that

$$\|u''\|_{L_2(t_L,t_R)} \leq \|u''\|_{L_2(I)} \leq 2\|f\|_{L_2(I)} \leq 2M,$$

and so

$$\|e\|^2_{H^1(t_L,t_R)} \leq (1+\pi^{-2})4M\pi^{-2}(t_R - t_L)^2.$$

Since $t_R - t_L = \Theta(2^{-l}) = \Theta(\gamma^{-n})$, (10) follows as claimed. Adding (8), (9), and (10), we find that

$$\|e\|_{H^1(I)} = O(\gamma^{-n}),$$

completing the proof. $\square$

Note that the optimal algorithms presented in Theorem 5.6.1 are all linear. Hence the results in §4.4 immediately yield

COROLLARY 5.7.1. *For* $p = 1$, *let* $\text{comp}^{\text{a}}(\varepsilon)$ *and* $\text{comp}^{\text{non}}(\varepsilon)$ *respectively denote the* ε-*complexity over the classes of algorithms using adaptive and nonadaptive information. Then*

$$\text{comp}^{\text{a}}(\varepsilon) = \Theta\big(\log(\varepsilon^{-1})\big) \qquad \text{as } \varepsilon \to 0$$

and

$$\text{comp}^{\text{non}}(\varepsilon) = \Theta(\varepsilon^{-1}) \qquad \text{as } \varepsilon \to 0. \qquad \square$$

Hence for one breakpoint, adaption gives an exponential improvement in both the nth minimal error and the ε-complexity. What happens if we allow more than one breakpoint?

THEOREM 5.7.3. *Let* $p \geq 2$. *Then*

$$r^{\text{a}}(n, \Lambda^{\text{std}}) = \Theta(n^{-1}) \qquad \text{as } n \to \infty$$

and

$$r^{\text{non}}(n, \Lambda^{\text{std}}) = \Theta(n^{-1}) \qquad \text{as } n \to \infty.$$

PROOF. We first show that

$$r^{\text{a}}(n, \Lambda^{\text{std}}) \geq \frac{M}{n}. \tag{11}$$

Let N be adaptive information of cardinality n. Apply N to the function $f_M \equiv M$, i.e. f_M is a constant. Let the points of evaluation chosen by N for f_M be $t_1, \ldots, t_n$, which we assume are increasing without loss of generality. We also let $t_0 = 0$ and $t_{n+1} = 1$. Choose an index $i = i(f_M)$ such that

$$t_{i+1} - t_i = \max_{0 \leq j \leq n} t_{j+1} - t_j.$$

For $\eta \in \big(0, (2n+1)^{-1}\big)$, we let

$$\tilde{f}_M(t) = \begin{cases} -M & \text{if } t \in [t_i + \eta, t_{i+1} - \eta], \\ M & \text{otherwise.} \end{cases}$$

Then $f_M, \tilde{f}_M \in F$, with $N f_M = N \tilde{f}_M$. Using Lemmas 5.7.1 and 5.7.2, we find that

$$d(N, y) \geq \|S f_M - S \tilde{f}_M\|_{H^1(I)} = \|f_M - \tilde{f}_M\|_{H^{-1}(I)}$$

$$= \sup_{\substack{v \in H^1(I) \\ \|v\|_{H^1(I)} \leq 1}} \left| \int_0^1 (f_M(x) - \tilde{f}_M(x)) v(x) \, dx \right|$$

$$= \sup_{\substack{v \in H^1(I) \\ \|v\|_{H^1(I)} \leq 1}} \left| \int_{t_i + \eta}^{t_{i+1} - \eta} v(x) \, dx \right|$$

$$\geq 2M(t_{i+1} - t_i - 2\eta).$$

Since $\eta > 0$ may be chosen arbitrarily small, we have

$$r(N, y) \geq \tfrac{1}{2} d(N, y) \geq M(t_{i+1} - t_i) \geq \frac{M}{n}.$$

Since $y \in N(F)$ and adaptive information N of cardinality n are arbitrary, the desired lower bound (11) follows immediately.

To complete the proof of this theorem, we describe nonadaptive information N_n of cardinality n and an algorithm ϕ using N such that

$$e(\phi, N) \leq \frac{M\beta p}{n-1}. \tag{12}$$

Let

$$t_i = \frac{i-1}{n-1} \quad (1 \leq i \leq n)$$

and choose information N of the form

$$Nf = \begin{bmatrix} f(t_1) \\ \vdots \\ f(t_n) \end{bmatrix} \qquad \forall f \in F.$$

Define functions $f_L, f_U \in F$ by

$$f_L\big|_{[t_i, t_{i+1}]} = \min\{y_i, y_{i+1}\} \quad \text{for } 1 \leq i \leq n-1$$

and

$$f_U\big|_{[t_i, t_{i+1}]} = \max\{y_i, y_{i+1}\} \qquad \text{for } 1 \leq i \leq n-1.$$

Define $f_M \in F$ to be

$$f_M = \tfrac{1}{2}(f_L + f_U),$$

so that

$$f(t) = \begin{cases} y_i & \text{if } y_i = y_{i+1} \\ y_{i+1} & \text{otherwise} \end{cases} \qquad \text{if } t \in [t_i, t_{i+1}] \text{ for } 1 \leq i \leq n-1.$$

We see that f_L and f_U form an envelope for $N^{-1}(y)$, and that f_M is its center. Define an algorithm ϕ by

$$\phi(y) = Sf_M.$$

Since f_M is a center of $N^{-1}(y)$, we have

$$e(\phi, N, y) = r(N, y) = \|Sf_M - Sf_U\|_{H^1(I)} = \|f_M - f_U\|_{H^{-1}(I)}.$$

Now f_M and f_U differ on r subintervals, where $r \leq p$. Each of these subintervals has length $1/(n-1)$. On the jth such subinterval (say, $[t_{s_j}, t_{s_j+1}]$), we have

$$|f_M(t) - f_U(t)| = \tfrac{1}{2}|y_{s_j+1} - y_{s_j}|.$$

Hence Lemmas 5.7.1 and 5.7.2 imply that

$$\|f_M - f_U\|_{H^{-1}(I)} = \sup_{\substack{v \in H^1(I) \\ \|v\|_{H^1(I)} \leq 1}} \left| \int_0^1 \big(f_M(t) - f_U(t)\big) v(t)\, dt \right|$$

$$= \sup_{\substack{v \in H^1(I) \\ \|v\|_{H^1(I)} \leq 1}} \sum_{j=1}^r \tfrac{1}{2}|y_{s_j+1} - y_{s_j}| \left| \int_{t_{s_j}}^{t_{s_j+1}} v(t)\, dt \right|$$

$$= M \sup_{\substack{v \in H^1(I) \\ \|v\|_{H^1(I)} \leq 1}} \sum_{j=1}^r \left| \int_{t_{s_j}}^{t_{s_j+1}} v(t)\, dt \right|$$

$$\leq \frac{M\beta r}{n-1} \leq \frac{M\beta p}{n-1}.$$

Thus

$$e(\phi, N) = \sup_{y \in N(F)} e(\phi, N, y) = \sup_{f \in F} \|f_M - f_U\|_{H^{-1}(I)} \leq \frac{M\beta p}{n-1},$$

establishing (12). $\square$

Note that the optimal algorithm ϕ in the proof of this theorem is (in principle) simple to compute. Indeed, since $\phi(y) = Sf_M$ and f_M is a piecewise constant with at most p pieces, one can find an explicit formula for Sf_M. However, if p is large, the formula for $\phi(y)$ becomes quite complicated. Furthermore, if S were to be replaced by a more complicated solution operator, computing Sf_M would become more difficult. As in the case $p = 1$, we want to find a less complicated algorithm that is almost optimal. Once again, we find that a finite element method will serve our purposes.

Let $\mathcal{S}_n$ be a space of continuous, piecewise linear polynomials over a uniform mesh. That is, letting

$$x_i = \frac{i-1}{n-1} \qquad (1 \leq i \leq n),$$

we say that $s \in \mathcal{S}_n$ iff s is continuous on I and

$$s\big|_{[x_i, x_{i+1}]} \in P_1([x_i, x_{i+1}]) \qquad (1 \leq i \leq n-1).$$

Let $s_1, \ldots, s_n$ be a basis for $\mathcal{S}_n$ such that

$$s_j(x_i) = \delta_{i,j} \qquad (1 \leq i, j \leq n).$$

For continuous $v \colon I \to \mathbb{R}$, we define the $\mathcal{S}_n$-interpolant $\Pi_n f$ to be

$$\Pi_n f = \sum_{j=1}^{n} f(x_j) s_j.$$

For $n \in \mathbb{P}$, we define finite element information N_n using $\mathcal{S}_n$ to be

$$N_n f = \begin{bmatrix} f(x_1) \\ \vdots \\ f(x_n) \end{bmatrix} \qquad \forall f \in F.$$

Then for $f \in F$, we choose $\tilde{u}_n \in \mathcal{S}_n$ satisfying

$$\begin{aligned}
\langle \tilde{u}_n, s_i \rangle_{H^1(I)} &= \langle \Pi_n f, s_i \rangle_{L_2(I)} \\
&= \sum_{j=1}^{n} f(x_j) \langle s_j, s_i \rangle_{L_2(I)} \qquad (1 \leq i \leq n).
\end{aligned} \tag{13}$$

Since $\tilde{u}_n$ depends on f only through $N_n f$, we may write

$$\tilde{u}_n = \phi_n(N_n f).$$

We see that ϕ_n is a (modified) FEM based on $\mathcal{S}_n$, using the finite element information N_n.

We now show that this FEM has almost minimal error:

THEOREM 5.7.4.

$$e(\phi_n, N_n) = \Theta(n^{-1}) \qquad \text{as } n \to \infty.$$

PROOF. Let $f \in F$. Define $u_n \in \mathcal{S}_n$ by

$$\langle u_n, s_i \rangle_{H^1(I)} = \langle f, s_i \rangle_{L_2(I)} \qquad (1 \leq i \leq n).$$

Hence u_n and $\tilde{u}_n$ respectively are the "pure" and "modified" finite element approximations to $u = Sf$. Subtracting this equation from (13), we find that

$$\langle u_n - \tilde{u}_n, s \rangle_{H^1(I)} = \langle f - \Pi_n f, s \rangle_{L_2(I)} \qquad \forall s \in \mathcal{S}_n.$$

In particular, let $s = u_n - \tilde{u}_n$ to find

$$\|u_n - \tilde{u}_n\|_{H^1(I)}^2 = \langle f - \Pi_n f, u_n - \tilde{u}_n \rangle_{L_2(I)} \leq \|f - \Pi_n f\|_{H^{-1}(I)} \|u_n - \tilde{u}_n\|_{H^1(I)},$$

and so

$$\|u_n - \tilde{u}_n\|_{H^1(I)} \le \|f - \Pi_n f\|_{H^{-1}(I)}.$$

Suppose we can show that

$$\|f - \Pi_n f\|_{H^{-1}(I)} \le CMn^{-1}. \tag{14}$$

Since we know that the error of the "pure" FEM satisfies

$$\|u - u_n\|_{H^1(I)} \le Cn^{-1}\|u\|_{H^2(I)} \le Cn^{-1}\|f\|_{L_\infty(I)} \le CMn^{-1},$$

we then will have

$$\|Sf - \phi_n(N_n f)\|_{H^1(I)} \le \|u - u_n\|_{H^1(I)} + \|u_n - \tilde{u}_n\|_{H^1(I)} \le CMn^{-1},$$

completing the proof.

To prove (14), let $v \in H^1(I)$ with $\|v\|_{H^1(I)} \le 1$. Then f and $\Pi_n f$ differ on r subintervals $[t_{s_i}, t_{s_i+1}]$, where $1 \le i \le r$. Writing $e = f - \Pi_n f$, we have

$$\left| \int_0^1 e(x)v(x)\,dx \right| \le \sum_{i=1}^r \left| \int_{t_{s_i}}^{t_{s_i+1}} e(x)v(x)\,dx \right|$$

$$\le \sum_{i=1}^r \|e\|_{L_\infty(t_{s_i}, t_{s_i+1})} \int_{t_{s_i}}^{t_{s_i+1}} |v(x)|\,dx$$

$$\le r\,\|e\|_{L_\infty(I)} \max_{1 \le r \le r} \int_{t_{s_i}}^{t_{s_i+1}} |v(x)|\,dx\,.$$

By Lemma 5.4.3, there exists a constant C (independent of f and n) such that

$$\|e\|_{L_\infty(I)} \le C\|f\|_{L_\infty(I)} \le CM.$$

Let $h = (n+1)^{-1}$. Since $r \le p$ and p is fixed, and the length of each subinterval $[t_{s_i}, t_{s_i+1}]$ is h, we see that there is a positive constant C such that

$$\left| \int_0^1 e(x)v(x)\,dx \right| \le CM \sup_{0 \le a \le 1-h} \int_a^{a+h} |v(x)|\,dx. \tag{15}$$

Let $\delta > 0$, and choose a polynomial v_δ such that

$$\|v - v_\delta\|_{H^1(I)} < \delta.$$

Define a second polynomial w_δ to be

$$w_\delta = \frac{v_\delta}{1+\delta}.$$

Since

$$\int_a^{a+h} |v(x) - v_\delta(x)| \, dx \le \|v - v_\delta\|_{L_1(I)} \le \|v - v_\delta\|_{L_2(I)} \le \|v - v_\delta\|_{H^1(I)} < \delta,$$

we find that

$$\int_a^{a+h} |v(x)| \, dx \le \int_a^{a+h} |v(x) - v_\delta(x)| \, dx + \frac{1}{1+\delta} \int_a^{a+h} |w_\delta(x)| \, dx$$

$$\le \delta + \frac{1}{1+\delta} \int_a^{a+h} |w_\delta(x)| \, dx. \tag{16}$$

Clearly $w_\delta \in H^1(I)$, with $\|w_\delta\|_{H^1(I)} \le 1$. Let $|w_\delta|: [0,1] \to \mathbb{R}$ be the function whose value at $x \in [0,1]$ is given by $|w_\delta|(x) = |w_\delta(x)|$. We claim that

$$|w_\delta| \in H^1(I) \qquad \text{and} \qquad \||w_\delta|\|_{H^1(I)} \le 1. \tag{17}$$

Indeed, we have

$$\big||w_\delta|'(x)\big| = |w_\delta'(x)| \qquad \forall x \notin w_\delta^{-1}(0).$$

Since w_δ is a polynomial, $w_\delta^{-1}(0)$ is finite, and so $\big||w_\delta|'\big| = |w_\delta'|$ almost everywhere. Since $w_\delta \in H^1(I)$ with $\|w_\delta\|_{H^1(I)} \le 1$, we thus see that $|w_\delta| \in H^1(I)$, with

$$\||w_\delta|\|_{H^1(I)}^2 = \||w_\delta|\|_{L_2(I)}^2 + \int_0^1 \big(|w_\delta(x)|'\big)^2 \, dx$$

$$= \|w_\delta\|_{L_2(I)}^2 + \int_0^1 \big(w_\delta'(x)\big)^2 \, dx$$

$$= \|w_\delta\|_{H^1(I)}^2 \le 1,$$

as claimed.

Now Lemma 5.6.2, (16), and (17) yield

$$\int_a^{a+h} |v(x)| \, dx \le \delta + \frac{1}{1+\delta} \sqrt{\frac{4}{3}} h.$$

Since $\delta > 0$ may be chosen arbitrarily small, we thus have

$$\int_a^{a+h} |v(x)| \, dx \le \sqrt{\frac{4}{3}} h.$$

Substituting this result into (15), we then have

$$\left| \int_0^1 e(x) v(x) \, dx \right| \le \frac{CM}{n}.$$

Since $v \in H^1(I)$ with $\|v\|_{H^1(I)} \leq 1$ is arbitrary, we have

$$\|e\|_{H^{-1}(I)} = \sup_{\substack{v \in H^1(I) \\ \|v\|_{H^1(I)} \leq 1}} \left| \int_0^1 e(x)v(x)\, dx \right| \leq \frac{CM}{n},$$

which establishes (14) and completes the proof of the theorem.   □

Since the optimal algorithm presented in the last two theorems is linear, we may once again use the results in §4.4 to find that

COROLLARY 5.7.2. *For $p \geq 2$, let* $\text{comp}^a(\varepsilon)$ *and* $\text{comp}^{non}(\varepsilon)$ *respectively denote the ε-complexity over the classes of algorithms using adaptive and nonadaptive information. Then*

$$\text{comp}^a(\varepsilon) = \Theta(\varepsilon^{-1}) \qquad \text{as } \varepsilon \to 0$$

and

$$\text{comp}^{non}(\varepsilon) = \Theta(\varepsilon^{-1}) \qquad \text{as } \varepsilon \to 0. \qquad \square$$

Thus adaptive information is no more powerful than nonadaptive information when there are more than two breakpoints.

Notes and Remarks

NR 5.7:1 The idea of using this particular nonconvex class F of problem elements was used by Huerta (1986). However, his paper was concerned with the integration problem (i.e. the problem of computing the definite integral of functions from F using standard information about such functions) rather than with boundary value problems.

NR 5.7:2 We cannot bring the upper bound β of Lemma 5.6.2 down to (say) 1 for all a and b. Indeed, choose

$$v_t(x) = \frac{\sqrt{3}(tx+1)}{\sqrt{4t^2 + 3t + 3}},$$

so that $\|v_t\|_{H^1(I)} = 1$. Suppose that $a = 0$. Then

$$\frac{1}{b-a} \int_a^b v_t(x)\, dx = \frac{1}{b} \int_0^b v_t(x)\, dx = \frac{\sqrt{3}(2+bt)}{2\sqrt{4t^2 + 3t + 3}}.$$

Chosing $t = t^* = -(6b-6)/(3b-16)$, we find that the right-hand side of this equation is maximized. Hence

$$\frac{1}{b-a} \int_a^b v_{t^*}(x)\, dx = \sqrt{\frac{16 - 6b + 3b^2}{13}}.$$

For small b (i.e. as $b \to 0$), we find

$$\frac{1}{b-a} \int_a^b v_{t^*}(x)\, dx \sim \sqrt{\frac{16}{13}} \doteq 1.10940.$$

Note that the ratio between the upper bound $\beta = \sqrt{4/3}$ of the lemma and the asymptotic lower bound $\sqrt{16/13}$ is about 1.0408. Hence, the upper bound of the lemma is off by at most about 4%.

NR 5.7:3 There are three ways to increase the accuracy of a finite element method. One can

(i)   decrease the mesh-size and keep the degree of the finite element space fixed (the h-method),

(ii) increase the degree and keep the mesh-size fixed (the p-method), or

(iii) simultaneously increase the degree and decrease the mesh-size (the (h, p)-method).

The "classical" FEM is an h-method, whereas the adaptive FEM used in the proof of Theorem 5.7.2 is an (h, p)-method. The p- and (h, p)-methods have become more widely studied over the last few years. For references, see Babuška (1987) and Babuška and Guo (1988).

6

Other problems in
the worst case setting

6.1 Introduction

In the previous chapter, we applied the techniques of information-based complexity to find algorithms that are optimal in the worst case setting for elliptic partial differential equations (PDE). In this chapter, we apply these techniques to find algorithms that are (worst case) optimal for several other important classes of problems. The first section deals with elliptic systems of PDE. The second section deals with well-posed integral equations (IE), namely the Fredholm problem of the second kind. The third section is a general treatment of linear ill-posed problems, with special emphasis on the Fredholm IE of the first kind. Finally, the fourth section deals with ordinary differential equations (ODE).

Although it is more common in many texts to discuss ODE before PDE, we have reversed the customary order in this book. This is because the elliptic PDE and the IE that we study are linear, and so we can use the results in §4.4 to help us find optimal algorithms if the class of problem elements is balanced and convex.

Recall our main results from the previous chapter:

(i) We found conditions that were necessary and sufficient for a finite element method (FEM or FEMQ) to be an almost optimal complexity algorithm.

(ii) If the error was measured in the most natural norm for the problem, and the problem elements had the minimal amount of smoothness to guarantee the existence and uniqueness of solutions, then the problem was non-convergent.

(iii) If the problem was convergent, it was intractable, in the sense that its complexity depended exponentially on the dimension of the domain.

In this chapter, we show that these results also hold for elliptic systems and for Fredholm problems of the second kind. This means that if we wish to solve these problems, avoiding either the non-convergence or the inherent intractability, we must abandon the worst case setting. We deal with other settings later in this book.

Our main result for ill-posed problems is also quite striking. Ill-posed problems cannot be solved with finite error in the worst case setting under the absolute error criterion; however, they *can* be solved under the residual error criterion.

We then discuss initial value problems for ordinary differential equations. The flavor of the results depends on what kind of information is permissible. If only standard information is available, we find that certain Taylor-series and Runge-Kutta methods are nearly optimal algorithms. However, if continuous linear information is available, a "Taylor-integral" method turns out to be nearly optimal. Moreover, standard information is weaker than continuous linear information, in the sense that the asymptotic penalty for using standard information (when Taylor-integral information is available) is unbounded. Finally we examine the complexity of the problem if mildly continuous nonlinear information is available, showing that such "finitely continuous" information is no stronger than continuous linear information.

Once again, we remind the reader that we will follow the standard notational convention of using the letter C in inequalities to denote a generic finite positive constant, independent of the functions and parameters appearing in that inequality.

6.2 Linear elliptic systems

In this section, we consider the complexity of linear elliptic systems. These systems typically arise in the mathematical formulation of boundary value problems arising in elasticity and fluid dynamics. For further background regarding theoretical properties of such systems, techniques for computing approximations to their solutions, and numerical test results of such techniques, see (among others) Agmon *et al.* (1959), Aziz and Leventhal (1978), Aziz and Werschulz (1980), Baker (1973), Bramble and Schatz (1971), Fix *et al.* (1981), Fix and Gurtin (1977), Hormander (1983), Lions and Magenes (1972), Roïtberg and Šeftel (1969), and Wendland (1979).

We briefly outline the contents of this section. First, we define tth-order elliptic systems over a region $\Omega \subset \mathbb{R}^d$ in the sense of Petrovsky, and then formulate our problem of finding a Sobolev l-norm ε-approximation of the solution, where $0 \leq l \leq t$. Let $r \geq 0$. We analyze the normed case $F = BH^r(\Omega)$ in detail, leaving the seminormed case for the Notes and Remarks. First, we assume that arbitrary linear information is permissible. We find that the ε-complexity of the problem is $\Theta(\varepsilon^{-d/(r+t-l)})$. Moreover, the least squares FEM of degree k is optimal iff $k \geq t - 1 + r$. Note that the problem is intractable if the dimension d is large, and is not convergent if $r + t = l$, i.e. if we are trying to solve in the problem's natural norm ($l = r$) and if the problem elements are only smooth enough to guarantee existence and uniqueness of solutions in that norm ($r = 0$).

Next, we suppose that only standard information is available. To guarantee that standard information is well-defined, we must assume that $r > \frac{1}{2}d$; for $r \leq \frac{1}{2}d$, the complexity is infinite for sufficiently-small ε. We then find that the ε-complexity is $\Theta(\varepsilon^{-d/r})$. Furthermore, the least squares FEM of degree k with quadrature is optimal iff $k \geq r + l - 1$. Hence, we see that there is an unavoidable penalty to be paid if only standard information is available. In all the cases mentioned, regardless of whether the least squares FEM is almost optimal, we find that the information used by the least squares FEM is always almost optimal, i.e. the spline algorithm using this information is always an optimal algorithm.

In this section, we will use Sobolev spaces $H^s(\Omega)^d$ of $\mathbb{R}^d$-valued functions, as well as the more familiar Sobolev spaces $H^s(\Omega)$ of $\mathbb{R}$-valued functions. The space $H^s(\Omega)^d$ is a Hilbert space under the inner product $\langle \cdot, \cdot \rangle_{H^s(\Omega)}$ defined by

$$\langle v, w \rangle_{H^s(\Omega)} = \sum_{j=1}^{d} \langle v_i, w_i \rangle_{H^s(\Omega)}$$

for $v = (v_1, \ldots, v_d)$ and $w = (w_1, \ldots, w_d)$. Since which space is intended can be inferred from context, we will use $H^s(\Omega)$ in the sequel to denote both the $\mathbb{R}$- and $\mathbb{R}^d$-valued Sobolev spaces (where this will cause no confusion).

6.2.1 Problem formulation

In this subsection, we define precisely what is meant by an elliptic system in the sense of Petrovsky. We use the standard notation for elliptic systems found in Agmon *et al.* (1959). However, we point out that this notation does conflict with the standard notation for the scalar elliptic problem discussed in the previous section.

Let $\Omega \subset \mathbb{R}^d$ be a bounded, simply connected, C^∞ region. Define the differential operator

$$\ell(x, \partial) = [\ell_{ij}(x, \partial)]_{1 \leq i, j \leq d},$$

with

$$\ell_{ij}(x, \xi) = \sum_{|\alpha| \leq t} a_\alpha^{ij}(x) \xi^\alpha \qquad (1 \leq i, j \leq d)$$

for coefficients $a_\alpha^{ij} \in C^\infty(\overline{\Omega})$. Here, t is a nonnegative integer. Let

$$\ell_{ij}^0(x, \xi) = \sum_{|\alpha| = t} a_\alpha^{ij}(x) \xi^\alpha$$

denote the *principal part* of ℓ_{ij}. We assume that ℓ is *elliptic*, i.e. that

$$L(x,\xi) \equiv \det[\ell_{ij}^0(x,\xi)] \neq 0 \qquad \forall\, x \in \overline{\Omega}, \forall \text{ nonzero } \xi \in \mathbb{R}^d.$$

We claim that td is even, i.e. $td = 2m$ for some integer m. Indeed, for $x \in \partial\Omega$, let τ_x and ν_x respectively denote the unit tangent and outer unit normal to $\partial\Omega$ at x, and set

$$L_x(\eta) = L(x, \tau_x + \eta\nu_x) \qquad \forall\, \eta \in \mathbb{C}.$$

Then L_x is a polynomial of degree td in the complex variable η. Since ℓ is elliptic, L_x has no real roots. Since L_x has only real coefficients, the degree td of L must be even, i.e. $td = \deg L_x = 2m$, as claimed. Moreover, we see that there is a factorization

$$L_x(\eta) = L_x^+(\eta)L_x^-(\eta),$$

where the zeros of L_x^+ and L_x^- have positive and negative real part, respectively, and $\deg L_x^+ = \deg L_x^- = m$.

We now specify a boundary operator

$$b(x, \partial) = [b_{ij}(x, \partial)]_{1 \leq i \leq m, 1 \leq j \leq d}$$

by

$$b_{ij}(x, \xi) = \sum_{|\alpha| \leq r_i} b_\alpha^{ij}(x)\xi^\alpha,$$

where $r_1, \ldots, r_n \in \mathbb{P}$ and the coefficients $b_\alpha^{ij} \in C^\infty(\partial\Omega)$.

We now need the boundary operator b to be compatible with the operator ℓ. Define the *principal part* b_{ij}^0 of b_{ij} to be

$$b_{ij}^0(x, \xi) = \sum_{|\alpha| = r_i} b_\alpha^{ij}(x)\xi^\alpha,$$

and let $L^{jk}(x, \xi)$ be the cofactor of ℓ_{jk}^0 in the matrix of principal parts. For $x \in \partial\Omega$ and $\eta \in \mathbb{C}$, let

$$C_x(\eta) = [c_{ij}(x, \eta)]_{1 \leq i \leq m, 1 \leq j \leq d}$$

with

$$c_{ij}(x, \eta) = \sum_{k=1}^d b_{ik}^0(x, \tau_x + \eta\nu_x)L^{jk}(x, \tau_x + \eta\nu_x).$$

Then b is *compatible* with ℓ if the row vectors of C_x, considered as polynomials in $\eta \in \mathbb{C}$, are linearly independent relative to the modulus of $L_x^+(\eta)$.

We say that ℓ and b are *elliptic* on $\overline{\Omega}$ if ℓ is elliptic and b is compatible with ℓ. For $s \geq 0$, let $H^s(\partial)$ denote the completion of

$$\{\, v \in C^\infty(\overline{\Omega}) : bu = 0 \text{ on } \partial\Omega \,\}$$

with respect to the usual Sobolev norm on $H^s(\Omega)$.

Since we will not require that our problems be self-adjoint, we must define the *formal adjoint* ℓ^+ of ℓ to be

$$\ell^+(x, \partial) = [\ell_{ij}^+(x, \partial)]_{1 \leq i,j \leq d},$$

with

$$\ell_{ij}^+(x, \partial)u_j(x) = \sum_{|\alpha| \leq t} \partial^\alpha \big(a_\alpha^{ij}(x)u_j(x)\big) \qquad (1 \leq i, j \leq d).$$

Integrating by parts, we may define an adjoint boundary operator b^+ such that

$$\langle \ell v, w \rangle_{L_2(\Omega)} = \langle v, \ell^+ w \rangle_{L_2(\Omega)} \qquad \forall\, v \in H^t(\partial),\ \forall\, w \in H^t(\partial)^+,$$

where for $s \geq 0$, we let $H^s(\partial)^+$ denote the completion of

$$\{\, v \in C^\infty(\overline{\Omega}) : b^+ u = 0 \text{ on } \partial\Omega \,\}$$

with respect to the usual Sobolev norm on $H^s(\Omega)$.

In the remainder of this section, we shall assume that

(i) ℓ and b are elliptic on $\overline{\Omega}$, and
(ii) ℓ^+ and b^+ are elliptic on $\overline{\Omega}$.

EXAMPLE: *An elliptic system for Poisson's equation.* Let $\Omega \subset \mathbb{R}^2$ be a smooth region, and let $\psi \colon \overline{\Omega} \to \mathbb{R}$ be the solution of

$$-\Delta\psi = f \qquad \text{in } \Omega,$$
$$\psi = 0 \qquad \text{on } \partial\Omega.$$

We are often more interested in the gradient of the solution than the solution itself. For instance, in fluid flow problems, ψ is a velocity potential, whereas $\nabla\psi$ is a velocity field. To this end, we may recast this problem as a first-order linear system. Let $u_1 = \partial_1\psi$ and $u_2 = \partial_2\psi$. Then $u = (u_1, u_2)$ satisfies the linear system

$$-(\partial_1 u_1 + \partial_2 u_2) = f \qquad \text{in } \Omega$$
$$\partial_2 u_1 - \partial_1 u_2 = 0$$

with boundary conditions

$$u(x)\cdot\tau_x = 0 \qquad \forall\, x \in \partial\Omega.$$

It is straightforward to check that this is an elliptic system, and that $\ell^+ = \ell$ and $b^+ = b$. Hence the conditions of this section are satisfied. $\square$

For future reference, we have the following result from Roĭtberg and Šeftel (1969):

THEOREM 6.2.1.1. *Let* $r \geq 0$.

(i) *For any* $f \in H^r(\Omega)$, *there exists a unique* $u \in H^{r+t}(\Omega)$ *such that*

$$\ell u = f \qquad \text{in } \Omega,$$
$$bu = 0 \qquad \text{on } \partial\Omega.$$

Moreover, there exists a constant $\sigma \geq 1$, *independent of* f *and* u, *such that*

$$\sigma^{-1}\|u\|_{H^{r+t}(\Omega)} \leq \|f\|_{H^r(\Omega)} \leq \sigma\|u\|_{H^{r+t}(\Omega)}.$$

(ii) *For any* $g \in H^r(\Omega)$, *there exists a unique* $v \in H^{r+t}(\Omega)$ *such that*

$$\ell^+ v = g \qquad \text{in } \Omega,$$
$$b^+ v = 0 \qquad \text{on } \partial\Omega.$$

Moreover, there exists a constant $\sigma \geq 1$, *independent of* g *and* v, *such that*

$$\sigma^{-1}\|v\|_{H^{r+t}(\Omega)} \leq \|g\|_{H^r(\Omega)} \leq \sigma\|v\|_{H^{r+t}(\Omega)}. \qquad \square$$

We are now ready to formulate our problem. Let l satisfy $0 \leq l \leq t$. We define our solution operator $S: H^r(\Omega) \to H^l(\Omega)$ by letting $u = Sf$ be the unique solution of

$$\ell u = f \qquad \text{in } \Omega,$$
$$bu = 0 \qquad \text{on } \partial\Omega.$$

Since $r \geq 0$ and $0 \leq l \leq t$, Theorem 6.2.1.1 implies that S is well-defined.

Next we choose our class Λ of permissible information operators as in the previous chapter, i.e. we will allow either $\Lambda = \Lambda'$ (arbitrary linear information), or $\Lambda = \Lambda^{\text{std}}$ (standard information).

For our choice of problem elements, we will mainly concentrate on the normed case $F = BH^r(\Omega)$. We leave a discussion of the seminormed case $F = \mathcal{B}H^r(\Omega)$ for the Notes and Remarks at the end of the appropriate subsections.

Notes and Remarks

NR 6.2.1:1 We only consider problems elliptic in the sense of Petrovsky in this section. This is mainly done to simplify the exposition. One could also consider systems that are elliptic in the sense of Agmon *et al.* (1959), which are more general than Petrovsky-elliptic systems.

NR 6.2.1:2 Roĭtberg and Šeftel (1969) give a *normality condition* such that ellipticity of ℓ and b over $\overline{\Omega}$ implies the ellipticity of ℓ^+ and b^+ over $\overline{\Omega}$.

NR 6.2.1:3 We deal only with the absolute error criterion in our treatment of elliptic systems. As in **NR 5.3:11**, we see that the following results hold for other error criteria:

(i) Using Theorem 6.2.1.1, we see that the results for the residual error criterion in the $\|\cdot\|_{H^{t+r}(\Omega)}$-norm are the same as those for the absolute error criterion under the $\|\cdot\|_{H^t(\Omega)}$-norm, to within a constant factor.

(ii) The results for the normalized error criterion are the same as for the absolute error criterion.

(iii) The problem is not solvable under the relative error criterion.

6.2.2 Arbitrary linear information

In this section, we assume that arbitrary linear information is permissible. Our general approach is analogous to that in §5.4. First, we give tight bounds on the nth minimal radius. Next, we consider minimal error algorithms for our problem. Since the spline algorithm using nth optimal information requires spectral information that is usually unavailable, we turn to a special class of finite element methods, called least squares finite element methods (LSFEM), hoping to determine whether such methods are optimal. We prove that the LSFEM of degree k is almost optimal iff $k \geq t - 1 + r$, where r is the regularity of the problem elements and t is the maximal order of the differential operators defining ℓ. Moreover, the least squares finite element information (LSFEI) used by the LSFEM is always almost optimal information. We finally determine the inherent problem complexity, comparing it to the complexity of using the LSFEM of degree k to solve an elliptic system.

Since we are studying a linear problem, nonadaptive information is as powerful as adaptive information. Hence, we first need to give tight bounds on the minimal radius of nonadaptive information of cardinality at most n.

THEOREM 6.2.2.1.

$$r(n) = \Theta(n^{-(r+t-l)/d}) \qquad \text{as } n \to \infty.$$

PROOF. The proof is analogous to that of Theorem 5.4.1, and hence we only sketch it. From Theorem 4.4.3.3, we have

$$r(n) = d^n\left(SF, H^l(\partial)\right).$$

For $\theta > 0$, let

$$X(\theta) = \theta BH^{r+t}(\partial).$$

From Theorem 6.2.1.1, we have

$$X(\sigma^{-1}) \subseteq SF \subseteq X(\sigma).$$

Since

$$d^n\left(X(\theta), H^l(\partial)\right) = \theta\, d^n\left(BH^{r+t}(\Omega), H^l(\partial)\right),$$

it then follows that

$$r(n) = \Theta\left(d^n\left(BH^{r+t}(\Omega), H^l(\partial)\right)\right).$$

Using Theorem 2.5.1 of Babuška and Aziz (1972), we find

$$d^n\left(BH^{r+t}(\Omega), H^l(\partial)\right) = \Theta\left(d^n\left(BH_0^1(\Omega), L_2(\Omega)\right)^{r+t-l}\right).$$

Since the results in Jerome (1968) yield that

$$d^n\left(BH_0^1(\Omega), L_2(\Omega)\right) = \Theta(n^{-1/d}),$$

the desired conclusion follows. □

As was the case for a scalar elliptic problem, the general theory tells how to construct an nth minimal error algorithm. Once again, this algorithm is useful in practice only if we can find the eigenvectors of S^*S. Since this seems to be impossible for our problem, we must consider other algorithms. If we can find a practical algorithm using information of cardinality n whose error is $\Theta\left(r(n)\right)$, then this algorithm will be an almost minimal error algorithm. Encouraged by our success in the scalar case, we will consider finite element methods of a special type.

Let $\{\mathcal{T}_n\}_{n=1}^\infty$ be a quasi-uniform family of triangulations of Ω. For each $n \in \mathbb{P}$, we let $\mathcal{S}_{n,k}$ be an n-dimensional subspace of $H^t(\partial)$ consisting of functions that are piecewise polynomial of degree k over a triangulation $\mathcal{T}_n$. From Lemma 5.4.2, we must have

$$k \geq t.$$

In what follows, we additionally assume that

$$\text{if } l < t, \text{ then } k \geq 2t - 1$$

(see §5.4 for discussion).

We now recall the definition of the least squares finite element method using $\{\mathcal{S}_{n,k}\}_{n=1}^\infty$, as applied to elliptic systems. Let $\{s_1, \dots, s_n\}$ be a basis for $\mathcal{S}_{n,k}$. For $f \in F$, we evaluate *least squares finite element information* (LSFEI)

$$N_{n,k}^{\text{LSFE}} f = \begin{bmatrix} \langle f, \ell s_1 \rangle_{L_2(\Omega)} \\ \vdots \\ \langle f, \ell s_n \rangle_{L_2(\Omega)} \end{bmatrix}.$$

Since ℓ and b form an elliptic system, the Gram matrix

$$G = [\langle \ell s_j, \ell s_i \rangle_{L_2(\Omega)}]_{1 \leq i,j \leq n}$$

is positive definite. This implies that there exists a unique vector

$$a = \begin{bmatrix} \alpha_1 \\ \vdots \\ \alpha_n \end{bmatrix}$$

such that

$$Ga = N_{n,k}^{\mathrm{LSFE}} f. \tag{1}$$

Let

$$u_{n,k} = \sum_{j=1}^{n} \alpha_j s_j.$$

Since $u_{n,k}$ depends on f only through the information $N_{n,k}^{\mathrm{LSFE}}$, we may write

$$u_{n,k} = \phi_{n,k}^{\mathrm{LSFE}}(N_{n,k}^{\mathrm{LSFE}} f).$$

We say that $\phi_{n,k}^{\mathrm{LSFE}}$ is the *least squares finite element method* (LSFEM) of cardinality n.

Clearly we have

$$\langle \ell u_{n,k}, \ell s \rangle_{L_2(\Omega)} = \langle f, \ell s \rangle_{L_2(\Omega)} \qquad \forall\, s \in \mathcal{S}_{n,k}.$$

It then follows that

$$\|f - \ell u_{n,k}\|_{L_2(\Omega)} = \min_{s \in \mathcal{S}_{n,k}} \|f - \ell s\|_{L_2(\Omega)},$$

explaining why this algorithm is called a "least squares method."

By analogy with the case of the FEM in Chapter 5, we can (in principle) compute the LSFEM $\phi_{n,k}^{\mathrm{LSFE}}(N_{n,k}^{\mathrm{LSFE}} f)$ using n information operations and $2n - 1$ arithmetic operations. Indeed, for $1 \leq j \leq n$, we first compute $u_j \in \mathcal{S}_{n,k}$ satisfying

$$\langle \ell u_j, \ell s_i \rangle_{L_2(\Omega)} = \delta_{i,j} \qquad (1 \leq i \leq n).$$

That is, we precompute $u_1, \ldots, u_n$, independently of the problem element f. Then for $f \in F$, we have

$$\phi_{n,k}^{\mathrm{LSFE}}(N_{n,k}^{\mathrm{LSFE}}) = u_{n,k} = \sum_{j=1}^{n} \langle f, s_j \rangle_{L_2(\Omega)} u_j.$$

Hence, the cost of computing $\phi_{n,k}^{\mathrm{LSFE}}(N_{n,k}^{\mathrm{LSFE}}f)$ is bounded by

$$\mathrm{cost}(\phi_{n,k}^{\mathrm{LSFE}}, N_{n,k}^{\mathrm{LSFE}}) \leq cn + 2n - 1,$$

as claimed.

Of course, this works only when the u_j are easy to obtain or if we wish to compute the LSFEM at a fixed n for many right-hand sides f. Otherwise, we have to solve the $n \times n$ linear system (1). See §5.4 for further discussion of this point, as well as pointers to the literature.

We are now ready to give tight bounds on the error of the LSFEM in the Sobolev l-norm.

THEOREM 6.2.2.2. *Let*

$$\mu = \min\{k + 1 - t, r\}.$$

Then

$$e(\phi_{n,k}^{\mathrm{LSFE}}, N_{n,k}^{\mathrm{LSFE}}) = \Theta(n^{-(\mu+t-l)/d}) \qquad \text{as } n \to \infty.$$

PROOF. We first show that

$$e(\phi_{n,k}^{\mathrm{LSFE}}, N_{n,k}^{\mathrm{LSFE}}) = O(n^{-(\mu+t-l)/d}) \qquad \text{as } n \to \infty. \qquad (2)$$

Suppose first that $l = t$. For $f \in F$, let $u = Sf$. Using Theorem 6.2.1.1 (twice) and the least squares property, we have

$$
\begin{aligned}
\|Sf - \phi_{n,k}^{\mathrm{LSFE}}(N_{n,k}^{\mathrm{LSFE}}f)\|_{H^t(\Omega)} &= \|Sf - S\ell u_{n,k}\|_{H^t(\Omega)} \\
&\leq \sigma \|f - \ell u_{n,k}\|_{L_2(\Omega)} \\
&\leq \sigma \|f - \ell \Pi_n Sf\|_{L_2(\Omega)} \\
&= \sigma \|\ell(Sf - \Pi_n Sf)\|_{L_2(\Omega)} \\
&\leq \sigma^2 \|Sf - \Pi_n Sf\|_{H^t(\Omega)},
\end{aligned}
$$

where Π_n is the $L_2(\Omega)$-orthoprojector onto $\mathcal{S}_{n,k}$. Since $\mu = \min\{r, k+1-t\}$, the basic approximation properties of $\{\mathcal{S}_{n,k}\}_{n=1}^{\infty}$ and Theorem 6.2.1.1 imply that

$$\|Sf - \Pi_n Sf\|_{H^t(\Omega)} \leq Cn^{-\mu/d}\|Sf\|_{H^{t+r}(\Omega)} \leq Cn^{-\mu/d}\|f\|_{H^r(\Omega)}.$$

Combining these results, we have

$$\|Sf - \phi_{n,k}^{\mathrm{LSFE}}(N_{n,k}^{\mathrm{LSFE}}f)\|_{H^t(\Omega)} \leq Cn^{-\mu/d}\|f\|_{H^r(\Omega)}. \qquad (3)$$

Taking the supremum over all $f \in F$, we find that the desired upper bound (2) holds when $l = t$.

Suppose now that $l < t$. Let $f \in F$. For $n \in \mathbb{P}$, let $e_n = Sf - \phi_{n,k}^{\mathrm{LSFE}}(N_{n,k}^{\mathrm{LSFE}} f)$. Choose $v_n = \Pi_n SS^+ e_n \in \mathcal{S}_{n,k}$, where $S^+ = (\ell^+)^{-1}$. Since $v_n \in \mathcal{S}_{n,k}$, we have $\langle \ell e_n, \ell v_n \rangle_{L_2(\Omega)} = 0$. Using this result along with Theorem 6.2.1.1, we have

$$
\begin{aligned}
\|e_n\|_{L_2(\Omega)}^2 &= \langle e_n, \ell^+ S^+ e_n \rangle_{L_2(\Omega)} \\
&= \langle \ell e_n, S^+ e_n \rangle_{L_2(\Omega)} \\
&= \langle \ell e_n, S^+ e_n - \ell v_n \rangle_{L_2(\Omega)} \\
&= \langle \ell e_n, \ell(SS^+ e_n - v_n) \rangle_{L_2(\Omega)} \\
&\leq \|\ell e_n\|_{L_2(\Omega)} \|\ell(SS^+ e_n - v_n)\|_{L_2(\Omega)} \\
&\leq \sigma^2 \|e_n\|_{H^t(\Omega)} \|SS^+ e_n - v_n\|_{H^t(\Omega)}.
\end{aligned}
\tag{4}
$$

Since $k \geq 2t - 1$, we see that $\min\{t, k + 1 - t\} = t$, and so

$$
\begin{aligned}
\|SS^+ e_n - v_n\|_{H^t(\Omega)} &\leq Cn^{-t/d} \|SS^+ e_n\|_{H^{2t}(\Omega)} \\
&\leq C\sigma n^{-t/d} \|S^+ e_n\|_{H^t(\Omega)} \leq C\sigma^2 n^{-t/d} \|e_n\|_{L_2(\Omega)}.
\end{aligned}
$$

(Here we have used Theorem 6.2.1.1 twice.) Substituting this inequality into (4), we have

$$
\|e_n\|_{L_2(\Omega)} \leq Cn^{-t/d} \|e_n\|_{H^t(\Omega)},
$$

which, along with (3), gives

$$
\|Sf - \phi_{n,k}^{\mathrm{LSFE}}(N_{n,k}^{\mathrm{LSFE}} f)\|_{L_2(\Omega)} \leq Cn^{-(\mu+t)/d} \|f\|_{H^r(\Omega)}.
\tag{5}
$$

Using (4), (5), and Hilbert-space interpolation Lemma A.2.2.1, we find that

$$
\begin{aligned}
\|Sf - \phi_{n,k}^{\mathrm{LSFE}}(N_{n,k}^{\mathrm{LSFE}} f)\|_{H^l(\Omega)} &\leq C \|Sf - \phi_{n,k}^{\mathrm{LSFE}}(N_{n,k}^{\mathrm{LSFE}} f)\|_{L_2(\Omega)}^{1-l/t} \times \\
&\quad \|Sf - \phi_{n,k}^{\mathrm{LSFE}}(N_{n,k}^{\mathrm{LSFE}} f)\|_{H^m(\Omega)}^{l/t} \\
&\leq Cn^{-(\mu+t-l)/d} \|f\|_{H^r(\Omega)}.
\end{aligned}
$$

Taking the supremum over all $f \in F$, the desired upper bound (2) holds. We next need to show the lower bound

$$
e(\phi_{n,k}^{\mathrm{LSFE}}, N_{n,k}^{\mathrm{LSFE}}) = \Omega(n^{-(\mu+t-l)/d}) \qquad \text{as } n \to \infty.
\tag{6}
$$

If $k + 1 - t \geq r$, then $\mu + t - l = r + t - l$, and our lower bound (6) follows from the Θ-estimate of $r(n)$ in the previous theorem. Otherwise, if $k + 1 - t < r$, then as in the proof of Theorem 5.4.2, we can find an open region Ω^1, independent of n, whose closure is contained in the interior of Ω, as well as $u \in H^{r+t}(\partial)$ such that

(i) there is a constant $C > 0$, independent of n, such that

$$\inf_{s \in \mathcal{S}_{n,k}} |u - s|_{H^l(\Omega)} \geq C n^{-(k+1-l)/d} |u|_{H^{k+1}(\Omega^1)},$$

(ii) $|u|_{H^{k+1}(\Omega^1)}$ is bounded from below by a positive constant, independent of n, and

(iii) $\|\ell u\|_{H^r(\Omega)}$ is bounded from above by a positive constant, independent of n.

Set

$$f^* = \frac{1}{\|\ell u\|_{H^r\Omega}} \ell u,$$

so that $f^* \in F$. Let

$$C' = C \inf_{n \in \mathbb{P}} \frac{|u|_{H^{k+1}(\Omega^1)}}{\|\ell u\|_{H^r(\Omega)}}$$

where C is the constant in (i) above. Since C' is a positive constant, we have

$$e(\phi_{n,k}^{\mathrm{LSFE}}, N_{n,k}^{\mathrm{LSFE}}) \geq \|Sf^* - \phi_{n,k}^{\mathrm{LSFE}}(N_{n,k}^{\mathrm{LSFE}} f^*)\|_{H^l(\Omega)}$$

$$\geq \frac{1}{\|\ell u\|_{H^r(\Omega)}} \inf_{s \in \mathcal{S}_{n,k}} |u - s|_{H^l(\Omega)}$$

$$\geq C' n^{-(k+1-l)/d},$$

which establishes (6) and completes the proof of the theorem. $\qquad\square$

Combining the previous two theorems, we immediately find that

COROLLARY 6.2.2.1. *The LSFEM of degree k is an almost minimal error algorithm iff $k \geq t - 1 + r$.* $\qquad\square$

As was the case for the scalar elliptic problem, we now ask why the LSFEM is not an almost minimal error algorithm. As before, the reason could be either of the following:

(i) The LSFEM does not use its information well, and there is a better algorithm using LSFEI whose error is almost minimal.

(ii) There is no algorithm using LSFEI whose error is almost minimal.

Given our experience in the previous chapter, it should come as no surprise that (i) is the reason the LSFEM does not have almost minimal error.

To prove this, we must first establish

LEMMA 6.2.2.1. *There exists $\sigma \geq 1$ such that*

$$\|\ell w\|_{H^{-r}(\Omega)} \leq \sigma \|w\|_{H^{t-r}(\Omega)} \qquad \forall w \in H^t(\partial).$$

PROOF. If $r = 0$, this result follows from Theorem 6.2.1.1. Suppose now that $r \geq t$. Let $w \in H^t(\partial)$. For any $v \in C_0^\infty(\Omega)$, Theorem 6.2.1.1 (with r replaced by the nonnegative real number $r - t$) implies that

$$|\langle \ell w, v\rangle_{L_2(\Omega)}| = |\langle w, \ell^+ v\rangle_{L_2(\Omega)}| \leq \|w\|_{H^{t-r}(\Omega)}\|\ell^+ v\|_{H^{r-t}(\Omega)}$$
$$\leq \sigma\|w\|_{H^{t-r}(\Omega)}\|v\|_{H^r(\Omega)},$$

and so

$$\|\ell w\|_{H^{-r}(\Omega)} = \sup_{\substack{v \in C_0^\infty(\Omega) \\ v \neq 0}} \frac{|\langle \ell w, v\rangle_{L_2(\Omega)}|}{\|v\|_{H^r(\Omega)}}$$
$$\leq \sigma\|w\|_{H^{t-r}(\Omega)},$$

as required. Finally, if $0 < r < t$, the result holds by Hilbert space interpolation of the results for $r = 0$ and $r = t$. $\qquad\square$

We now show that FEI is almost optimal information, regardless of whether $k \geq t - 1 + r$ holds.

THEOREM 6.2.2.3. *The spline algorithm $\phi_{n,k}^s$ using FEI $N_{n,k}^{\mathrm{LSFE}}$ is an optimal error algorithm using FEI and is an almost minimal error algorithm. That is,*

$$e(\phi_{n,k}^s, N_{n,k}^{\mathrm{LSFE}}) = r(N_{n,k}^{\mathrm{LSFE}}) = \Theta\big(r(n)\big) = \Theta(n^{-(r+t-l)/d}) \qquad \text{as } n \to \infty.$$

PROOF. Since the spline algorithm using any information N is always an optimal error algorithm using N, the first equality follows immediately. From Theorem 6.2.2.1, we have the lower bound

$$r(N_{n,k}^{\mathrm{LSFE}}) \geq r(n) = \Theta(n^{-(r+t-l)/d}) \qquad \text{as } n \to \infty.$$

It only remains to prove the upper bound

$$r(N_{n,k}^{\mathrm{LSFE}}) = O(n^{-(r+t-l)/d}) \qquad \text{as } n \to \infty. \tag{7}$$

To prove (7), note that if $z \in F \cap \ker N_{n,k}^{\mathrm{LSFE}}$, then

$$\langle z, \ell s\rangle_{L_2(\Omega)} = 0 \quad \forall s \in \mathcal{S}_{n,k} \qquad \text{and} \qquad \|z\|_{H^r(\Omega)} \leq 1.$$

Suppose first that $l = t$. Let $z \in \ker N_{n,k}^{\mathrm{LSFE}} \cap F$. Then for any $s \in \mathcal{S}_{n,k}$, we have

$$\sigma^{-2}\|Sz\|_{H^t(\Omega)}^2 \leq \|z\|_{L_2(\Omega)}^2 = \langle \ell Sz, \ell Sz\rangle_{L_2(\Omega)}$$
$$= \langle z, \ell Sz\rangle_{L_2(\Omega)} = \langle z, \ell(Sz - s)\rangle_{L_2(\Omega)}$$
$$\leq \|\ell(Sz - s)\|_{H^{-r}(\Omega)} \leq \sigma\|Sz - s\|_{H^{t-r}(\Omega)},$$

where we have used the previous lemma in the last step. Since $k + 1 \geq t$, Lemma 5.4.3 yields that there exists $s \in \mathcal{S}_{n,k}$ such that

$$\|Sz - s\|_{H^{t-r}(\Omega)} \leq C n^{-r/d} \|Sz\|_{H^t(\Omega)}$$

for some positive constant C, independent of n, z, and s. Combining these last two inequalities, we have

$$\|Sz\|_{H^t(\Omega)} \leq C n^{-r/d}. \tag{8}$$

Since $z \in \ker N^{\mathrm{LSFE}}_{n,k} \cap F$ is arbitrary, this implies

$$r(N^{\mathrm{LSFE}}_{n,k}) = \sup_{z \in \ker N^{\mathrm{LSFE}}_{n,k} \cap F} \|Sz\|_{H^t(\Omega)} \leq C n^{-r/d},$$

which establishes (7) when $l = t$.

We now consider the case $l < t$. Let $z \in \ker N^{\mathrm{LSFE}}_{n,k} \cap F$. Then for any $s \in \mathcal{S}_{n,k}$, we may use the previous lemma to find that

$$
\begin{aligned}
\|Sz\|^2_{L_2(\Omega)} &= \langle Sz, Sz \rangle_{L_2(\Omega)} = \langle z, S^+ Sz \rangle_{L_2(\Omega)} = \langle z, S^+ Sz - ls \rangle_{L_2(\Omega)} \\
&\leq \|z\|_{H^r(\Omega)} \|S^+ Sz - ls\|_{H^{-r}(\Omega)} \leq \|\ell(SS^+ Sz - s)\|_{H^{-r}(\Omega)} \\
&\leq \sigma \|SS^+ Sz - s\|_{H^{t-r}(\Omega)}.
\end{aligned}
$$

Since $k + 1 - t \geq t$, we find that $t + r \leq k + 1 - (t - r)$, and so Lemma 5.4.3 yields that there exists $s \in \mathcal{S}_{n,k}$ such that

$$
\begin{aligned}
\|SS^+ Sz - s\|_{H^{t-r}(\Omega)} &\leq C n^{-(r+t)/d} \|SS^+ Sz\|_{H^{2t}(\Omega)} \\
&\leq C \sigma^2 n^{-(r+t)/d} \|Sz\|_{L_2(\Omega)}.
\end{aligned}
$$

Combining the previous two inequalities, we have

$$\|Sz\|_{L_2(\Omega)} \leq C n^{-(r+t)/d}.$$

Using Lemma A.2.2.1 on this inequality and (8), we find that

$$\|Sz\|_{H^l(\Omega)} \leq C \|Sz\|^{1-l/t}_{L_2(\Omega)} \|Sz\|^{l/t}_{H^t(\Omega)} \leq C n^{-(r+t-l)/d}$$

for some constant C, independent of n and z. Hence

$$r(N) = \sup_{z \in \ker N^{\mathrm{LSFE}}_{n,k} \cap F} \|Sz\|_{H^l(\Omega)} \leq C n^{-(r+t-l)/d},$$

and so (7) holds, as required. $\square$

For $\varepsilon > 0$, we now wish to find comp(ε), the ε-complexity of our problem. We also wish to determine conditions under which the LSFEM is an almost optimal complexity algorithm for our problem. Let

$$\text{cost}_k^{\text{LSFE}}(\varepsilon) = \inf\{\,\text{cost}(\phi_{n,k}^{\text{LSFE}}, N_{n,k}^{\text{LSFE}}) : e(\phi_{n,k}^{\text{LSFE}}, N_{n,k}^{\text{LSFE}}) \leq \varepsilon\,\}$$

denote the minimal cost of using the LSFEM of degree k to compute an ε-approximation. We also let

$$\text{cost}_k^{\text{spline}}(\varepsilon) = \inf\{\,\text{cost}(\phi_{n,k}^{\text{s}}, N_{n,k}^{\text{LSFE}}) : e(\phi_{n,k}^{\text{LSFE}}, N_{n,k}^{\text{LSFE}}) \leq \varepsilon\,\}$$

be the minimal cost of computing an ε-approximation with the spline algorithm using LSFEI of degree k. Using the previous three theorems, along with the results in Chapter 4, we have

THEOREM 6.2.2.4.

(i) *The ε-complexity is given by*

$$\text{comp}(\varepsilon) = \Theta(\varepsilon^{-d/(r+t-l)}) \qquad \text{as } \varepsilon \to 0.$$

(ii) *Let $\mu = \min\{k+1-t, r\}$. Then*

$$\text{cost}_k^{\text{LSFE}}(\varepsilon) = \Theta(\varepsilon^{-d/(\mu+t-l)}) \qquad \text{as } \varepsilon \to 0.$$

Hence, if $k \geq t-1+r$, then the LSFEM of degree k is an almost optimal complexity algorithm, i.e.

$$\text{cost}_k^{\text{LSFE}}(\varepsilon) = \Theta\big(\text{comp}(\varepsilon)\big) = \Theta(\varepsilon^{-d/(r+t-l)}) \qquad \text{as } \varepsilon \to 0.$$

(iii) *The spline algorithm using LSFEI of degree k is an almost optimal complexity algorithm, i.e.*

$$\text{cost}_k^{\text{spline}}(\varepsilon) = \Theta\big(\text{comp}(\varepsilon)\big) = \Theta(\varepsilon^{-d/(r+t-l)}) \qquad \text{as } \varepsilon \to 0. \qquad \square$$

Hence if $k \geq t-1+r$, the LSFEM of degree k is an almost optimal complexity algorithm. Suppose that $k < t-1+r$. Then

$$\frac{\text{cost}_k^{\text{LSFE}}(\varepsilon)}{\text{comp}(\varepsilon)} = \Theta(\varepsilon^{-\delta d}) \qquad \text{as } \varepsilon \to 0,$$

where

$$\delta = \frac{1}{k+1-l} - \frac{1}{r+t-l} > 0.$$

Thus

$$\lim_{\varepsilon \to 0} \frac{\text{cost}_k^{\text{LSFE}}(\varepsilon)}{\text{comp}(\varepsilon)} = \infty,$$

and so the asymptotic penalty for using an LSFEM of too low a degree is unbounded.

Note that the results of this subsection are analogous to those for a scalar elliptic PDE, as described in §5.4. This implies that many of the same remarks about that problem hold for the problem of solving elliptic systems. In particular, the following hold:

(i) If the problem elements are only smooth enough to guarantee existence and uniqueness of the solution (i.e. $r = 0$ and $l = t$), then the problem is not convergent. That is, for some positive ε_0, we have comp$(\varepsilon) = \infty$ for $\varepsilon < \varepsilon_0$.

(ii) Solving elliptic systems is inherently intractable for problems of large dimension.

(iii) The inherent complexity of solving small-dimensional problems may be prohibitively expensive. For example, consider the first-order system for the Poisson equation in a three-dimensional region. If we have problem elements in $L_2(\Omega)$ and we seek an $L_2(\Omega)$-norm ε-approximation, then the complexity is $\Theta(\varepsilon^{-3})$. As in Chapter 5, this implies that four-place accuracy will require about two weeks' worth of computation on a megaflop machine.

(iv) If we don't know the exact value of the regularity r, but only know an upper bound r^* on r, then the LSFEM of degree $k \geq t - 1 + r^*$ is an almost optimal algorithm.

(v) The condition for the LSFEM to be a spline algorithm is quite restrictive, namely that the LSFEM is a spline algorithm iff $\mathcal{S}_{n,k} = SS^*\ell\mathcal{S}_{n,k}$.

Notes and Remarks

NR 6.2.2:1 The results in this section are based on Werschulz (1985a).

NR 6.2.2:2 As mentioned previously, we deal with Petrovsky-elliptic systems rather than Agmon *et al.* (1959)-elliptic systems mainly for ease of exposition. For a discussion of LSFEMs for Agmon *et al.* (1959)-elliptic systems, see Aziz *et al.* (1985).

NR 6.2.2:3 We briefly discuss the seminormed case for arbitrary linear information, in which $F = \mathcal{B}H^r(\Omega)$. We find that that the nth minimal error and the ε-complexity are the same for the normed and seminormed cases. Furthermore, the LSFEM has infinite error unless $SP_{r-1}(\Omega) \subseteq \mathcal{S}_{n,k}$. However, when this inclusion holds, the results for the LSFEM in the normed case carry over to the seminormed case.

We describe these results more precisely. Let

$$n^* = d\binom{d + r - 1}{r - 1}.$$

Then:

(i) $r(n) = \infty$ if $n < n^*$.

(ii) $r(n) = \Theta(n^{-(r+t-l)/d})$ as $n \to \infty$.

(iii) If $SP_{r-1}(\Omega) \not\subseteq \mathcal{S}_{n,k}$, then

$$e(\phi_{n,k}^{\text{LSFE}}, N_{n,k}^{\text{LSFE}}) = \infty.$$

(iv) If $SP_{r-1}(\Omega) \subseteq \mathcal{S}_{n,k}$ for all sufficiently large n, then

$$e(\phi_{n,k}^{\text{LSFE}}, N_{n,k}^{\text{LSFE}}) = \Theta(n^{-(\mu+t-l)/d}) \qquad \text{as } n \to \infty,$$

where

$$\mu = \min\{k+1-t, r\}.$$

(v) The spline algorithm $\phi_{n,k}^{\text{s}}$ using LSFEI $N_{n,k}^{\text{LSFE}}$ is an almost minimal error algorithm, i.e.

$$e(\phi_{n,k}^{\text{s}}, N_{n,k}^{\text{LSFE}}) = r(N_{n,k}^{\text{LSFE}}) = \Theta(r(n)) = \Theta(n^{-(r+t-l)/d}) \qquad \text{as } n \to \infty.$$

(vi) $\text{comp}(\varepsilon) = \Theta(\varepsilon^{-d/(r+t-l)})$ as $\varepsilon \to 0$.

(vii) If there exists no integer n for which $SP_{r-1}(\Omega) \subseteq \mathcal{S}_{n,k}$, then

$$\text{cost}_k^{\text{LSFE}}(\varepsilon) = \infty \qquad \forall \varepsilon > 0.$$

(viii) Suppose that there exists an integer $n_0 \in \mathbb{N}$ such that $SP_{r-1}(\Omega) \subseteq \mathcal{S}_{n,k}$ for $n \geq n_0$. Then

$$\text{cost}_k^{\text{LSFE}}(\varepsilon) = \Theta(\varepsilon^{-d/(\mu+t-l)}) \qquad \text{as } \varepsilon \to 0.$$

Hence, if $SP_{r-1}(\Omega) \subseteq \mathcal{S}_{n,k}$ for $n \geq n_0$ and $k \geq t-1+r$, then the LSFEM of degree k is an almost optimal complexity algorithm, i.e.

$$\text{cost}_k^{\text{FE}}(\varepsilon) = \Theta(\text{comp}(\varepsilon)) = \Theta(\varepsilon^{-d/(r+t-l)}) \qquad \text{as } \varepsilon \to 0.$$

(ix) The spline algorithm is an almost optimal complexity algorithm, i.e.

$$\text{cost}_k^{\text{spline}}(\varepsilon) = \Theta(\text{comp}(\varepsilon)) = \Theta(\varepsilon^{-d/(r+t-l)}) \qquad \text{as } \varepsilon \to 0.$$

These results may be proved by following the analysis in the first part of §5.6, making the obvious changes.

6.2.3   Standard information

We now assume that $\Lambda = \Lambda^{\text{std}}$. That is, we suppose that only standard information is permissible, so that we can only evaluate a function (or any of its derivatives) at a point.

Our results are analogous to those of §5.5. We find that standard information involving only function values is roughly as powerful as standard information involving function values and derivatives. Moreover, there is a loss in changing our class Λ of permissible information operations from Λ' to Λ^{std}. That is, if $0 \leq l \leq t$, then the nth minimal radius of standard information is $\Theta(n^{-r/d})$ as $n \to \infty$, as compared with the nth minimal radius $\Theta(n^{-(r+t-l)/d})$ or arbitrary linear information. Similarly, the ε-complexity changes from $\Theta(\varepsilon^{-d/(r+t-l)})$ as $\varepsilon \to 0$ to $\Theta(\varepsilon^{-d/r})$ when going from arbitrary linear information to standard information.

Again, note that the nth minimal radius and the ε-complexity are independent of l when we use standard information. Hence we do not get any improvement by measuring error in a lower-order Sobolev norm when we are restricted to using standard information.

In this subsection, we shall assume that $r > \frac{1}{2}d$, since standard information consisting of function evaluations is ill-defined on $H^r(\Omega)$, and the problem complexity is infinite, if this inequality is not satisfied.

Since our problem is linear, there is no loss of generality in considering only nonadaptive information in what follows. We first show a lower bound on the nth minimal radius of (nonadaptive) standard information, remembering that we are permitting the evaluation of a function or one of its derivatives at a point. Recall that we are measuring error in the Sobolev $\| \cdot \|_{H^l(\Omega)}$-norm, our problem is a tth-order elliptic system, and that the unit ball of $H^r(\Omega)$ is our set of problem elements.

THEOREM 6.2.3.1. *There exists a positive constant C, independent of n and l, such that*

$$r(n, \Lambda^{\mathrm{std}}) \geq Cn^{-r/d} \qquad \forall \, n \in \mathbb{P}.$$

PROOF. Let Ω^0 be a C^∞ region whose closure is a subset of int Ω. Choose a nonnegative function $v \in C_0^\infty(\Omega)$ such that $v \equiv 1$ on Ω^0. Let N be standard information of cardinality n. As in the proof of Theorem 5.5.1, there exists a nonzero function $z \in \ker N \cap F$ such that

$$\int_{\Omega^0} z(x) \, dx \geq \kappa n^{-r/d},$$

where $\kappa > 0$ is independent of n and z. Since $C_0^\infty(\Omega) \subset H^t(\partial)$, we may use Theorem 6.2.1.1 to find that

$$\int_{\Omega^0} z(x) \, dx \leq \langle z, v \rangle_{L_2(\Omega)} = \langle \ell Sz, v \rangle_{L_2(\Omega)}$$

$$= \langle Sz, \ell^+ v \rangle_{L_2(\Omega)} \leq \| Sz \|_{L_2(\Omega)} \| \ell^+ v \|_{L_2(\Omega)}$$

$$\leq \sigma \| Sz \|_{H^l(\Omega)} \| v \|_{H^t(\Omega)}.$$

Combining the last two inequalities, we find

$$\| Sz \|_{H^l(\Omega)} \geq \frac{\kappa}{\sigma \| v \|_{H^t(\Omega)}} n^{-r/d}.$$

Setting $C = \kappa \sigma^{-1} \| v \|_{H^t(\Omega)}^{-1}$, we then have

$$r(N) = \sup_{h \in \ker N \cap F} \| Sh \|_{H^l(\Omega)} \geq \| Sz \|_{H^l(\Omega)} \geq Cn^{-r/d},$$

as required. □

We now need to find an almost minimal error algorithm using standard information. Following the approach of the previous chapter, this algorithm will be a least squares finite element method with quadrature, in which all the integrals appearing in the LSFEM are approximated using quadrature rules.

In what follows, we will be using the notation of §5.5. We let $\hat{I}$ denote a (scalar) quadrature rule over the reference element $\hat{K}$, with nodes $\hat{b}_1, \ldots, \hat{b}_J$ and weights $\hat{\omega}_1, \ldots, \hat{\omega}_J$. That is, we have

$$\hat{I}\hat{v} = \sum_{j=1}^{J} \hat{\omega}_j \hat{v}(\hat{b}_j) \qquad \forall \hat{v} \in C\left(\overline{\hat{K}}\right).$$

The reference quadrature rule $\hat{I}$ induces a local quadrature rule I_K over each element K in $\mathcal{T}_{\tilde{n}}$, the nodes and weights of I_K being respectively denoted by $b_{1,K}, \ldots, b_{J,K}$ and $\omega_{1,K}, \ldots, \omega_{J,K}$, with

$$\omega_{j,K} = \det B_K \cdot \hat{\omega}_j \quad \text{and} \quad b_{j,K} = F_K(\hat{b}_j) \qquad (1 \le j \le J),$$

with $F_K : \hat{K} \to K$ being the affine mapping that maps the reference element $\hat{K}$ onto the local element K. Thus

$$I_K v = \sum_{j=1}^{J} \omega_{j,K} v(b_{j,K}) \qquad \forall v \in C(\overline{K}).$$

We let $\hat{E}$ and E_K respectively denote the error functionals of $\hat{I}$ and I_K.

Let $\tilde{n} \in \mathbb{P}$. Then

$$\mathcal{N}_{\tilde{n}} = \bigcup_{K \in \mathcal{T}_{\tilde{n}}} \bigcup_{j=1}^{J} \{b_{j,K}\}$$

denotes the set of all nodes of the triangulation $\mathcal{T}_{\tilde{n}}$. We define a new bilinear form $B_{\tilde{n}}$ on $\mathcal{S}_{\tilde{n},k}$ by

$$B_{\tilde{n}}(v, w) = \sum_{\substack{K \in \mathcal{T}_{\tilde{n}} \\ 1 \le i, s, s' \le d \\ |\alpha|, |\alpha'| \le t}} I_K(a_\alpha^{is} a_{\alpha'}^{is'} \partial^\alpha v_s \partial^{\alpha'} w_{s'}) \qquad \forall v, w \in \mathcal{S}_{\tilde{n},k}.$$

Note that $B_{\tilde{n}}(v, w)$ approximates $\langle \ell v, \ell w \rangle_{L_2(\Omega)}$ on $\mathcal{S}_{\tilde{n},k}$. For $f \in F$, we define a linear functional $f_{\tilde{n}}$ on $\mathcal{S}_{\tilde{n},k}$ by

$$f_{\tilde{n}}(w) = \sum_{\substack{K \in \mathcal{T}_{\tilde{n}} \\ 1 \le i, s \le d \\ |\alpha| \le t}} I_K(f_i a_\alpha^{is} \partial^\alpha w_s) \qquad \forall v \in \mathcal{S}_{\tilde{n},k}.$$

That is, $f_{\tilde{n}}(w)$ approximates $\langle f, \ell w \rangle_{L_2(\Omega)}$ for $w \in \mathcal{S}_{\tilde{n},k}$.

We are now ready to define our least squares finite element method with quadrature. Let $n \in \mathbb{P}$. Choose $\tilde{n}$ to be the largest integer such that

$$d \cdot \operatorname{card} \mathcal{N}_{\tilde{n}} \leq n.$$

(Remember that f has d components.) As before, $\tilde{n} = \Theta(n)$ as $n \to \infty$. Let $x_1, \ldots, x_{\tilde{n}}$ denote the elements of $\mathcal{N}_{\tilde{n}}$, and choose points $x_{\tilde{n}+1}, \ldots, x_n$ not belonging to $\mathcal{N}_{\tilde{n}}$. For $f \in F$, define information $N_{n,k}^Q$ to be

$$N_{n,k}^Q f = \begin{bmatrix} f(x_1) \\ \vdots \\ f(x_n) \end{bmatrix} \qquad \forall f \in F.$$

Then $N_{n,k}^Q$ is standard information of cardinality n. Suppose we can find $\tilde{u}_{\tilde{n},k} \in \mathcal{S}_{\tilde{n},k}$ satisfying

$$B_{\tilde{n}}(\tilde{u}_{\tilde{n},k}, s) = f_{\tilde{n}}(s) \qquad \forall s \in \mathcal{S}_{\tilde{n},k}. \tag{1}$$

Since $\tilde{u}_{\tilde{n},k}$ depends on f only through $N_{n,k}^Q f$, we write $\tilde{u}_{\tilde{n}} = \phi_{n,k}^Q(N_{n,k}^Q f)$. We say that $\phi_{n,k}^Q$ is the *least squares finite element method with quadrature* (LSFEMQ).

In what follows, let

$$\nu = \min\{k+1, r\}.$$

As in §5.5, we assume that

(i) The smoothness r of the problem elements F satisfies $r \geq 1$ (as well as our previous requirement $r > \frac{1}{2}d$).

(ii) The degree k of the finite element subspaces $\mathcal{S}_{\tilde{n},k}$ satisfies $k > \frac{1}{2}d - 1$.

(iii) $\hat{I}$ is exact of degree $2k + \nu - 1$ over the reference element $\hat{K}$, i.e. $\hat{I}(\hat{p}) = \int_{\hat{f}} \hat{p}$ for any $\hat{p} \in P_{2k+\nu-1}(\hat{K})$.

We now show that the LSFEMQ is well-defined, and give tight bounds on its error.

THEOREM 6.2.3.2. *There exists $n_1 \in \mathbb{P}$ such that $\phi_{n,k}^Q$ is well-defined for $n \geq n_1$. Furthermore,*

$$e(\phi_{n,k}^Q, N_{n,k}^Q) = \Theta(n^{-\mu/d}) \qquad \text{as } n \to \infty,$$

where

$$\mu = \min\{k+1-l, r\}.$$

PROOF. First, we show that there exists $n_1 \in \mathbb{P}$ such that the LSFEMQ is well-defined for all $n \geq n_1$. In what follows, let us write

$$A(r) = \sum_{\substack{1 \leq i,s \leq d \\ |\alpha| \leq t}} \|a_\alpha^{is}\|_{W^{r,\infty}(\Omega)},$$

$$B(v,w) = \langle \ell v, \ell w \rangle_{L_2(\Omega)},$$

$$f(w) = \langle f, \ell w \rangle_{L_2(\Omega)}.$$

For any $\tilde{n} \in \mathbb{P}$, we may sum the result of Lemma 5.5.1 over all $K \in \mathcal{T}_{\tilde{n}}$, finding that

$$|B(v,w) - B_n(v,w)| \leq C n^{-\nu/d} A(r)^2 \|v\|_{H^t(\Omega)} \|w\|_{H^t(\Omega)} \tag{2}$$
$$\forall\, v, w \in \mathcal{S}_{\tilde{n},k}$$

and

$$|f(w) - f_n(w)| \leq C n^{-\nu/d} A(r) \|f\|_{H^r(\Omega)} \|w\|_{H^t(\Omega)} \tag{3}$$
$$\forall f \in H^r(\Omega), w \in \mathcal{S}_{\tilde{n},k}.$$

By Strang's Lemma (Lemma A.3.2), we see that (2) implies that there exists $n_1 \in \mathbb{P}$ such that for any $f \in F$ and any $n \geq n_1$, there exists a unique $u_{\tilde{n},k} \in \mathcal{S}_{\tilde{n},k}$ such that (1) holds. Hence the LSFEMQ $\phi_{n,k}^Q$ is well-defined for $n \geq n_1$.

We next prove the lower bound

$$e(\phi_{n,k}^Q, N_{n,k}^Q) = \Omega(n^{-\mu/d}) \qquad \text{as } n \to \infty.$$

Since $\phi_{n,k}^Q$ uses standard information $N_{n,k}^Q$ of cardinality n, Theorem 6.2.3.1 implies that

$$e(\phi_{n,k}^Q, N_{n,k}^Q) \geq r(n, \Lambda^{\text{std}}) = \Omega(n^{-r/d}).$$

Define f^* as in the proof of Theorem 6.2.2.2. Since $\phi_{n,k}^Q(N_{n,k}^Q f^*) \in \mathcal{S}_{\tilde{n},k}$, we see that

$$e(\phi_{n,k}^Q, N_{n,k}^Q) \geq \|Sf^* - \phi_{n,k}^Q(N_{n,k}^Q f^*)\|_{H^l(\Omega)} \geq C n^{-(k+1-l)/d}.$$

The desired lower bound for the LSFEMQ follows immediately from these last two lower bounds.

It remains to establish the upper bound

$$e(\phi_{n,k}^Q, N_{n,k}^Q) \leq C n^{-\mu/d} \qquad \forall n \geq n_1. \tag{4}$$

We first suppose that $l = t$. For $n \geq n_1$ and $f \in F$, Lemma 5.4.3 and Theorem 6.2.1.1 yield that

$$\|Sf - \Pi_{\tilde{n}} Sf\|_{H^t(\Omega)} \leq C n^{-\min\{k+1-t,r\}/d} \|Sf\|_{H^{r+t}(\Omega)} \tag{5}$$
$$\leq C\sigma n^{-\min\{k+1-t,r\}/d} \|f\|_{H^r(\Omega)}.$$

Using Lemma 5.4.3 and Theorem 6.2.1.1, we have

$$\|\Pi_{\tilde{n}} Sf\|_{H^t(\Omega)} \leq \|Sf\|_{H^t(\Omega)} + \|Sf - \Pi_{\tilde{n}} Sf\|_{H^t(\Omega)}$$
$$\leq C\|Sf\|_{H^t(\Omega)} \leq C\sigma \|f\|_{L_2(\Omega)}$$
$$\leq C\sigma \|f\|_{H^r(\Omega)},$$

and so (2) implies that

$$|B(\Pi_{\tilde{n}} Sf, w) - B_n(\Pi_{\tilde{n}} Sf, w)| \leq CA(r)^2 n^{-\nu/d} \|f\|_{H^r(\Omega)} \|w\|_{H^t(\Omega)} \tag{6}$$
$$\leq CA(r)^2 n^{-\min\{k+1-t,r\}/d} \|f\|_{H^r(\Omega)} \|w\|_{H^t(\Omega)}$$

for any $w \in \mathcal{S}_{\tilde{n},k}$. Using (3), (5), and (6) along with Strang's Lemma, we see that there is a positive constant C, such that for $n \geq n_1$ and $f \in F$,

$$\|Sf - \phi_{n,k}^Q(N_{n,k}^Q f)\|_{H^t(\Omega)} \leq CA(r)^2 n^{-\min\{k+1-t,r\}/d} \|f\|_{H^r(\Omega)}. \tag{7}$$

Taking the supremum over all $f \in F$, we find that (4) holds for the case $l = t$.

We now suppose that $l < t$. First, we bound the $L_2(\Omega)$-error. Let $f \in F$ and let $n \geq n_1$. Write $\tilde{u}_{\tilde{n},k} = \phi_{n,k}^Q(N_{n,k}^Q f)$. For $g \in L_2(\Omega)$, let $\tilde{S}g \in H^t(\partial)$ satisfy

$$B(v, \tilde{S}g) = \langle g, v \rangle_{L_2(\Omega)} \qquad \forall v \in H^t(\partial).$$

Since B is symmetric, we have

$$\langle \ell S g, \ell v \rangle_{L_2(\Omega)} = B(\tilde{S}g, v) = B(v, \tilde{S}g) = \langle g, v \rangle_{L_2(\Omega)} = \langle g, S\ell v \rangle_{L_2(\Omega)}$$
$$= \langle S^+ g, \ell v \rangle_{L_2(\Omega)}.$$

Hence $\ell \tilde{S} g = S^+ g$, which means that $\tilde{S}g = S S^+ g$. Thus Theorem 6.2.1.1 yields that

$$\|\tilde{S}g\|_{H^{2t}(\Omega)} \leq \sigma^2 \|g\|_{L_2(\Omega)}. \tag{8}$$

Since $l < t$, we have $k + 1 - t \geq t$. Hence

$$\|\tilde{S}g - \Pi_{\tilde{n}} \tilde{S}g\|_{H^t(\Omega)} \leq C n^{-t/d} \|\tilde{S}g\|_{H^{2t}(\Omega)} \leq C\sigma^2 n^{-t/d} \|g\|_{L_2(\Omega)}. \tag{9}$$

Using (7) and Theorem 6.2.1.1, we have

$$\|\tilde{u}_{\tilde{n},k}\|_{H^t(\Omega)} \leq \|u - \tilde{u}_{\tilde{n},k}\|_{H^t(\Omega)} + \|u\|_{H^t(\Omega)}$$
$$\leq CA(r)^2 \left(\|f\|_{H^r(\Omega)} + \sigma \|f\|_{L_2(\Omega)} \right) \leq CA(r)^2 \|f\|_{H^r(\Omega)}. \tag{10}$$

But (8) and (9) imply that

$$\|\Pi_{\tilde{n}}\tilde{S}g\|_{H^t(\Omega)} \le \|\tilde{S}g\|_{H^t(\Omega)} + \|\tilde{S}g - \Pi_{\tilde{n}}\tilde{S}g\|_{H^t(\Omega)}$$
$$\le C\|\tilde{S}g\|_{H^t(\Omega)} \le C\|\tilde{S}g\|_{H^{2t}(\Omega)} \le C\sigma^2\|g\|_{L_2(\Omega)}. \tag{11}$$

So (2), (10), and (11) imply that

$$|B(\tilde{u}_{\tilde{n},k},\Pi_{\tilde{n}}\tilde{S}g) - B_n(\tilde{u}_{\tilde{n},k},\Pi_{\tilde{n}}\tilde{S}g)| \le CA(r)^2 n^{-\nu/d}\|f\|_{H^r(\Omega)}\|g\|_{L_2(\Omega)}. \tag{12}$$

Using (3) with $w = \Pi_{\tilde{n}}\tilde{S}g$, (9) and (12), along with the Aubin-Ciarlet-Nitsche Duality Argument (Lemma A.3.3), we find that

$$\|Sf - \phi_{n,k}^{Q}(N_{n,k}^{Q}f)\|_{L_2(\Omega)} \le CA(r)^2 n^{-\zeta/d}\|f\|_{H^r(\Omega)},$$

where

$$\zeta = \min\{\min\{k+1-t,r\}+t, \min\{k+1,r\}\}$$
$$= \min\{\min\{k+1,r+t\}, \min\{k+1,r\}\}$$
$$= \min\{k+1,r\}.$$

Thus

$$\|Sf - \phi_{n,k}^{Q}(N_{n,k}^{Q}f)\|_{L_2(\Omega)} \le CA(r)^2 n^{-\min\{k+1,r\}/d}\|f\|_{H^r(\Omega)}, \tag{13}$$

our desired bound on the $L_2(\Omega)$-error.

Finally, since $0 \le l \le m$, we may use Hilbert space interpolation, (7), and (13) to find that if $n \ge n_1$ and $f \in F$, then

$$\|Sf - \phi_{n,k}^{Q}(N_{n,k}^{Q}f)\|_{H^l(\Omega)} \le C\,\|Sf - \phi_{n,k}^{Q}(N_{n,k}^{Q}f)\|_{L_2(\Omega)}^{1-l/t}$$
$$\times \|Sf - \phi_{n,k}^{Q}(N_{n,k}^{Q}f)\|_{H^t(\Omega)}^{l/t}$$
$$\le CA(r)n^{-\min\{k+1-l,r\}/d}\|f\|_{H^r(\Omega)}.$$

Taking the supremum over all $f \in F$, we get our desired result. □

Comparing the previous two theorems, we immediately find that

COROLLARY 6.2.3.1. *The LSFEMQ of degree k is an almost minimal error algorithm iff $k \ge r + l - 1$.* □

For $\varepsilon > 0$, we now wish to find $\mathrm{comp}(\varepsilon)$, the ε-complexity of our problem, which we will compare to

$$\mathrm{cost}_k^{Q}(\varepsilon) = \inf\{\,\mathrm{cost}(\phi_{n,k}^{Q}, N_{n,k}^{Q}) : e(\phi_{n,k}^{Q}, N_{n,k}^{Q}) \le \varepsilon\,\},$$

which is the minimal cost of computing an ε-approximation with the LS-FEMQ of degree k. Using the last two theorems, along with the results in Chapter 4, we have

THEOREM 6.2.3.3.

(i) *The ε-complexity is given by*

$$\text{comp}(\varepsilon) = \Theta(\varepsilon^{-d/r}) \qquad \text{as } \varepsilon \to 0.$$

(ii) *Let $\mu = \min\{k + 1 - l, r\}$. Then*

$$\text{cost}_k^Q(\varepsilon) = \Theta(\varepsilon^{-d/\mu}) \qquad \text{as } \varepsilon \to 0.$$

Hence, if $k \geq r + l - 1$, then the LSFEMQ of degree k is an almost optimal complexity algorithm, i.e.

$$\text{cost}_k^Q(\varepsilon) = \Theta\big(\text{comp}(\varepsilon)\big) = \Theta(\varepsilon^{-d/r}) \qquad \text{as } \varepsilon \to 0. \qquad \square$$

Comparing this result with that in the previous subsection, we see that standard information is not as powerful as arbitrary linear information when solving elliptic systems with $t > 0$. Compare with the analogous result about elliptic PDE in Chapter 5.

As with the pure least squares finite element method, we pay a large asymptotic penalty for using an LSFEMQ of too low a degree. That is, for $k < r + l - 1$,

$$\frac{\text{cost}_k^Q(\varepsilon)}{\text{comp}(\varepsilon)} = \Theta(\varepsilon^{-\delta d}) \qquad \text{as } \varepsilon \to 0,$$

where

$$\delta = \frac{1}{k + 1 - l} - \frac{1}{r} > 0.$$

Thus,

$$\lim_{\varepsilon \to 0} \frac{\text{cost}_k^Q(\varepsilon)}{\text{comp}(\varepsilon)} = \infty.$$

Notes and Remarks

NR 6.2.3:1 We briefly describe results for the seminormed case $F = \mathcal{B}H^r(\Omega)$ with standard information. Basically, we find that the the nth minimal error and the ε-complexity are the same as for the normed case. However, the LSFEMQ has infinite error unless $SP_{r-1}(\Omega) \subseteq \mathcal{S}_{n,k}$. When this inclusion holds, the results for the LSFEMQ in the normed case carry over to the seminormed case.

We describe these results more precisely. Let

$$n^* = d\binom{d + r - 1}{r - 1}.$$

Then:

(i) $r(n, \Lambda^{\text{std}}) = \infty$ if $n < n^*$.
(ii) $r(n, \Lambda^{\text{std}}) = \Theta(n^{-(r+t-l)/d})$ as $n \to \infty$.

(iii) If $SP_{r-1}(\Omega) \not\subseteq \mathcal{S}_{n,k}$, then

$$e(\phi_{n,k}^Q, N_{n,k}^Q) = \infty.$$

(iv) If $SP_{r-1}(\Omega) \subseteq \mathcal{S}_{n,k}$ for all sufficiently large n, then

$$e(\phi_{n,k}^Q, N_{n,k}^Q) = \Theta(n^{-\mu)/d}) \qquad \text{as } n \to \infty,$$

where

$$\mu = \min\{k+1-l, r\}.$$

(v) $\mathrm{comp}(\varepsilon) = \Theta(\varepsilon^{-d/r})$ as $\varepsilon \to 0$.

(vi) If there exists no integer n for which $SP_{r-1}(\Omega) \subseteq \mathcal{S}_{n,k}$, then

$$\mathrm{cost}_k^Q(\varepsilon) = \infty \qquad \forall \varepsilon > 0.$$

(vii) Suppose that there exists a nonnegative integer n_0 such that $SP_{r-1}(\Omega) \subseteq \mathcal{S}_{n,k}$ for $n \geq n_0$. Then

$$\mathrm{cost}_k^Q(\varepsilon) = \Theta(\varepsilon^{-d/\mu}) \qquad \text{as } \varepsilon \to 0.$$

Hence, if $SP_{r-1}(\Omega) \subseteq \mathcal{S}_{n,k}$ for $n \geq n_0$ and $k \geq r+l-1$, then the LSFEMQ of degree k is an almost optimal complexity algorithm, i.e.

$$\mathrm{cost}_k^{FE}(\varepsilon) = \Theta(\mathrm{comp}(\varepsilon)) = \Theta(\varepsilon^{-d/r}) \qquad \text{as } \varepsilon \to 0.$$

These results may be proved by following the analysis in the last part of §5.6, making the obvious changes.

6.3   Fredholm problems of the second kind

In this section, we consider the complexity of linear Fredholm integral equations of the second kind over a region $\Omega \in \mathbb{R}^d$. These problems arise in many areas of mathematical physics and engineering. For example, let $\Omega \subset \mathbb{R}^2$, and let $u \colon \bar{\Omega} \to \mathbb{R}$ be the solution to

$$\Delta u = 0 \qquad \text{in } \Omega,$$
$$u = g \qquad \text{on } \partial\Omega.$$

Let ds_y denote the arc length measure at $y \in \partial\Omega$. Then

$$u(x) + \frac{1}{\pi} \int_{\partial\Omega} \frac{\partial}{\partial\nu} \ln \frac{1}{|x-y|} u(y) \, ds_y = \frac{1}{\pi} f(x) \qquad \forall x \in \partial\Omega,$$

see Riesz and Sz.-Nagy (1955). Hence u is the solution of a linear Fredholm integral equation of the second kind.

For further background regarding the Fredholm integral equation of the second kind (including applications, standard techniques for computing approximations, and numerical results of tests of such techniques), see Anderssen *et al.* (1980), Atkinson (1976), Baker (1977), Baker and Miller (1982), Chow and Tsitsiklis (1989*a*), Chow and Tsitsiklis (1989*b*), Delves and Walsh (1974), Emelyanov and Ilin (1967), Heinrich (1985), Pereverzev (1989), te Riele (1979), Schock (1982), and Tsitsiklis (1989).

We briefly outline the results of this section. Error will be measured in the $L_p(\Omega)$-norm, for some $p \in (1, \infty]$. We will concentrate on the normed case, so that our class F of problem elements will be the unit ball of $W^{r,p}(\Omega)$. Permissible information will either be arbitrary linear information or standard information. For the case of arbitrary linear information, we find that the ε-complexity of the problem is $\Theta(\varepsilon^{-d/r})$. Moreover, the FEM of degree k is optimal iff $k \geq r - 1$. If only standard information is available, we find that the ε-complexity of the problem is still $\Theta(\varepsilon^{-d/r})$, and that a FEM with quadrature is optimal iff $k \geq r - 1$. In short, we find that there is no penalty if only standard information is available, instead of arbitrary linear information. This should be contrasted with what we learned for elliptic PDE and elliptic systems, in which there was an unbounded asymptotic penalty in going from arbitrary linear information to standard information.

Note once again that our problem is not convergent if we have the minimal amount of smoothness necessary to solve the problem, i.e. if $r = 0$. Furthermore, if $r > 0$, then the problem complexity is exponential in the dimension d, i.e. the problem is intractable. Hence we must go to other settings if we really wish to solve these problems.

Notes and Remarks

NR 6.3:1 The results in this section extend the results of Werschulz (1985*b*), which dealt only with the one-dimensional case. The paper Werschulz (1985*b*), in turn, extended the results of Emelyanov and Ilin (1967) (applied to the scalar case $d = 1$) from the case $p = \infty$ to the case $1 < p \leq \infty$.

6.3.1 Problem formulation

Let $\Omega \subset \mathbb{R}^d$ be a bounded, simply connected, C^∞ region. Let $r \in \mathbb{N}$, and let $p \in (1, \infty]$. Let $k \colon \Omega \times \Omega \to \mathbb{R}$, with $\partial_x^\alpha k(x, y)$ a continuous function of $(x, y) \in \Omega \times \Omega$ for $|\alpha| \leq r$. Define a linear operator $K \colon L_p(\Omega) \to L_p(\Omega)$ by

$$(Kv)(x) = \int_\Omega k(x, y) v(y) \, dy.$$

Then K is compact. We assume that 1 is not an eigenvalue of K in what follows. Set

$$L = I - K.$$

By the usual Fredholm alternative theory of functional analysis, L is an invertible bounded operator on $L_p(\Omega)$; see Friedman (1970), Riesz and Sz.-Nagy (1955) or Schechter (1971) for further discussion.

For future reference, we now state and prove a useful regularity theorem.

THEOREM 6.3.1.1.

(i) *There exists a constant $\sigma > 0$, such that*

$$\sigma^{-1}\|u\|_{W^{r,p}(\Omega)} \le \|Lu\|_{W^{r,p}(\Omega)} \le \sigma\|u\|_{W^{r,p}(\Omega)} \qquad \forall\, u \in W^{r,p}(\Omega).$$

(ii) *For any $f \in W^{r,p}(\Omega)$, there exists a unique $u \in W^{r,p}(\Omega)$ such that*

$$Lu = (I - K)u = f.$$

PROOF. We need only prove (i), since (ii) follows immediately from (i), the bounded inverse theorem, and the Fredholm alternative theorem.

Suppose first that $r = 0$. Then (i) holds with

$$\sigma = \sigma_0 \equiv \{\max\{\|L\|, \|L^{-1}\|\}\},$$

the norm denoting the operator norm on $L_p(\Omega)$. We now suppose that $r \in \mathbb{P}$. Let $u \in W^{r,p}(\Omega)$. Set

$$\theta_\alpha(x) = \left(\int_\Omega |\partial_x^\alpha k(x,y)|\, dy \right)^{1/p'},$$

where

$$p' = \frac{p}{p-1}$$

denotes the exponent conjugate to p'. We then have

$$|(\partial_x^\alpha Ku)(x)| = \left| \int_\Omega |\partial_x^\alpha k(x,y)u(y)|\, dy \right| \le \theta_\alpha(x)\|u\|_{L_p(\Omega)}.$$

Let

$$\gamma = \left(\sum_{|\alpha| \le r} \|\theta\|_{L_p(\Omega)}^p \right)^{1/p},$$

so that

$$\|Ku\|_{W^{r,p}(\Omega)} \le \gamma\|u\|_{L_p(\Omega)}. \tag{1}$$

We then have

$$\|Ku\|_{W^{r,p}(\Omega)} \le \gamma\|u\|_{L_p(\Omega)} \le \sigma_0\gamma\|Lu\|_{L_p(\Omega)} \le \sigma_0\gamma\|Lu\|_{W^{r,p}(\Omega)}.$$

Set $\sigma = 1 + \sigma_0\gamma$. Then

$$\|u\|_{W^{r,p}(\Omega)} = \|(L+K)u\|_{W^{r,p}(\Omega)} \leq \|Lu\|_{W^{r,p}(\Omega)} + \|Ku\|_{W^{r,p}(\Omega)}$$
$$\leq \sigma\|Lu\|_{W^{r,p}(\Omega)},$$

proving the first half of the inequality in (i). To prove the other half, note that $\sigma_0 \geq 1$, and so (1) implies that

$$\|Lu\|_{W^{r,p}(\Omega)} \leq \|u\|_{W^{r,p}(\Omega)} + \|Ku\|_{W^{r,p}(\Omega)} \leq \|u\|_{W^{r,p}(\Omega)} + \gamma\|u\|_{L_p(\Omega)}$$
$$\leq (1+\gamma)\|u\|_{W^{r,p}(\Omega)} \leq \sigma\|u\|_{W^{r,p}(\Omega)},$$

completing the proof of (i) and of Theorem 6.3.1.1.   □

We are now ready to formulate our problem. Our solution operator $S\colon W^{r,p}(\Omega) \to L_p(\Omega)$ is defined by letting

$$u = Sf \qquad \text{iff} \qquad Lu = (I-K)u = f.$$

By Theorem 6.3.1.1, S is well-defined.

Next we choose our class Λ of permissible information operators as in the previous section, i.e. we will allow either $\Lambda = \Lambda'$ (arbitrary linear information), or $\Lambda = \Lambda^{\mathrm{std}}$ (standard information).

For our choice of problem elements, we will mainly concentrate on the normed case $F = BH^r(\Omega)$. We leave a discussion of the seminormed case $F = \mathcal{B}H^r(\Omega)$ for the Notes and Remarks.

Notes and Remarks

NR 6.3.1:1 Our conditions on the kernel function k are really somewhat restrictive, but chosen for the sake of exposition. For the one-dimensional case, V. Rokhlin (private communication) has extended our results to k which are piecewise analytic, the only discontinuity being on the diagonal.

NR 6.3.1:2 Other conditions establishing the compactness of K are given on p. 518 of Dunford and Schwartz (1963).

NR 6.3.1:3 We deal only with the absolute error criterion in our treatment of the Fredholm problem of the second kind. As in **NR 5.3:1** of the previous chapter, we see that the following results hold for other error criteria:

(i) Using Theorem 6.3.1.1, the results for the residual error criterion are (to within a constant factor) the same as those for the absolute error criterion.
(ii) The results for the normalized error criterion are the same as for the absolute error criterion.
(iii) The problem is not solvable under the relative error criterion.

6.3.2   Arbitrary linear information

We now study the complexity of our problem, assuming that arbitrary linear information is permissible. As we stated in the introduction to this section, it will turn out that arbitrary linear information is no more powerful than standard information. However, it will be easier to analyze the arbitrary linear case first, showing that there is an almost optimal FEM using finite element information. In the next subsection, we will show how to approximate this FEM with an FEMQ using standard information, showing conditions that are necessary and sufficient for said FEMQ to be almost optimal.

Our approach is similar to that of problems that we have studied previously. First, we determine lower bounds by computing the nth minimal radius of information. Next, we give sharp bounds on the FEM, which allow us to conclude that the FEM of degree k is an almost minimal error algorithm iff $k \geq r - 1$. Using these results, we then easily conclude that the FEM of degree k is an almost optimal algorithm iff $k \geq r - 1$.

Since we are studying a linear problem, nonadaptive information is as powerful as adaptive information. Hence, we first need to give tight bounds on the minimal radius of nonadaptive information of cardinality at most n.

THEOREM 6.3.2.1.

$$r(n) = \Theta(n^{-r/d}) \qquad \text{as } n \to \infty.$$

PROOF. From Theorem 4.4.3.3, we have

$$r(n) = d^n\left(SF, L_p(\Omega)\right).$$

For $\theta > 0$, let

$$X(\theta) = \theta BW^{r,p}(\Omega).$$

From Theorem 6.3.1.1, we have

$$X(\sigma^{-1}) \subseteq SF \subseteq X(\sigma).$$

Since

$$d^n\left(X(\theta), L_p(\Omega)\right) = \theta \, d^n\left(BW^{r,p}(\Omega), L_p(\Omega)\right),$$

it then follows that

$$r(n) = \Theta\left(d^n\left(BW^{r,p}(\Omega), L_p(\Omega)\right)\right) \qquad \text{as } n \to \infty.$$

From the results in Chapter VII of Pinkus (1985), we find that

$$d^n\left(BW^{r,p}(\Omega), L_p(\Omega)\right) = \Theta(n^{-r/d}) \qquad \text{as } n \to \infty,$$

completing the proof of the theorem. ☐

As in previous problems that we have studied, we can (in principle) use the general theory of Chapter 4 to construct optimal algorithms. Of course, we can only follow this prescription if we know eigenvectors of S^*S. Since this information is not generally available, we follow the path laid out previously. Namely, we will look at finite element methods for our problem, hoping to find conditions under which FEMs are almost optimal.

Let $\{\mathfrak{T}_n\}_{n=1}^\infty$ be a quasi-uniform family of triangulations of Ω. For each $n \in \mathbb{P}$, we let $\mathcal{S}_{n,k}$ be an n-dimensional space of $\mathfrak{T}_n$-piecewise polynomials of degree k. That is, we say that

$$s \in \mathcal{S}_{n,k} \quad \text{iff} \quad s\big|_K \in P_k(K) \quad \forall K \in \mathfrak{T}_n.$$

Note that although we require no inter-element continuity for elements of $\mathcal{S}_{n,k}$, we have $\mathcal{S}_{n,k} \subset L_q(\Omega)$ for any $q \in [1, \infty]$.

We define a bilinear form $B : L_p(\Omega) \times L_{p'}(\Omega) \to \mathbb{R}$ by

$$B(v, w) = \langle Lv, w \rangle = \langle (I - K)v, w \rangle \quad \forall v \in L_p(\Omega), w \in L_{p'}(\Omega).$$

Here, and in what follows, we let $\langle \cdot, \cdot \rangle$ denote the duality pairing of $L_p(\Omega)$ with $L_{p'}(\Omega)$ given by

$$\langle v, w \rangle = \int_\Omega v(x)w(x) \, dx \quad \forall v \in L_p(\Omega), w \in L_{p'}(\Omega).$$

Now we are ready to define our FEM using $\{\mathcal{S}_{n,k}\}_{n=1}^\infty$. For $n \in \mathbb{P}$, let $\{s_1, \ldots, s_n\}$ be a basis for $\mathcal{S}_{n,k}$. Define *finite element information* (FEI) $N_{n,k}^{\mathrm{FE}}$ of cardinality n by

$$N_{n,k}^{\mathrm{FE}} f = \begin{bmatrix} \langle f, s_1 \rangle \\ \vdots \\ \langle f, s_n \rangle \end{bmatrix} \quad \forall f \in F.$$

Let $u_{n,k} \in \mathcal{S}_{n,k}$ satisfy

$$B(u_{n,k}, s_i) = \langle f, s_i \rangle \quad (1 \le i \le n). \tag{1}$$

Since $u_{n,k}$ depends on f only through $N_{n,k}^{\mathrm{FE}}$, we write $u_{n,k} = \phi_{n,k}^{\mathrm{FE}}(N_{n,k}^{\mathrm{FE}} f)$, and we say that $\phi_{n,k}^{\mathrm{FE}}$ is the *finite element method* (FEM) using $N_{n,k}^{\mathrm{FE}}$.

Note that we have not yet shown that $\phi_{n,k}^{\mathrm{FE}}(N_{n,k}^{\mathrm{FE}} f)$ is well-defined, i.e. that there always exists $u_{n,k} \in \mathcal{S}_{n,k}$ satisfying (1). We momentarily defer this issue, preferring to first discuss the cost of computing the FEM.

What is $\mathrm{cost}(\phi_{n,k}^{\mathrm{FE}}, N_{n,k}^{\mathrm{FE}})$? Under the model of computation given previously, we would simply say that $\phi_{n,k}^{\mathrm{FE}}$ is a linear algorithm using linear information of cardinality n. More choose $u_1, \ldots, u_n \in \mathcal{S}_{n,k}$ so that

$$B(u_j, s_i) = \delta_{i,j} \qquad (1 \le i, j \le n).$$

Then

$$\phi_{n,k}^{\mathrm{FE}}(N_{n,k}^{\mathrm{FE}}f) = \sum_{j=1}^{n} \langle f, s_j \rangle u_j,$$

Hence, we find that

$$\mathrm{cost}(\phi_{n,k}^{\mathrm{FE}}, N_{n,k}^{\mathrm{FE}}) \le (c+2)n - 1.$$

Note that this estimate assumes that we disregard the cost of precomputing $u_1, \ldots, u_n$.

In practice, we are usually unwilling to do ignore these precomputations. Hence, we usually treat the problem of computing the FEM as a problem of numerical linear algebra. Define a matrix

$$G = [g_{i,j}]_{1 \le i,j \le n}$$

by

$$g_{i,j} = B(s_j, s_i) \qquad (1 \le i, j \le n).$$

For $f \in F$, let

$$b = N_{n,k}^{\mathrm{FE}}f.$$

Let

$$a = \begin{bmatrix} \alpha_1 \\ \vdots \\ \alpha_n \end{bmatrix}$$

be the solution to

$$Ga = b.$$

Then

$$\phi_{n,k}^{\mathrm{FE}}(N_{n,k}^{\mathrm{FE}}f) = u_{n,k} = \sum_{j=1}^{n} \alpha_j s_j.$$

Note that, unlike for elliptic PDEs or elliptic systems, the matrix G is generally a full matrix. However, the efficient solution of this linear system by multigrid methods has been well-investigated. For further details, see Dellwo (1988), Friedman and Dellwo (1982), and Schippers (1979).

In the remainder of this section, we assume that the FEM $\phi_{n,k}^{\mathrm{FE}}$ can be computed in $\Theta(n)$ operations.

Of course, all this presupposes that the FEM is well-defined. We need to show that (1) has a unique solution for sufficiently large n. To do this, we need to lay a little groundwork.

Recall that the mapping $P_n \colon L_2(\Omega) \to L_2(\Omega)$ defined by

$$\langle P_n h, s \rangle = \langle h, s \rangle \qquad \forall h \in L_2(\Omega), s \in \mathcal{S}_{n,k}$$

is the *orthogonal projector* of $L_2(\Omega)$ onto $\mathcal{S}_{n,k}$. It is well-known that P_n is a self-adjoint operator with range $\mathcal{S}_{n,k}$ and unit norm, with

$$\|v - P_n v\|_{L_2(\Omega)} = \inf_{s \in \mathcal{S}_{n,k}} \|v - s\|_{L_2(\Omega)}.$$

Thus the orthogonal projector gives a best approximation in the L_2-norm. The next result shows that P_n also gives an almost-best approximation in the L_q-norm.

LEMMA 6.3.2.1. *For any $q \in [1, \infty]$, there exists $\pi_q > 0$ such that for any $n \in \mathbb{P}$,*

$$\|P_n v\|_{L_q(\Omega)} \le \pi_q \|v\|_{L_q(\Omega)} \qquad \forall v \in L_q(\Omega),$$

and so

$$\|v - P_n v\|_{L_q(\Omega)} \le (1 + \pi_q) \inf_{s \in \mathcal{S}_{n,k}} \|v - s\|_{L_q(\Omega)} \qquad \forall v \in L_q(\Omega).$$

PROOF. To establish the first inequality, let $v \in L_q(\Omega)$. Let $K \in \mathcal{T}_n$. From Lemma A.2.3.3, there exists $C > 0$, independent of v and n, such that

$$\|P_n v\|_{L_q(K)} \le C (\operatorname{vol} K)^{1/2 - 1/q} \|P_n v\|_{L_2(K)}$$

and

$$\|P_n v\|_{L_{q'}(K)} \le C (\operatorname{vol} K)^{1/2 - 1/q'} \|P_n v\|_{L_2(K)},$$

where $1/q + 1/q' = 1$. Let

$$w_K = \begin{cases} P_n v & \text{on } K, \\ 0 & \text{otherwise.} \end{cases}$$

Since $P_n v = w_K$ on K, $w_K = 0$ outside K, P_n is self-adjoint, and $P_n w_K = w_K$ (because $w_K \in \mathcal{S}_{n,k}$), we have

$$\begin{aligned} \|P_n v\|_{L_2(K)}^2 &= \langle P_n, v, w_K \rangle_{L_2(K)} = \langle P_n v, w_K \rangle = \langle v, P_n w_K \rangle \\ &= \langle v, w_K \rangle = \langle v, w_K \rangle_{L_2(K)} = \langle v, P_n v \rangle_{L_2(K)} \\ &\le \|v\|_{L_q(K)} \|P_n v\|_{L_{q'}(K)}, \end{aligned}$$

the latter by the Hölder inequality. Hence we have

$$\|P_n v\|_{L_q(K)} \le C(\operatorname{vol} K)^{1/2-1/q} \|P_n v\|_{L_2(K)}$$

$$\le C(\operatorname{vol} K)^{1/2-1/q} \frac{\|v\|_{L_q(K)} \|P_n v\|_{L_{q'}(K)}}{\|P_n v\|_{L_2(K)}}$$

$$\le C(\operatorname{vol} K)^{1/2-1/q} (\operatorname{vol} K)^{1/2-1/q'} \|v\|_{L_q(K)} \le C\|v\|_{L_q(K)}.$$

Using the discrete version of Hölder's inequality, the desired result follows.

To prove the remainder of the lemma, let $v \in L_q(\Omega)$. Let $n \in \mathbb{P}$ and $s \in \mathcal{S}_{n,k}$. Then $P_n s = s$, and so

$$\|v - P_n s\|_{L_q(\Omega)} \le \|v - s\|_{L_q(\Omega)} + \|P_n(v - s)\|_{L_q(\Omega)} \le (1 + \pi_q)\|v - s\|_{L_q(\Omega)},$$

concluding the proof of the lemma. $\square$

Using this lemma, we then have

LEMMA 6.3.2.2.

 (i) *Let $s \ge 0$ and $q \in [1, \infty]$. There is a positive constant C such that for $v \in W^{s,q}(\Omega)$ and $n \in \mathbb{P}$,*

$$\|v - P_n v\|_{L_q(\Omega)} \le C n^{-\mu/d} \|v\|_{W^{s,q}(\Omega)},$$

   where

$$\mu = \min\{k + 1, s\}.$$

 (ii) *For any $v \in L_q(\Omega)$,*

$$\lim_{n \to \infty} \|v - P_n v\|_{L_q(\Omega)} = 0.$$

PROOF. Using the previous lemma and Lemma 5.4.3, we immediately have (i). To see (ii), let $v \in L_q(\Omega)$ and $\varepsilon > 0$. Choose $v_\varepsilon \in C^1(\Omega)$ such that

$$\|v - v_\varepsilon\|_{L_q(\Omega)} \le \frac{\varepsilon}{2(1 + \pi_q)}.$$

Set $n_0(\varepsilon) = \lceil (2\varepsilon^{-1} C \|v_\varepsilon\|_{W^{1,q}(\Omega)})^d \rceil$, where C is as in (i). Then for any $n \ge n_0(\varepsilon)$, we have

$$\|v_\varepsilon - P_n v_\varepsilon\|_{L_q(\Omega)} \le C n^{-1/d} \|v_\varepsilon\|_{W^{1,q}(\Omega)} \le \tfrac{1}{2}\varepsilon.$$

But the previous lemma also yields

$$\|(v - v_\varepsilon) - P_n(v - v_\varepsilon)\|_{L_q(\Omega)} \le \|v - v_\varepsilon\|_{L_q(\Omega)} + \|(P_n(v - v_\varepsilon)\|_{L_q(\Omega)}$$

$$\le (1 + \pi_q)\|v - v_\varepsilon\|_{L_q(\Omega)}.$$

So

$$\|v - P_n v\|_{L_q(\Omega)} \le \|v_\varepsilon - P_n v_\varepsilon\|_{L_q(\Omega)} + \|(v - v_\varepsilon) - P_n(v - v_\varepsilon)\|_{L_q(\Omega)} < \varepsilon,$$

completing the proof of the lemma. $\square$

We are now able to prove that the FEM is well-defined, and to give a tight estimate of its error.

THEOREM 6.3.2.2.

(i) *There exists $n_0 \in \mathbb{P}$ such that the FEM $\phi_{n,k}^{\mathrm{FE}}$ is well-defined for all $n \geq n_0$. That is, for any $f \in F$ and $n \geq n_0$, there exists a unique $u_{n,k} \in \mathcal{S}_{n,k}$ satisfying (1).*

(ii) *Let*

$$\mu = \min\{k+1, r\}.$$

Then

$$e(\phi_{n,k}^{\mathrm{FE}}, N_{n,k}^{\mathrm{FE}}) = \Theta(n^{-\mu/d}) \qquad \text{as } n \to \infty.$$

PROOF. We first prove (i). Since $u_{n,k} \in \mathcal{S}_{n,k}$ and $P_n^* = P_n$, we find that if $w \in \mathcal{S}_{n,k}$, then

$$\langle (I - P_n K)u_{n,k}, w \rangle = \langle u_{n,k}, w \rangle - \langle Ku_{n,k}, P_n w \rangle = \langle (I - K)u_{n,k}, w \rangle$$
$$= \langle f, w \rangle = \langle f, P_n w \rangle = \langle P_n f, w \rangle,$$

and so

$$(I - P_n K)u_{n,k} = P_n f.$$

Let $\kappa_n = \|(P_n - I)K\|$. By the previous lemma, $\lim_{n \to \infty} \|P_n v - v\|_{L_p(\Omega)} = 0$ for all $v \in L_p(\Omega)$. Since K is compact, this implies that $\lim_{n \to \infty} \kappa_n = 0$, and so there exists $n_0 \in \mathbb{P}$ such that $\kappa_n < \frac{1}{2}\|(I - K)^{-1}\|^{-1}$ for $n \geq n_0$. Hence, $\|(I - K)^{-1}(P_n K - K)\| \leq \frac{1}{2}$, so that the operator $I - (I - K)^{-1}(P_n K - K)$ is invertible for $n \geq n_0$, with

$$\left\| \left(I - (I - K)^{-1}(P_n K - K) \right)^{-1} \right\| \leq 2.$$

Since

$$[I - (I - K)^{-1}(P_n K - K)]^{-1}(I - K)^{-1}$$
$$= \left((I - K)[I - (I - K)^{-1}(P_n K - K)] \right)^{-1}$$
$$= (I - P_n K)^{-1},$$

we see that $(I - P_n K)$ is invertible, and so the FEM is well-defined, whenever $n \geq n_0$.

We next establish that

$$e(\phi_{n,k}^{\mathrm{FE}}, N_{n,k}^{\mathrm{FE}}) = O(n^{-\mu/d}) \qquad \text{as } n \to \infty. \qquad (2)$$

For $f \in F$, let $u = Sf$. Then

$$(I - P_n K)(u - u_{n,k}) = (u - Ku) - (I - P_n K)u_{n,k} + (I - P_n)Ku$$
$$= f - P_n f + (I - P_n)Ku.$$

Since

$$\|(I - P_n K)^{-1}\| \le \|I - (I - K)^{-1}(P_n K - K)\| \, \|(I - K)^{-1}\|$$
$$\le 2\|(I - K)^{-1}\|,$$

we have

$$\|u - u_{n,k}\|_{L_p(\Omega)} \le 2\|(I-K)^{-1}\| \left(\|(I - P_n)f\|_{L_p(\Omega)} + \|(I - P_n)Ku\|_{L_p(\Omega)} \right).$$

Using the previous lemma and Theorem 6.3.1.1, we find

$$\|(I - P_n)f\|_{L_p(\Omega)} \le Cn^{-\mu/d}\|f\|_{W^{r,p}(\Omega)}$$

and

$$\|(I - P_n)Ku\|_{L_p(\Omega)} \le Cn^{-\mu/d}\|Ku\|_{W^{r,p}(\Omega)} \le C\|K\|n^{-\mu/d}\|K\|\|u\|_{W^{r,p}(\Omega)}$$
$$\le C\|K\|\sigma n^{-\mu/d}\|f\|_{W^{r,p}(\Omega)}.$$

Combining these last three inequalities, we have

$$\|Sf - \phi_{n,k}^{\mathrm{FE}}(N_{n,k}^{\mathrm{FE}}f)\|_{L_p(\Omega)} = \|u - u_{n,k}\|_{L_p(\Omega)} \le Cn^{-\mu/d}\|f\|_{W^{r,p}(\Omega)}$$
$$\le Cn^{-\mu/d},$$

the latter since $f \in F$. Taking the supremum over all $f \in F$, the desired upper bound (2) follows.

We need only prove that

$$e(\phi_{n,k}^{\mathrm{FE}}, N_{n,k}^{\mathrm{FE}}) = \Omega(n^{-\mu/d}) \qquad \text{as } n \to \infty. \tag{3}$$

If $k + 1 \ge r$, then $\mu = r$ and our lower bound (3) follows from the Θ-estimate of $r(n)$ given in Theorem 6.3.2.1. Suppose now that $k + 1 < r$, so that $\mu = k + 1$. As in the proof of Theorem 5.4.2, we can find an open region Ω^1, independent of n, whose closure is contained in the interior of Ω, as well as $u \in W^{r,p}(\Omega)$, such that

(i) there is a constant $C > 0$, independent of n, such that

$$\inf_{s \in \mathcal{S}_{n,k}} |u - s|_{L_p(\Omega)} \ge Cn^{-(k+1)/d}|u|_{W^{r,p}(\Omega^1)},$$

(ii) $|u|_{W^{r,p}(\Omega^1)}$ is bounded from below by a positive constant, independent of n, and

(iii) $\|Lu\|_{W^{r,p}(\Omega)}$ is bounded from above by a positive constant, independent of n.

Set

$$f^* = \frac{1}{\|Lu\|_{W^{r,p}(\Omega)}} Lu,$$

so that $f^* \in F$. Let

$$C' = C \inf_{n \in \mathbb{P}} \frac{|u|_{W^{r,p}(\Omega^1)}}{\|Lu\|_{W^{r,p}(\Omega)}},$$

where C is the constant in (i) above. Since C' is a positive constant, we have

$$e(\phi_{n,k}^{\mathrm{LSFE}}, N_{n,k}^{\mathrm{LSFE}}) \geq \|Sf^* - \phi_{n,k}^{\mathrm{FE}}(N_{n,k}^{\mathrm{FE}}f^*)\|_{L_p(\Omega)}$$

$$\geq \frac{1}{\|Lu\|_{W^{r,p}(\Omega)}} \inf_{s \in \mathcal{S}_{n,k}} |u - s|_{L_p(\Omega)}$$

$$\geq C'n^{-(k+1)/d},$$

which establishes (3) and completes the proof of the theorem. $\square$

Combining the previous two theorems, we have

COROLLARY 6.3.2.1. *The FEM of degree k is an almost minimal error algorithm iff $k \geq r - 1$.* $\square$

As with the problems we have already studied, there can be two explanations of why the error of the FEM is not almost minimal when $k < r - 1$. Either

(i) the FEM $\phi_{n,k}^{\mathrm{FE}}$ does use its information $N_{n,k}^{\mathrm{FE}}$ as well as possible, and there is another method using $N_{n,k}^{\mathrm{FE}}$ whose error is almost minimal, regardless of whether $k \geq r - 1$, or,

(ii) the FEM is an almost optimal error algorithm using FEI, and FEI is not strong enough information to allow an nth almost minimal error algorithm.

We now show that in the Hilbert case $p = 2$, possibility (i) above holds, explaining why the FEM of too low a degree does not have almost minimal error.

THEOREM 6.3.2.3. *For the Hilbert case $p = 2$, the spline algorithm $\phi_{n,k}^{\mathrm{s}}$ using FEI $N_{n,k}^{\mathrm{FE}}$ is an optimal error algorithm using FEI and is an almost minimal error algorithm. That is,*

$$e(\phi_{n,k}^{\mathrm{s}}, N_{n,k}^{\mathrm{FE}}) = r(N_{n,k}^{\mathrm{FE}}) = \Theta(r(n)) = \Theta(n^{-r/d}) \qquad \text{as } n \to \infty.$$

PROOF. As with previous problems, it suffices to show that

$$r(N_{n,k}^{\mathrm{FE}}) = O(n^{-r/d}) \qquad \text{as } n \to \infty.$$

Let $z \in F \cap \ker N_{n,k}^{\mathrm{FE}}$. By Lemma 5.4.3, there exists $s \in \mathcal{S}_{n,k}$ such that

$$\|z - s\|_{H^{-r}(\Omega)} \leq Cn^{-r/d}\|z\|_{H^r(\Omega)} \leq Cn^{-r/d},$$

the latter because $z \in F$. Since $z \in \ker N_{n,k}^{\mathrm{FE}}$, we have $\langle z, s \rangle = 0$. Using Theorem 6.3.1.1, we find that

$$\|Sz\|_{L_2(\Omega)} \leq \sigma\|z\|_{L_2(\Omega)} = \frac{\sigma|\langle z, z - s \rangle|}{\|z\|_{L_2(\Omega)}}$$

$$\leq \frac{\sigma\|z\|_{H^r(\Omega)}\|z - s\|_{H^{-r}(\Omega)}}{\|z\|_{L_2(\Omega)}} \leq Cn^{-r/d}.$$

Taking the supremum over all $z \in F \cap \ker N_{n,k}^{\mathrm{FE}}$, the result follows.   □

So the information $N_{n,k}^{\mathrm{FE}}$ is always nth almost-optimal information in the Hilbert case. However, the FEM of degree k is an almost-optimal error algorithm using FEI iff $k \geq r - 1$.

For $\varepsilon > 0$, we now wish to find comp(ε), the ε-complexity of our problem. We also wish to determine conditions under which the FEM is an almost optimal complexity algorithm for our problem. Let

$$\mathrm{cost}_k^{\mathrm{FE}}(\varepsilon) = \inf\{\, \mathrm{cost}(\phi_{n,k}^{\mathrm{FE}}, N_{n,k}^{\mathrm{FE}}) : e(\phi_{n,k}^{\mathrm{FE}}, N_{n,k}^{\mathrm{FE}}) \leq \varepsilon \,\}$$

denote the minimal cost of using the FEM of degree k to compute an ε-approximation. We also let

$$\mathrm{cost}_k^{\mathrm{spline}}(\varepsilon) = \inf\{\, \mathrm{cost}(\phi_{n,k}^{\mathrm{s}}, N_{n,k}^{\mathrm{FE}}) : e(\phi_{n,k}^{\mathrm{FE}}, N_{n,k}^{\mathrm{FE}}) \leq \varepsilon \,\}$$

be the minimal cost of computing an ε-approximation with the spline algorithm using FEI of degree k. Using the previous three theorems, along with the results in §4.4, we have

THEOREM 6.3.2.4.

(i) *The ε-complexity is given by*

$$\mathrm{comp}(\varepsilon) = \Theta(\varepsilon^{-d/r}) \qquad \text{as } \varepsilon \to 0.$$

(ii) *Let $\mu = \min\{k + 1, r\}$. Then*

$$\mathrm{cost}_k^{\mathrm{FE}}(\varepsilon) = \Theta(\varepsilon^{-d/\mu}) \qquad \text{as } \varepsilon \to 0.$$

Hence, if $k \geq r - 1$, then the FEM of degree k is an almost optimal complexity algorithm, i.e.

$$\mathrm{cost}_k^{\mathrm{FE}}(\varepsilon) = \Theta\big(\mathrm{comp}(\varepsilon)\big) = \Theta(\varepsilon^{-d/r}) \qquad \text{as } \varepsilon \to 0.$$

(iii) *The spline algorithm using FEI of degree k is an almost optimal complexity algorithm in the Hilbert case. That is, if $p = 2$, then*

$$\mathrm{cost}_k^{\mathrm{spline}}(\varepsilon) = \Theta\big(\mathrm{comp}(\varepsilon)\big) = \Theta(\varepsilon^{-d/r}) \qquad \text{as } \varepsilon \to 0. \qquad □$$

Hence if $k \geq r - 1$, the FEM of degree k is an almost optimal complexity algorithm. Suppose that $k < r - 1$. Then

$$\frac{\text{cost}_k^{\text{FE}}(\varepsilon)}{\text{comp}(\varepsilon)} = \Theta(\varepsilon^{-\delta d}) \qquad \text{as } \varepsilon \to 0,$$

where

$$\delta = \frac{1}{k + 1 - l} - \frac{1}{r + t - l} > 0.$$

Thus

$$\lim_{\varepsilon \to 0} \frac{\text{cost}_k^{\text{FE}}(\varepsilon)}{\text{comp}(\varepsilon)} = \infty,$$

and so the asymptotic penalty for using an FEM of too low a degree is unbounded.

Note that the results of this subsection are analogous to those for a scalar elliptic PDE and for elliptic systems. This implies that many of the same remarks about those problems hold for solving Fredholm problems of the second kind. In particular, the following hold:

 (i) If the problem elements are only smooth enough to guarantee existence and uniqueness of the solution (i.e. $r = 0$), then the problem is not convergent.

 (ii) Solving Fredholm problems of the second kind is inherently intractable for problems of large dimension.

 (iii) The inherent complexity of solving small-dimensional problems may be prohibitively expensive. For example, suppose that $d = 3$. If we have problem elements in $W^{1,p}(\Omega)$ and we seek an $L_p(\Omega)$-norm ε-approximation, then the complexity is $\Theta(\varepsilon^{-3})$. As in Chapter 5, this implies that four-place accuracy will require about two weeks' worth of computation on a megaflop machine.

 (iv) If we don't know the exact value of the regularity r, but only know an upper bound r^* on r, then the FEM of degree $k \geq r^* - 1$ is an almost optimal algorithm.

 (v) The condition for the FEM to be a spline algorithm is quite restrictive, namely that the FEM is a spline algorithm iff $\mathcal{S}_{n,k} = SS^*\mathcal{S}_{n,k}$.

Notes and Remarks

NR 6.3.2:1 Lemma 6.3.2.1 is due to Douglass *et al.* (1975).

NR 6.3.2:2 The proofs that the FEM is well-defined and of the upper bound on its error (in Theorem 6.3.2.2) are based on Ikebe (1972).

NR 6.3.2:3 We have only been able to prove Theorem 6.3.2.3 in the Hilbert case $p = 2$. If we could prove an estimate of the form

$$\inf_{s \in \mathcal{S}_{n,k}} \|z - s\|_{W^{-r,p'}(\Omega)} \leq Cn^{-r/d}\|z\|_{L_p(\Omega)},$$

(which is an extension of (ii) in Lemma 5.4.3) then it is easy to see that the proof of Theorem 6.3.2.3 would carry over to the case of arbitrary p. Most of the proof in Babuška and Aziz (1972) of (ii) in Lemma 5.4.3 *do* appear to hold in the general case, with one exception: we need to be able establish that the $H^{-s}(\Omega)$-orthogonal projection onto $S_{n,k}$ is bounded in the $W^{-s,q}(\Omega)$-norm. That is, we need to extend Lemma 6.3.2.1 from the $L_q(\Omega)$-norm to the $W^{-s,q}(\Omega)$-norm. It is not clear whether this extension holds.

NR 6.3.2:4 We briefly discuss the seminormed case for arbitrary linear information, in which $F = \mathcal{B}H^r(\Omega)$. We find that that the nth minimal error and the ε-complexity are the same for the normed and seminormed cases. Furthermore, the FEM has infinite error unless $SP_{r-1}(\Omega) \subseteq S_{n,k}$. However, when this inclusion holds, the results for the FEM in the normed case carry over to the seminormed case.

We describe these results more precisely. Let

$$n^* = \binom{d + r - 1}{r - 1}.$$

Then:

- (i) $r(n) = \infty$ if $n < n^*$.
- (ii) $r(n) = \Theta(n^{-r/d})$ as $n \to \infty$.
- (iii) If $SP_{r-1}(\Omega) \not\subseteq S_{n,k}$, then

$$e(\phi_{n,k}^{\mathrm{FE}}, N_{n,k}^{\mathrm{FE}}) = \infty.$$

- (iv) If $SP_{r-1}(\Omega) \subseteq S_{n,k}$ for all sufficiently large n, then

$$e(\phi_{n,k}^{\mathrm{FE}}, N_{n,k}^{\mathrm{FE}}) = \Theta(n^{-\mu/d}) \qquad \text{as } n \to \infty,$$

  where

$$\mu = \min\{k + 1 - t, r\}.$$

- (v) The spline algorithm $\phi_{n,k}^{\mathrm{s}}$ using FEI $N_{n,k}^{\mathrm{FE}}$ is an almost minimal error algorithm, i.e.

$$e(\phi_{n,k}^{\mathrm{s}}, N_{n,k}^{\mathrm{FE}}) = r(N_{n,k}^{\mathrm{FE}}) = \Theta\bigl(r(n)\bigr) = \Theta(n^{-r/d}) \qquad \text{as } n \to \infty.$$

- (vi) $\mathrm{comp}(\varepsilon) = \Theta(\varepsilon^{-d/r})$ as $\varepsilon \to 0$.
- (vii) If there exists no integer n for which $SP_{r-1}(\Omega) \subseteq S_{n,k}$, then

$$\mathrm{cost}_k^{\mathrm{FE}}(\varepsilon) = \infty \qquad \forall\, \varepsilon > 0.$$

- (viii) Suppose that there exists a nonnegative integer n_0 such that $SP_{r-1}(\Omega) \subseteq S_{n,k}$ for $n \ge n_0$. Then

$$\mathrm{cost}_k^{\mathrm{FE}}(\varepsilon) = \Theta(\varepsilon^{-d/(\mu+t-l)}) \qquad \text{as } \varepsilon \to 0.$$

  Hence, if $SP_{r-1}(\Omega) \subseteq S_{n,k}$ for $n \ge n_0$ and $k \ge r-1$, then the FEM of degree k is an almost optimal complexity algorithm, i.e.

$$\mathrm{cost}_k^{\mathrm{FE}}(\varepsilon) = \Theta\bigl(\mathrm{comp}(\varepsilon)\bigr) = \Theta(\varepsilon^{-d/r}) \qquad \text{as } \varepsilon \to 0.$$

- (ix) The spline algorithm using FEI of degree k is an almost optimal complexity algorithm, i.e.

$$\mathrm{cost}_k^{\mathrm{spline}}(\varepsilon) = \Theta\bigl(\mathrm{comp}(\varepsilon)\bigr) = \Theta(\varepsilon^{-d/r}) \qquad \text{as } \varepsilon \to 0.$$

Once again, these results may be proved by following the analysis in the first part of §5.6, making the obvious changes.

6.3.3 Standard information

We now suppose that $\Lambda = \Lambda^{\text{std}}$. That is, we assume that only standard information (i.e. the value of f or one of its derivatives at some point in Ω) is permissible information about f. Our goal is to show that the minimal error and complexity remain essentially the same as for the case $\Lambda = \Lambda'$ of arbitrary linear information.

Since our problem is linear, we need only consider nonadaptive information. As in previous problems, we first need a lower bound on $r(n, \Lambda^{\text{std}})$. However, since any standard information is an instance of arbitrary linear information, we have $r(n, \Lambda^{\text{std}}) \geq r(n)$. Since $r(n) = \Theta(n^{-r/d})$ by Theorem 6.3.2.1, we immediately have

THEOREM 6.3.3.1.   *There exists a positive constant C, independent of n, such that*

$$r(n, \Lambda^{\text{std}}) \geq C n^{-r/d} \qquad \forall\, n \in \mathbb{P}. \qquad \Box$$

Now we need to find an almost minimal error algorithm using standard information. Based on our past experiences, it should be no surprise that a finite element method with quadrature, using standard information consisting of function evaluations alone (i.e. no derivative evaluations), will be an almost minimal error algorithm.

In what follows, we assume that $r > d/p$, since standard information is not well-defined (and the complexity is infinite) if this inequality is violated. By the Sobolev embedding theorem, this means that $W^{r,p}(\Omega)$ is embedded in $C(\bar{\Omega})$, and so standard information consisting of only function values is a well-defined, bounded linear operator on our class of problem elements.

Once again, we let $\hat{I}$ denote a quadrature rule over the reference element $\hat{K}$, with nodes $\hat{b}_1, \ldots, \hat{b}_J$ and weights $\hat{\omega}_1, \ldots, \hat{\omega}_J$. That is, we have

$$\hat{I}\hat{v} = \sum_{j=1}^{J} \hat{\omega}_j \hat{v}(\hat{b}_j) \qquad \forall\, \hat{v} \in C\left(\overline{\hat{K}}\right).$$

The reference quadrature rule $\hat{I}$ induces a local quadrature rule I_K over each element K in $\mathcal{T}_{\bar{n}}$, the nodes and weights of I_K being respectively denoted by $b_{1,K}, \ldots, b_{J,K}$ and $\omega_{1,K}, \ldots, \omega_{J,K}$, with

$$\omega_{j,K} = \det B_K \cdot \omega_K \quad \text{and} \quad b_{j,K} = F_K(b_K) \qquad (1 \leq j \leq J),$$

with $F_K \hat{x} \equiv B_K \hat{x} + b_K$ being the affine mapping that maps the reference element $\hat{K}$ onto the local element K. Thus

$$I_K v = \sum_{j=1}^{J} \omega_{j,K} v(b_{j,K}) \qquad \forall\, v \in C(\overline{K}).$$

We let $\hat{E}$ and E_K respectively denote the error functionals of $\hat{I}$ and I_K.
   Let $\tilde{n} \in \mathbb{P}$. Then

$$\mathcal{N}_{\tilde{n}} = \bigcup_{K \in \mathcal{T}_{\tilde{n}}} \bigcup_{j=1}^{J} \{b_{j,K}\}$$

denotes the set of all nodes of the triangulation $\mathcal{T}_{\tilde{n}}$. Recall that the bilinear form B is defined by

$$B(v,w) = \langle (I - K)v, w \rangle = \int_{\Omega} \left[v(x) - \int_{\Omega} k(x,y)v(y)\,dy \right] w(x)\,dx.$$

We define a bilinear form $B_{\tilde{n}}$ approximating B on $\mathcal{S}_{n,k}$ by replacing the integrals above by quadratures. Thus we have

$$B_{\tilde{n}}(v,w) = \sum_{\substack{K \in \mathcal{T}_{\tilde{n}} \\ 1 \le i \le J}} \omega_{i,K} \left[v(b_{i,K}) - \sum_{\substack{K' \in \mathcal{T}_{\tilde{n}} \\ 1 \le j \le J}} \omega_{j,K'} k(b_{i,K}, b_{j,K'}) v(b_{j,K'}) \right] w(b_{i,K})$$

for $v,\, w \in \mathcal{S}_{n,k}$. Analogously, for $f \in F$, we define a linear functional $f_{\tilde{n}}$ approximating $\langle f, \cdot \rangle$ on $\mathcal{S}_{n,k}$ by

$$f_{\tilde{n}}(v) = \sum_{\substack{K \in \mathcal{T}_{\tilde{n}} \\ 1 \le j \le J}} \omega_{j,K} f(b_{j,K}) v(b_{j,K})$$

for $w \in \mathcal{S}_{n,k}$.
   We are now ready to define our finite element method with quadrature. Let $n \in \mathbb{P}$. Choose $\tilde{n}$ to be the largest integer such that

$$\operatorname{card} \mathcal{N}_{\tilde{n}} \le n.$$

As in Chapter 5, $\tilde{n} = \Theta(n)$ as $n \to \infty$. Let $x_1, \dots, x_{\tilde{n}}$ denote the elements of $\mathcal{N}_{\tilde{n}}$, and choose points $x_{\tilde{n}+1}, \dots, x_n$ not belonging to $\mathcal{N}_{\tilde{n}}$. For $f \in F$, define information $N_{n,k}^{Q}$ to be

$$N_{n,k}^{Q} f = \begin{bmatrix} f(x_1) \\ \vdots \\ f(x_n) \end{bmatrix} \qquad \forall f \in F.$$

Then $N_{n,k}^Q$ is standard information of cardinality n. Suppose we can find $\tilde{u}_{\tilde{n},k} \in \mathcal{S}_{\tilde{n},k}$ satisfying

$$B_{\tilde{n}}(\tilde{u}_{\tilde{n},k}, s) = f_{\tilde{n}}(s) \qquad \forall\, s \in \mathcal{S}_{\tilde{n},k}. \tag{1}$$

Then $\tilde{u}_{\tilde{n},k}$ depends on f only through $N_{n,k}^Q f$, so we write $\tilde{u}_{\tilde{n}} = \phi_{n,k}^Q(N_{n,k}^Q f)$. We say that $\phi_{n,k}^Q$ is the *finite element method with quadrature* (FEMQ).

As in the previous subsection, we let

$$\mu = \min\{k+1, r\}.$$

Analogously with §5.5, we assume that

   (i) The smoothness r of the problem elements F satisfies $r \geq 1$ (as well as our previous requirement $r > d/p$).

   (ii) The degree k of the finite element subspaces $\mathcal{S}_{\tilde{n},k}$ satisfies $k > d/p - 1$.

   (iii) $\hat{I}$ is exact of degree $2k$ over the reference element $\hat{K}$, i.e. $\hat{I}(\hat{p}) = \int_{\hat{I}} \hat{p}$ for any $\hat{p} \in P_{2k}(\hat{K})$.

Before analyzing the FEMQ, we need to do some preliminary work. For $n \in \mathbb{P}$ and $K, K' \in \mathcal{T}_{\tilde{n}}$, let

$$E_{K \times K'}(kvw) = \int_K \left[\int_{K'} k(x,y)v(y)\,dy \right] w(x)\,dx$$
$$- \sum_{i=1}^J \omega_{i,K} \left[\sum_{j=1}^J \omega_{j,K'} k(b_{i,K}, b_{j,K'}) v(b_{j,K'}) \right] w(b_{i,K}).$$

LEMMA 6.3.3.1. *There is a positive constant C such that for any $n \in \mathbb{P}$, any $v, w \in \mathcal{S}_{n,k}$, any $f \in W^{r,p}(\Omega)$, and any $K, K' \in \mathcal{T}_{\tilde{n}}$, the following hold:*

   (i) $|E_{K \times K'}(kvw)| \leq Cn^{-(1+2/d)}|k|_{W^{1,\infty}(K \times K')}\|v\|_{L_p(K')}\|w\|_{L_{p'}(K)}.$

   (ii) $|E_{K \times K'}(kvw)| \leq Cn^{-(1+\mu/d)}|k|_{W^{r,\infty}(K \times K')}\|v\|_{W^{r,p}(K')}\|w\|_{L_{p'}(K)}.$

   (iii) $|E_K(fw)| \leq Cn^{-\mu/d}\|f\|_{W^{r,p}(K)}\|w\|_{L_{p'}(K)}.$

PROOF. Before proving (i) and (ii), we write

$$\hat{E}_{\hat{K} \times \hat{K}}(\hat{z}) = \int_{\hat{K}} \left[\int_{\hat{K}} \hat{z}(\hat{x}, \hat{y})\,d\hat{y} \right] d\hat{x} - \sum_{i=1}^J \hat{\omega}_i \left[\sum_{j=1}^J \hat{\omega}_j \hat{z}(\hat{b}_i, \hat{b}_j) \right]$$

for $\hat{z}: \hat{K} \times \hat{K} \to \mathbb{R}$. Since $\{\mathcal{T}_n\}_{n=1}^\infty$ is quasi-uniform,

$$|\det B_K| = \frac{\mathrm{vol}\,K}{\mathrm{vol}\,\hat{K}} = \Theta(h_K^d) = \Theta(n^{-1}) \qquad \forall K \in \mathcal{T}_{\tilde{n}}.$$

Hence,

$$|E_{K \times K'}(kvw)| = |\det B_K||\det B_{K'}||\hat{E}_{\hat{K} \times K}(\hat{k}\hat{v}\hat{w})| \qquad (2)$$
$$\leq Cn^{-2}|\hat{E}_{\hat{K} \times K}(\hat{k}\hat{v}\hat{w})|.$$

We first prove (i). For $\hat{v}, \hat{w} \in P_k(\hat{K})$, define $\lambda_{\hat{v},\hat{w}} \colon W^{1,\infty}(\hat{K}) \to \mathbb{R}$ by

$$\lambda_{\hat{v},\hat{w}}(\hat{k}) = \hat{E}_{\hat{K} \times K}(\hat{k}\hat{v}\hat{w}).$$

Since all norms are equivalent on the finite-dimensional space $P_k(\hat{K})$, there exist positive constants C, independent of k, v, and w, such that

$$|\lambda_{\hat{v},\hat{w}}(\hat{k})| \leq C\|\hat{k}\hat{v}\hat{w}\|_{L_\infty(\hat{K} \times \hat{K})} \leq C\|\hat{k}\|_{L_\infty(\hat{K} \times \hat{K})}\|\hat{v}\|_{L_\infty(\hat{K})}\|\hat{w}\|_{L_\infty(\hat{K})}$$
$$\leq C\|\hat{k}\|_{W^{1,\infty}(\hat{K} \times \hat{K})}\|\hat{v}\|_{L_p(\hat{K})}\|\hat{w}\|_{L_{p'}(\hat{K})},$$

and hence $\lambda_{\hat{v},\hat{w}}$ is a bounded linear functional on $W^{1,\infty}(\hat{K})$ with norm

$$\|\lambda_{\hat{v},\hat{w}}\|^* \leq C\|\hat{v}\|_{L_p(\hat{K})}\|\hat{w}\|_{L_{p'}(\hat{K})}.$$

Since $\hat{I}$ is exact of degree $2k \geq k$ and $\hat{v}, \hat{w} \in P_k(\hat{K})$, we have $\lambda_{\hat{v},\hat{w}} \equiv 0$ on $P_0(\hat{K} \times \hat{K})$. By the Bramble-Hilbert Lemma (Lemma A.2.3.2), we thus have

$$|\lambda_{\hat{v},\hat{w}}(\hat{k})| \leq C|\hat{k}|_{W^{1,\infty}(\hat{K} \times \hat{K})}\|\hat{v}\|_{L_p(\hat{K})}\|\hat{w}\|_{L_{p'}(\hat{K})}.$$

From Lemma A.2.2.3 and quasi-uniformity, we have

$$|\hat{k}|_{W^{1,\infty}(\hat{K} \times \hat{K})} \leq Ch_{K \times K'}|k|_{W^{1,\infty}(K \times K')} \leq Cn^{-2/d}|k|_{W^{1,\infty}(K \times K')}$$
$$\|\hat{v}\|_{L_p(\hat{K})} \leq C|\det B'_K|^{-1/p}\|v\|_{L_p(K')} \leq Cn^{1/p}\|v\|_{L_p(K')}$$
$$\|\hat{w}\|_{L_{p'}(\hat{K})} \leq C|\det B_K|^{-1/p'}\|w\|_{L_{p'}(K)} \leq Cn^{1/p'}\|w\|_{L_{p'}(K)}.$$

Since $1/p + 1/p' = 1$, we thus have

$$|\hat{E}_{\hat{K} \times K}(\hat{k}\hat{v}\hat{w})| \leq Cn^{1-2/d}|k|_{W^{1,\infty}(K \times K')}\|v\|_{L_p(K')}\|w\|_{L_{p'}(K)},$$

which, when combined with (2), yields (i).

We next show (ii). For $\hat{w} \in P_k(\hat{K})$, define $\lambda_{\hat{w}} \colon W^{\mu,\infty}(\hat{K} \times \hat{K}) \to \mathbb{R}$ by

$$\lambda_{\hat{w}}(\hat{z}) = \hat{E}_{\hat{K} \times K}(\hat{z}\hat{w}).$$

Then there exist positive constants C, independent of z and w, such that

$$|\lambda_{\hat{w}}(\hat{z})| = |\hat{E}_{\hat{K} \times K}(\hat{z}\hat{w})| \leq C\|\hat{z}\hat{w}\|_{L_\infty(\hat{K} \times \hat{K})} \leq C\|\hat{z}\|_{L_\infty(\hat{K} \times \hat{K})}\|\hat{w}\|_{L_\infty(\hat{K})}$$
$$\leq C\|\hat{z}\|_{W^{\mu,\infty}(\hat{K} \times \hat{K})}\|\hat{w}\|_{L_{p'}(\hat{K})},$$

and so $\lambda_{\hat{w}}$ is a bounded linear functional on $W^{\mu,\infty}(\hat{K} \times \hat{K})$, with norm

$$\|\lambda_{\hat{w}}\|^* \le C\|\hat{w}\|_{L_{p'}(\hat{K})}.$$

Since $\hat{I}$ is exact of degree $2k \ge k + \mu - 1$, it is easy to see that $\lambda_{\hat{w}} \equiv 0$ on $P_{\mu-1}(\hat{K})$. Hence the Bramble-Hilbert Lemma (Lemma A.2.3.2) yields

$$|\hat{E}_{\hat{K} \times K}(\hat{z}\hat{w})| = |\lambda_{\hat{w}}(\hat{z})| \le C|\hat{z}|_{W^{\mu,\infty}(\hat{K} \times \hat{K})}\|\hat{w}\|_{L_{p'}(\hat{K})},$$

and so

$$|\hat{E}_{\hat{K} \times K}(\hat{k}\hat{v}\hat{w})| \le C|\hat{k}\hat{v}|_{W^{\mu,\infty}(\hat{K} \times \hat{K})}\|\hat{w}\|_{L_{p'}(\hat{K})}.$$

Now for $|\alpha| = j$, norm equivalence on $P_k(\hat{K})$ yields

$$\|D^\alpha \hat{w}\|_{L_\infty(\hat{K})} \le C\|D^\alpha \hat{w}\|_{L_p(\hat{K})},$$

and so

$$|\hat{v}|_{W^{j,\infty}(\hat{K})} \le C|\hat{v}|_{W^{j,p}(\hat{K})}.$$

Using the Leibnitz rule, we thus find

$$|\hat{k}\hat{w}|_{W^{\mu,\infty}(\hat{K} \times \hat{K})} \le C \sum_{j=0}^\mu |\hat{k}|_{W^{\mu-j,\infty}(\hat{K} \times \hat{K})}|\hat{v}|_{W^{j,\infty}(\hat{K})}$$

$$\le C \sum_{j=0}^\mu |\hat{k}|_{W^{\mu-j,\infty}(\hat{K} \times \hat{K})}|\hat{v}|_{W^{j,p}(\hat{K})},$$

and so

$$|\hat{E}_{\hat{K} \times K}(\hat{k}\hat{v}\hat{w})| \le C \sum_{j=0}^\mu |\hat{k}|_{W^{\mu-j,\infty}(\hat{K} \times \hat{K})}|\hat{v}|_{W^{j,p}(\hat{K})}\|\hat{w}\|_{L_{p'}(\hat{K})}.$$

From Lemma A.2.3.3 and quasi-uniformity, we find

$$|\hat{k}|_{W^{\mu-j,\infty}(\hat{K} \times \hat{K})} \le Ch_{K \times K'}^{\mu-j}|k|_{W^{\mu-j,\infty}(K \times K')}$$

$$\le Cn^{-2(\mu-j)/d}|k|_{W^{\mu-j,\infty}(K \times K')},$$

$$|\hat{v}|_{W^{j,p}(\hat{K})} \le Ch_K^j|\det B_K|^{-1/p}|v|_{W^{j,p}(K')} \le Cn^{1/p-j/d}|v|_{W^{j,p}(K')},$$

$$\|\hat{w}\|_{L_{p'}(\hat{K})} \le C|\det B_K|^{-1/p'}\|w\|_{L_{p'}(K)} \le Cn^{1/p'}\|w\|_{L_{p'}(K)}.$$

Hence

$$|\hat{E}_{\hat{K} \times K}(\hat{k}\hat{v}\hat{w})| \le C \sum_{j=0}^\mu n^{-(2\mu-j)/d+1}|k|_{W^{\mu-j,\infty}(K \times K')}|v|_{W^{j,p}(K')}\|w\|_{L_{p'}(K)}$$

$$\le Cn \cdot n^{-\mu/d}|k|_{W^{\mu,\infty}(K \times K')}|v|_{W^{\mu,p}(K')}\|w\|_{L_{p'}(K)},$$

Since $\mu \leq r$, the desired result (ii) now follows from this inequality and (2). To prove (iii), let $f \in W^{r,\infty}(K)$ and $v \in P_k(K)$. Then

$$|E_K(fv)| = |\det B_K| |\hat{E}(\hat{f}\hat{v})|.$$

For $\hat{v} \in P_k(\hat{K})$, define $\lambda_{\hat{v}} \colon W^{\mu,\infty}(\hat{K}) \to \mathbb{R}$ by

$$\lambda_{\hat{v}}(\hat{f}) = \hat{E}(\hat{f}\hat{v}).$$

Since $\mu = \min\{k+1,r\} > d/p$, we find that $W^{\mu,p}(\hat{K})$ is embedded in $C\left(\overline{\hat{K}}\right)$. Thus

$$|\lambda_{\hat{v}}(\hat{f})| \leq C\|\hat{f}\hat{v}\|_{L_\infty(\hat{K})} \leq C\|\hat{f}\|_{L_\infty(\hat{K})}\|\hat{v}\|_{L_\infty(\hat{K})}$$
$$\leq C\|\hat{f}\|_{W^{\mu,p}(\hat{K})}\|\hat{v}\|_{L_{p'}(\hat{K})},$$

where we have used norm-equivalence on $P_k(\hat{K})$. So $\lambda_{\hat{v}}$ is a bounded linear functional on $W^{\mu,p}(\hat{K})$, with norm bounded by

$$\|\lambda_{\hat{v}}\|^* \leq C\|\hat{v}\|_{L_{p'}(\hat{K})}.$$

Since $\hat{I}$ is exact of degree $2k \geq k + \mu - 1$, we see that $\lambda_{\hat{v}} = 0$ on $P_{\mu-1}(\hat{K})$. Hence the Bramble-Hilbert Lemma implies that

$$|\lambda_{\hat{v}}(\hat{f})| \leq C|\hat{f}|_{W^{\mu,\infty}(\hat{K})}\|\hat{v}\|_{L_{p'}(K)}.$$

Hence

$$|E_K(fv)| = |\det B_K| |\hat{E}(\hat{f}\hat{v})| \leq C|\det B_K| |f|_{W^{\mu,p}(\hat{K})}\|\hat{v}\|_{L_{p'}(K)}.$$

Since $\mu \leq r$, Lemma A.2.3.3 yields that

$$|\hat{f}|_{W^{\mu,p}(\hat{K})} \leq Ch_K^\mu|\det B_K|^{-1/p}|f|_{W^{\mu,\infty}(K)}$$
$$\leq Cn^{-\mu/d}|\det B_K|^{-1/p}\|f\|_{W^{r,p}(K)},$$

and

$$\|\hat{v}\|_{L_{p'}(\hat{K})} \leq C|\det B_K|^{-1/p'}\|v\|_{L_{p'}(K)}.$$

Using these last two inequalities, (iii) follows immediately . $\square$

We now translate the local estimates of the previous lemma to global results. Recall that $\Pi_{\tilde{n}}v$ is the $\mathcal{S}_{\tilde{n},k}$-interpolant of v.

LEMMA 6.3.3.2. *There is a positive constant C such that for any $n \in \mathbb{P}$, any $v, w \in \mathcal{S}_{n,k}$, and any $f \in W^{r,p}(\Omega)$, the following hold:*

(i) $|B(v, w) - B_{\tilde{n}}(v, w)|$
$$\leq Cn^{-(1+2/d)}|k|_{W^{1,\infty}(\Omega \times \Omega')}\|v\|_{L_p(\Omega)}\|w\|_{L_{p'}(\Omega)}.$$

(ii) $|B(P_{\tilde{n}}Sf, w) - B_{\tilde{n}}(P_{\tilde{n}}Sf, w)|$
$$\leq Cn^{-(1+\mu/d)}|k|_{W^{r,\infty}(\Omega \times \Omega)}\|v\|_{W^{r,p}(\Omega)}\|w\|_{L_{p'}(\Omega)}.$$

(iii) $|\langle f, w \rangle - f_{\tilde{n}}(w)| \leq Cn^{-\mu/d}\|f\|_{W^{r,p}(\Omega)}\|w\|_{L_{p'}(\Omega)}.$

PROOF. First, note that since $\hat{I}$ is exact of degree $2k$, we have

$$\int_\Omega vw = \sum_{K \in \mathcal{T}_{\tilde{n}}} I_K(vw).$$

Hence we find that

$$|B(v, w) - B_n(v, w)| \leq \sum_{K, K' \in \mathcal{T}_{\tilde{n}}} |E_{K \times K'}(kvw)|$$
$$\leq Cn^{-(1+\alpha/d)} \sum_{K, K' \in \mathcal{T}_{\tilde{n}}} a_{K,K'}\|v\|_{W^{\beta,p}(K')}\|w\|_{L_{p'}(K)}, \tag{3}$$

where either

$$\alpha = 2, \qquad a_{K,K'} = |k|_{W^{1,\infty}(K \times K')}, \qquad \beta = 0 \tag{4}$$

or

$$\alpha = \mu, \qquad a_{K,K'} = \|k\|_{W^{r,\infty}(K \times K')}, \qquad \beta = r. \tag{5}$$

Using the discrete Hölder inequality, we have

$$\sum_{K, K' \in \mathcal{T}_{\tilde{n}}} a_{K,K'}|v|_{W^{\beta,p}(K)}\|w\|_{L_{p'}(K}$$
$$\leq C\left[\max_{K,K' \in \mathcal{T}_{\tilde{n}}} a_{K,K'}\right]$$
$$\times \left[\sum_{K' \in \mathcal{T}_{\tilde{n}}} \|v\|^p_{W^{\beta,p}(K')}\right]^{1/p} \left[\sum_{K \in \mathcal{T}_{\tilde{n}}} \|w\|^{p'}_{L_{p'}(K')}\right]^{1/p'}$$
$$= C\left[\max_{K,K' \in \mathcal{T}_{\tilde{n}}} a_{K,K'}\right]\left[\sum_{K' \in \mathcal{T}_{\tilde{n}}} \|v\|^p_{W^{\beta,p}(K')}\right]^{1/p} \|w\|_{L_{p'}},$$

and so we have

$$|B(v, w) - B_n(v, w)| \leq Cn^{-(1+\alpha/d)}\left[\max_{K,K' \in \mathcal{T}_{\tilde{n}}} a_{K,K'}\right]$$
$$\times \left[\sum_{K' \in \mathcal{T}_{\tilde{n}}} \|v\|^p_{W^{\beta,p}(K')}\right]^{1/p} \|w\|_{L_{p'}(\Omega)}. \tag{6}$$

The result (i) follows immediately from (4) and from (6). To prove (ii), let $v = \Pi_{\tilde{n}} Sf$ in (3). From Lemma A.2.3.3, we see that there exists a constant C such that

$$\|\Pi_{\tilde{n}} Sf\|_{W^{r,p}(K')} \leq \|Sf\|_{W^{r,p}(K')} + \|Sf - \Pi_{\tilde{n}} Sf\|_{W^{r,p}(K')} \leq C\|Sf\|_{W^{r,p}(K')}.$$

Summing over all elements $K' \in \mathcal{T}_{\tilde{n}}$ and using Theorem 6.3.1.1, we see that there exist constants C such that

$$\left[\sum_{K' \in \mathcal{T}_{\tilde{n}}} \|\Pi_{\tilde{n}} Sf\|_{W^{r,p}(K')}^p \right]^{1/p} \leq C\|Sf\|_{W^{r,p}(\Omega)} \leq C\|f\|_{W^{r,p}(\Omega)},$$

which proves (ii).

To prove (iii), note that we may use the discrete version of Hölder's inequality and the previous lemma to find that

$$|\langle f, w \rangle - f_{\tilde{n}}(w)| \leq \sum_{K \in \mathcal{T}_{\tilde{n}}} |E_K(fw)| \leq Cn^{-\mu/d} \sum_{K \in \mathcal{T}_{\tilde{n}}} \|f\|_{W^{r,p}(K)} \|w\|_{L_{p'}(K)}$$

$$\leq Cn^{-\mu/d} \left[\sum_{K \in \mathcal{T}_{\tilde{n}}} \|f\|_{W^{r,p}(K)}^p \right]^{1/p} \left[\sum_{K \in \mathcal{T}_{\tilde{n}}} \|w\|_{L_{p'}(K)}^{p'} \right]^{1/p'}$$

$$= Cn^{-\mu/d} \|f\|_{W^{r,p}(\Omega)} \|w\|_{L_{p'}(\Omega)},$$

completing the proof of the lemma. $\qquad\square$

We now show that the FEMQ is well-defined, and give tight bounds on its error.

THEOREM 6.3.3.2.   *There exists $n_1 \in \mathbb{P}$ such that $\phi_{n,k}^Q$ is well-defined for $n \geq n_1$. Furthermore,*

$$e(\phi_{n,k}^Q, N_{n,k}^Q) = \Theta(n^{-\mu/d}) \qquad \text{as } n \to \infty,$$

where

$$\mu = \min\{k+1, r\}.$$

PROOF.   We first claim that B is weakly coercive over $\{\mathcal{S}_{n,k}\}_{n=n_0}^\infty$ for some $n_0 \in \mathbb{P}$. That is, there exists $n_0 \in \mathbb{P}$ and a constant $\gamma' > 0$, such that for any $n \geq n_0$ and any $v \in \mathcal{S}_{n,k}$, there is a nonzero $w \in \mathcal{S}_{n,k}$ such that

$$|B(v, w)| \geq \gamma' \|v\|_{L_p(\Omega)} \|w\|_{L_{p'}(\Omega)}.$$

If $v = 0$, then this inequality holds for any nonzero $w \in \mathcal{S}_{n,k}$, so we need only consider the case of nonzero v. Since $1 < p \leq \infty$, we know that

$$\|v\|_{L_p(\Omega)} = \sup_{\substack{w \in L_{p'}(\Omega) \\ w \neq 0}} \frac{|\langle v, w \rangle|}{\|w\|_{L_{p'}(\Omega)}};$$

see (4.14.3) and (4.14.8) of Friedman (1970). So, there exists nonzero $g \in L_{p'}(\Omega)$ such that

$$|\langle v, g \rangle| \geq \tfrac{1}{2} \|v\|_{L_p(\Omega)} \|g\|_{L_{p'}(\Omega)}.$$

Now choose $w \in \mathcal{S}_{n,k}$ to be the finite element approximation of $(L^*)^{-1}g$, i.e.

$$B(s, w) = \langle s, g \rangle \qquad \forall s \in \mathcal{S}_{n,k}.$$

By (the adjoint version of) Theorem 6.3.2.2, there exists $n_0 \in \mathbb{P}$ such that w is well-defined for all $n \geq n_0$, and

$$\|w\|_{L_{p'}(\Omega)} \leq C \|g\|_{L_{p'}(\Omega)}$$

for a positive constant C, independent of n, g, and w. Letting $\gamma' = 1/(2C)$, we have

$$|B(v, w)| = |\langle v, g \rangle| \geq \tfrac{1}{2} \|v\|_{L_p(\Omega)} \|g\|_{L_{p'}(\Omega)} \geq \gamma' \|v\|_{L_p(\Omega)} \|w\|_{L_{p'}(\Omega)}.$$

Note that since g and v are nonzero, this inequality implies that $B(v, w) \neq 0$. Since B is linear, we see that $w \neq 0$, and so B is weakly coercive, as claimed.

Since B is weakly coercive, part (i) of the previous lemma and Strang's Lemma (Lemma A.3.2) imply that there exists $n_1 \in \mathbb{P}$ such that for $n \geq n_1$ and $f \in F$, there exists a unique $\tilde{u}_{\tilde{n},k} \in \mathcal{S}_{n,k}$ satisfying (1). Hence the FEMQ is well-defined for $n \geq n_1$.

We next need to show the lower bound

$$e(\phi^Q_{n,k}, N^Q_{n,k}) = \Omega(n^{-\mu/d}) \qquad \text{as } n \to \infty.$$

If $k + 1 \geq r$, then $\mu = r$ and we have

$$e(\phi^Q_{n,k}, N^Q_{n,k}) \geq r(n, \Lambda^{\mathrm{std}}) = \Theta(n^{-r/d}) = \Theta(n^{-\mu/d}) \qquad \text{as } n \to \infty.$$

Suppose now that $k + 1 < r$. Define f^* as in the proof of Theorem 6.3.2.2. Since $\phi^Q_{n,k}(N^Q_{n,k}f) \in \mathcal{S}_{\tilde{n},k}$, we see that

$$e(\phi^Q_{n,k}, N^Q_{n,k}) \geq \|Sf^* - \phi^Q_{n,k}(N^Q_{n,k}f^*)\|_{L_p(\Omega)} \geq Cn^{-(k+1)/d}.$$

The desired lower bound for the FEMQ follows immediately from the last two inequalities.

Finally, we need to show the analogous upper bound

$$e(\phi^Q_{n,k}, N^Q_{n,k}) = O(n^{-\mu/d}) \qquad \text{as } n \to \infty.$$

From Lemma 5.4.3 and Theorem 6.3.1.1, we have

$$\|Sf - \Pi_{\tilde{n}}Sf\|_{L_p(\Omega)} \le Cn^{-\mu/d}\|Sf\|_{W^{r,p}(\Omega)} \le Cn^{-\mu/d}\|f\|_{W^{r,p}(\Omega)}.$$

Since B is weakly coercive, we may use the estimates in the previous lemma and Strang's Lemma to see that

$$\|Sf - \phi_{n,k}^Q(N_{n,k}^Q f)\|_{L_p(\Omega)} \le Cn^{-\mu/d}\|f\|_{W^{r,p}(\Omega)} \qquad \forall f \in W^{r,p}(\Omega),$$

and so

$$e(\phi_{n,k}^Q, N_{n,k}^Q) = \sup_{\substack{f \in W^{r,p}(\Omega) \\ \|f\|_{W^{r,p}(\Omega)} \le 1}} \|Sf - \phi_{n,k}^Q(N_{n,k}^Q f)\|_{L_p(\Omega)} = O(n^{-\mu/d}),$$

as required.   □

Comparing the previous two theorems, we immediately find that

COROLLARY 6.3.3.1. *The FEMQ of degree k is an almost minimal error algorithm iff $k \ge r - 1$.*   □

For $\varepsilon > 0$, we now wish to find comp(ε), the ε-complexity of our problem, which we will compare to

$$\text{cost}_k^Q(\varepsilon) = \inf\{\,\text{cost}(\phi_{n,k}^Q, N_{n,k}^Q) : e(\phi_{n,k}^Q, N_{n,k}^Q) \le \varepsilon\,\},$$

which is the minimal cost of computing an ε-approximation with the FEMQ of degree k. Using the last two theorems, along with the results in Chapter 4, we have

THEOREM 6.3.3.3.

(i) *The ε-complexity is given by*

$$\text{comp}(\varepsilon) = \Theta(\varepsilon^{-d/r}) \qquad \text{as } \varepsilon \to 0.$$

(ii) *Let $\mu = \min\{k+1, r\}$. Then*

$$\text{cost}_k^Q(\varepsilon) = \Theta(\varepsilon^{-d/\mu}) \qquad \text{as } \varepsilon \to 0.$$

Hence, if $k \ge r - 1$, then the FEMQ of degree k is an almost optimal complexity algorithm, i.e.

$$\text{cost}_k^Q(\varepsilon) = \Theta\big(\text{comp}(\varepsilon)\big) = \Theta(\varepsilon^{-d/r}) \qquad \text{as } \varepsilon \to 0.$$ □

Hence we see that standard information is as powerful as arbitrary linear information for our problem. In either case, the nth minimal error is $\Theta(n^{-r/d})$ as $n \to \infty$, and the ε-complexity is $\Theta(\varepsilon^{-d/r})$ as $\varepsilon \to 0$. This should be contrasted with what we have learned for elliptic PDE and systems.

As with the pure least squares finite element method, we pay a large asymptotic penalty for using an FEMQ of too low a degree. That is, for $k < r - 1$,

$$\frac{\text{cost}_k^Q(\varepsilon)}{\text{comp}(\varepsilon)} = \Theta(\varepsilon^{-\delta d}) \qquad \text{as } \varepsilon \to 0,$$

where

$$\delta = \frac{1}{k+1} - \frac{1}{r} > 0.$$

Thus,

$$\lim_{\varepsilon \to 0} \frac{\text{cost}_k^Q(\varepsilon)}{\text{comp}(\varepsilon)} = \infty.$$

Notes and Remarks

NR 6.3.3:1 We briefly describe results for the seminormed case $F = \mathcal{B}H^r(\Omega)$ with standard information. Basically, we find that the the nth minimal error and the ε-complexity are the same as for the normed case. However, the FEMQ has infinite error unless $SP_{r-1}(\Omega) \subseteq \mathcal{S}_{n,k}$. When this inclusion holds, the results for the FEMQ in the normed case carry over to the seminormed case.

We describe these results more precisely. Let

$$n^* = \binom{d+r-1}{r-1}.$$

Then:

    (i) $r(n, \Lambda^{\text{std}}) = \infty$ if $n < n^*$.
    (ii) $r(n, \Lambda^{\text{std}}) = \Theta(n^{-r/d})$ as $n \to \infty$.
    (iii)   If $SP_{r-1}(\Omega) \not\subseteq \mathcal{S}_{n,k}$, then

$$e(\phi_{n,k}^Q, N_{n,k}^Q) = \infty.$$

    (iv)   If $SP_{r-1}(\Omega) \subseteq \mathcal{S}_{n,k}$ for all sufficiently large n, then

$$e(\phi_{n,k}^Q, N_{n,k}^Q) = \Theta(n^{-\mu/d}) \qquad \text{as } n \to \infty,$$

where

$$\mu = \min\{k+1, r\}.$$

    (v) $\text{comp}(\varepsilon) = \Theta(\varepsilon^{-d/r})$ as $\varepsilon \to 0$.
    (vi)   If there exists no integer n for which $SP_{r-1}(\Omega) \subseteq \mathcal{S}_{n,k}$, then

$$\text{cost}_k^Q(\varepsilon) = \infty \qquad \forall \varepsilon > 0.$$

(vii) Suppose that there exists a nonnegative integer n_0 such that $SP_{r-1}(\Omega) \subseteq \mathcal{S}_{n,k}$ for $n \geq n_0$. Then

$$\mathrm{cost}_k^Q(\varepsilon) = \Theta(\varepsilon^{-d/\mu}) \qquad \text{as } \varepsilon \to 0.$$

Hence, if $SP_{r-1}(\Omega) \subseteq \mathcal{S}_{n,k}$ for $n \geq n_0$ and $k \geq r + l - 1$, then the FEMQ of degree k is an almost optimal complexity algorithm, i.e.

$$\mathrm{cost}_k^{FE}(\varepsilon) = \Theta\big(\mathrm{comp}(\varepsilon)\big) = \Theta(\varepsilon^{-d/r}) \qquad \text{as } \varepsilon \to 0.$$

These results may be proved by following the analysis in the last part of §5.6, making the obvious changes.

6.4   Ill-posed problems

The notion of a well-posed (or "correctly set") problem was introduced in Hadamard (1952). A problem is said to be well-posed if it its solution exists, is unique, and depends continuously on its data; a problem that is not well-posed is said to be ill-posed. The problems that we have studied so far have all been well-posed (mainly because shift theorems exist for these problems).

Hadamard gave the impression that any well-formulated physical problem must be well-posed, and that ill-posed problems were merely problems that had not been formulated correctly. (This explains his choice of terminology.) However, in the years since the appearance of Hadamard's treatise, many important practical problems have been found to be ill-posed, such as the following:

(i)   inversion of the Laplace transform, whether the "usual" transform defined over $[0, \infty)$ or the finite transform whose inversion is discussed in Dunn (1967),

(ii) Fujita's equation relating molecular weight distribution to steady-state concentration or optical density in a centrifuged sample (see Gehatia and Wiff 1970),

(iii) problems in computational vision, such as edge detection, optical flow, surface reconstruction, and determining shape from shading (see Poggio *et al.* 1985),

(iv) the backwards heat equation (see Lavrentjev 1955 and Nedelkov 1972), and

(v)   problems in remote sensing (see Twomey 1977).

Ill-posed problems are notoriously hard to solve. They have been the subject of research since the 1950s, the canonical problem being the backwards heat equation. The monograph of Morozov (1985) contains a substantial bibliography, giving about one hundred pointers to the Soviet literature alone.

Most of the ill-posed problems arising in practice are Fredholm problems of the first kind. That is, for a compact operator L and a problem element f, we wish to find u such that $Lu = f$. Regularization is often used to solve such problems; for further discussion, see Tikhonov (1963) and Tikhonov and Arsenin (1977). Here, we choose a regularization parameter $\lambda > 0$ and a quadratic penalty functional J, and seek u_λ minimizing $\|Lu_\lambda - f\| + \lambda J(f)$. Of course, the success of such a technique depends on a good choice of λ and J. While generalized cross-validation (see Wahba 1984, 1985) is a good technique for picking the regularization parameter, determining the right quadratic penalty functional is still difficult. Furthermore, good error bounds appear to be available only if the solution satisfies (hard-to-verify) a priori conditions.

Since we have been quite successful in finding optimal algorithms for PDE, elliptic systems, and Fredholm problems of the second kind, it is only natural to try applying the techniques of IBC to ill-posed problems. Recall that our techniques yield (almost) optimal complexity algorithms. If these techniques are at all usable for ill-posed problems, we will either find good algorithms for such problems, or prove that good algorithms do not exist.

We now give an overview of the results in this section. First, we precisely describe the class of ill-posed problems that we study, our motivating example being the inversion of the finite Laplace transform. Next, we study ill-posed problems under the absolute error criterion. We find that the radius of information for an ill-posed problem is always infinite. This means that there is no algorithm with finite error, and so the ε-complexity of ill-posed problems is infinite for *any* finite ε, no matter how large.

Thus we cannot solve ill-posed problems in the worst case setting under the absolute error criterion. If we need to solve such problems, we must change either the setting or the error criterion. Since we defer the study of other settings until later in this book, we consider a different error criterion, namely, the residual error criterion. Using the results in §4.5, we show that it *is* possible to solve ill-posed problems under the residual error criterion.

6.4.1 Definition

We now describe the problem to be solved, i.e. the solution operator, the class of problem elements, and the permissible information. Let F_1 and G be normed linear spaces (where F_1 is infinite-dimensional). Let D be a linear subspace of F_1. Without loss of generality, we will assume that D is dense in F_1. Then the solution operator for our ill-posed problem will be an unbounded linear operator

$$S: D \subseteq F_1 \to G,$$

Since S is unbounded with domain D, this means that there exist linearly independent $x_1, x_2, \ldots \in D$ of unit norm such that $\lim_{n \to \infty} \|Sx_n\| = \infty$.

EXAMPLE: *The Fredholm problem of the first kind.* The most common source of such ill-posed problems is the Fredholm problem of the first kind, i.e. the inversion of a compact operator. Let $L \colon G \to F_1$ be a compact linear injection, with $\dim G = \infty$. We assume that the range $D = L(G)$ of L is dense in F_1. Define $S \colon D \subseteq F_1 \to G$ by

$$u = Sf \quad \text{iff} \quad Lu = f \qquad \forall f \in D.$$

Since L is compact with infinite-dimensional range, the solution operator S is unbounded.   □

We now give an example of a Fredholm problem of the first kind. This example arises in the "measurement of the distribution of an absorbing gas (such as ozone in the earth's atmosphere) from the spectrum of scattered light;" see pp. 12–13 of Twomey (1977) for details.

EXAMPLE: *Inverting the finite Laplace transform.* For the sake of normalization, we work with functions defined over the unit interval $I = [0, 1]$. We define an operator $L \colon L_2(I) \to H^r(I)$ by

$$(Lu)(s) = \int_0^1 e^{-st} u(t)\, dt \qquad (0 \le s \le 1)$$

for $u \in L_2(I)$. That is, Lu is the finite Laplace transform of u.

Marti (1983) shows that the operator L is injective. We claim that L is compact. Indeed, since the mapping $(s, t) \mapsto e^{-st}$ is infinitely differentiable, there is a positive constant C such that

$$\|Lu\|_{H^{r+1}(I)} \le C\|u\|_{L_2(I)} \qquad \forall u \in L_2(I).$$

Since the injection of $H^{r+1}(I)$ into $H^r(I)$ is compact by the Rellich-Kondrasov theorem (Lemma A.2.2.3), the mapping $L \colon L_2(I) \to H^r(I)$ is compact, as claimed.

Hence the problem of inverting the finite Laplace transform is a Fredholm problem of the first kind from $H^r(I)$ to $L_2(I)$, and is thus ill-posed.

Of course, in our formulation, we require that the operator L have dense range. Hence, we now show that the range D of L is dense in $H^r(I)$. In what follows, we write $L_r \colon L_2(I) \to H^r(I)$ for the finite Laplace transform to emphasize dependence on the codomain. Since range L_r is dense in $(\ker L_r^*)^\perp$, it suffices to show that the adjoint L_r^* of L_r is injective. Let E denote the standard embedding of $H^r(I)$ into $L_2(I)$. Clearly $EL_r = L_0$, so that $L_r^* E^* = (EL_r)^* = L_0$. From the result of Marti (1983), it follows that L_0 is a dense, self-adjoint injection of $L_2(I)$ into $L_2(I)$. Hence, there exists an orthonormal basis $\{g_i\}_{i=1}^{\infty}$ of $L_2(I)$ and a nonincreasing sequence $\{\gamma_i\}_{i=1}^{\infty}$ of positive numbers such that

$$L_r^* E^* g_i = \gamma_i g_i \qquad (i = 1, 2, \dots).$$

Since E is a dense injection, E^* is also a dense injection. Hence, $\{E^* g_i\}_{i=1}^{\infty}$ is a basis for $H^r(I)$. Now let $v \in \ker L_r^*$. Writing

$$v = \sum_{i=1}^{\infty} \nu_i E^* g_i,$$

we find

$$\sum_{i=1}^{\infty} \gamma_i \nu_i g_i = \sum_{i=1}^{\infty} \nu_i L_r^* E^* g_i = L_r^* v = 0.$$

Since $\{g_i\}_{i=1}^{\infty}$ is a basis for $L_2(I)$ and the γ_i are all non-zero, the ν_i must all be zero, so that $v = 0$. Hence $\ker L_r^* = 0$, as required. $\square$

We define our class F of *problem elements* to be

$$F = BF_1 \cap D = \{\, f \in D : \|f\| \le 1 \,\}.$$

That is, our problem elements will be those elements f of the unit ball of F_1 for which Sf is defined.

We will study several classes of permissible information. For the most part, we will consider algorithms using continuous linear information Λ^*. We will also discuss algorithms using arbitrary linear information Λ' and various kinds of nonlinear information in our treatment of the absolute error criterion.

6.4.2   The absolute error criterion

Our main result will be that there is no algorithm using "reasonable" information to approximate the solution of an ill-posed problem with finite error under the absolute error criterion. We show this by successively considering more general classes of information. As a corollary, it then follows that the ε-complexity of ill-posed problems is infinite for any finite positive ε, no matter how large.

We first consider the case of continuous linear information.

THEOREM 6.4.2.1. *There is no algorithm using (adaptive or nonadaptive) linear continuous information of finite cardinality whose error is finite.*

PROOF. Note that S is a linear operator, and the problem elements F form a convex, balanced subset of D. Since our information is linear, the results of §4.4.2 imply that nonadaptive information is as powerful as adaptive information (to within a factor of two). Hence, it suffices to prove the result for nonadaptive information.

Suppose now that the theorem is false. Then there exists information N of the form

$$Nf = \begin{bmatrix} \lambda_1(f) \\ \vdots \\ \lambda_n(f) \end{bmatrix} \qquad \forall f \in F,$$

with linear continuous $\lambda_1, \ldots, \lambda_n$, whose radius is finite. Hence, the diameter

$$d(N) = 2 \sup_{h \in \ker N \cap F} \|Sh\|$$

is also finite. Without loss of generality, we assume that the functionals $\lambda_1, \ldots, \lambda_n$ are linearly independent. As in Edwards (1965), we may decompose $D = \ker N \oplus (\ker N)^{\perp}$, where $(\ker N)^{\perp}$ is any algebraic complement of $\ker N$ in D. Hence, there exist $e_1, \ldots, e_n \in D$ such that

$$\lambda_i(e_j) = \delta_{ij} .$$

Define

$$\beta = \sum_{j=1}^{n} \|\lambda_j\| \, \|e_j\|$$

and

$$\sigma = \max \left\{ \frac{\|Se_j\|}{\|e_j\|} : 1 \leq j \leq n \right\} .$$

Now let $f \in D$. Then

$$\left\| \sum_{j=1}^{n} \lambda_j(f) \, e_j \right\| \leq \beta \|f\| .$$

Let

$$f_0 = f - \sum_{j=1}^{n} \lambda_j(f) \, e_j .$$

Since $f_0 \in \ker N$, the previous inequality implies that

$$\|Sf_0\| \leq \tfrac{1}{2} d(N) \|f_0\| \leq \tfrac{1}{2} d(N) \left(\|f\| + \left\| \sum_{j=1}^{n} \lambda_j(f) \, e_j \right\| \right) \leq \tfrac{1}{2} d(N)(1 + \beta) \|f\| ,$$

while

$$\|S(f - f_0)\| \leq \sum_{j=1}^{n} |\lambda_j(f)| \, \|Se_j\| \leq \beta \sigma \|f\| .$$

Hence the triangle inequality yields

$$\|Sf\| \le \left(\tfrac{1}{2}d(N)(1+\beta) + \beta\sigma\right)\|f\| \qquad \forall f \in D,$$

so that S is bounded, which is a contradiction. $\qquad\qquad\square$

Hence linear continuous information (whether adaptive or nonadaptive) is useless for solving ill-posed problems under the absolute error criterion.

Things are even worse than suggested by this theorem. There is no algorithm using linear continuous adaptive information N of finite cardinality that is substantially better than the *zero algorithm* ϕ^0 which is defined by

$$\phi^0(Nf) = 0 \qquad \forall f \in F_0.$$

To state this result precisely, recall that for any subset $A \subseteq F_0$, we say that $w \in F_0$ belongs to the *relative interior* of A in F_0 iff w is the center of an open F_0-ball that is contained in A.

THEOREM 6.4.2.2. *Let $q \in [0,1)$. For any algorithm ϕ using (adaptive or nonadaptive) linear continuous information N of finite cardinality, let*

$$A_q = \left\{ f \in F_0 : \frac{\|Sf - \phi(Nf)\|}{\|Sf - \phi^0(Nf)\|} \le q \right\}.$$

Then the relative interior of A_q in F_0 is empty.

PROOF. Suppose not. Choose information N, an algorithm ϕ using N, and $q \in [0,1)$ such that the relative interior of A_q in F_0 is nonempty. Then there exist $w \in F_0$ and $\varepsilon > 0$ such that the ball

$$B = \{ f \in F_0 : \|f - w\| < \varepsilon \},$$

centered at w with radius ε, is contained in A_q. We

Since N is (possibly adaptive) linear continuous information, there exist functionals

$$\lambda_i \colon D \times \mathbb{R}^{i-1} \to \mathbb{R} \qquad (i = 1, 2, \dots),$$

which are linear in their first variable, such that

$$Nf = \begin{bmatrix} \lambda_1(f) \\ \lambda_2(f; y_1) \\ \vdots \\ \lambda_{n(f)}(f; y_1, \dots, y_{n(f)-1}) \end{bmatrix},$$

where we set

$$y_i = \lambda_i(f; y_1, \dots, y_{i-1}) \qquad (1 \le i \le n(f)).$$

Here, $n(f)$ satisfies

$$n(f) = \min\{\, i : \text{ter}_i(y_1, \dots, y_i) = 1 \,\} < \infty\,,$$

where ter_i is the ith termination function, see §3.3.

We now define a new linear continuous (non-adaptive) information operator $\tilde{N}$ of fixed cardinality $n(w)$ as follows: Apply N to w. For $1 \le i \le n(w)$, let

$$\tilde{y}_i = \lambda_i(w; \tilde{y}_1, \dots, \tilde{y}_{i-1})\,.$$

Since the real numbers $\tilde{y}_1, \dots, \tilde{y}_{n(w)}$ have been defined, we can now define continuous linear functionals $\tilde{\lambda}_1, \dots, \tilde{\lambda}_{n(w)}$ on F_1 by

$$\tilde{\lambda}_i = \lambda_i(\,\cdot\,; \tilde{y}_1, \dots, \tilde{y}_{i-1}) \qquad \big(1 \le i \le n(w)\big)\,.$$

Finally, we let

$$\tilde{N} = \begin{bmatrix} \tilde{\lambda}_1 \\ \vdots \\ \tilde{\lambda}_{n(w)} \end{bmatrix}\,.$$

Note that $\tilde{N}$ is continuous (non-adaptive) linear information of cardinality $n(w)$.

From §4.2, we have $d(\tilde{N}) \ge r(\tilde{N})$. But $r(\tilde{N}) = \infty$ by the previous theorem, and so $d(\tilde{N}) = \infty$. Hence there is a sequence $\{h_m\}_{m=1}^{\infty}$ in $\ker \tilde{N}$ such that

$$\|h_m\| = 1 \qquad \forall\, m \ge 1$$
$$\lim_{m \to \infty} \|Sh_m\| = \infty\,.$$

For each m, let

$$f_m = w + \varepsilon h_m\,,$$

and define

$$y_i = \lambda_i(f_m; y_1, \dots, y_{i-1}) \qquad \big(1 \le i \le n(w)\big)\,.$$

A straightforward induction establishes that

$$y_i = \tilde{y}_i \qquad \big(1 \le i \le n(w)\big)\,.$$

Hence, for each index m, we see that

$$n(f_m) = n(w)$$

and that

$$Nf_m = \begin{bmatrix} y_1 \\ \vdots \\ y_{n(f_m)} \end{bmatrix} = \begin{bmatrix} \tilde{y}_1 \\ \vdots \\ \tilde{y}_{n(w)} \end{bmatrix} = \tilde{N}w = Nw.$$

Thus,

$$\phi(Nf_m) = \phi(Nw).$$

By construction, $f_m \in B$. Since ϕ^0 is the zero algorithm and $B \subseteq A_q$, we find that

$$\|Sf_m - \phi(Nw)\| = \|Sf_m - \phi(Nf_m)\| \le q\|Sf_m - \phi^0(Nf_m)\|$$
$$= q\|Sf_m\| \le q\|Sw\| + q\varepsilon\|Sh_m\|.$$

On the other hand,

$$\|Sf_m - \phi(Nw)\| \ge \varepsilon\|Sh_m\| - \|Sw - \phi(Nw)\|.$$

Combining the last two inequalities, we find

$$\|Sh_m\| \le \frac{1}{(1-q)\varepsilon}\left(\|Sw - \phi(Nw)\| + q\|Sw\|\right),$$

contradicting the fact that $\lim_{m\to\infty}\|Sh_m\| = \infty$. □

Hence there is no algorithm using linear continuous adaptive information of finite cardinality that can do better than the zero algorithm (except possibly on a set having empty relative interior) when solving an ill-posed problem. In other words, the set of functions for which a given algorithm beats the zero algorithm is small.

So, if we are to find an algorithm to solve ill-posed problems with finite error, we must once again extend the permissible class of information that algorithms may use.

To this end, we now consider the use of general (i.e. not necessarily linear) information. By *general information* of cardinality n, we mean any mapping $N: F \to \mathbb{R}^n$. Using the results in Chapter 7 of GTOA, we find that arbitrary general information is too powerful: if either F_1 or G is a separable space, there exists (horribly discontinuous) general information of cardinality one whose radius is arbitrarily close to zero. (See also **NR 6.4.2:3.**)

So, we impose a mild smoothness requirement on the general information to be considered. We say that general information N of cardinality n is *finitely continuous information* if N, when restricted to the intersection of F with any $(n+1)$-dimensional subspace of D, is continuous.

We now consider the solution of the Fredholm problem of the first kind, using finitely continuous information. We restrict our attention to the

Hilbert case, i.e. we assume that the spaces F_1 and G are infinite-dimensional Hilbert spaces. Hence, we are given a compact operator $L: G \to F_1$ with dense range $D = L(G)$, and we define our solution operator $S: D \subset F_1 \to G$ by

$$u = Sf \quad \text{iff} \quad Lu = f \qquad \forall f \in D.$$

Adapting the ideas found in Kacewicz and Wasilkowski (1986), we use the Borsuk-Ulam theorem (see, for example, Dugundji 1966) to establish

THEOREM 6.4.2.3. *In the Hilbert case of the Fredholm problem of the first kind, there is no algorithm using finitely continuous information of finite cardinality having finite error.*

PROOF. Since L is a compact injection, there is an orthonormal basis $\{ e_i \}_{i=1}^{\infty}$ for G consisting of eigenvectors of L^*L, i.e.

$$L^*Le_i = \lambda_i^2 e_i \qquad i = 1, 2, \ldots$$

with

$$\lambda_1 \geq \lambda_2 \geq \ldots > 0 \qquad \text{and} \qquad \lim_{i \to \infty} \lambda_i = 0.$$

We let

$$g_i = \frac{1}{\lambda_i} Le_i \in D,$$

so that $\{g_i\}_{i=1}^{\infty}$ is orthonormal in G.

Let N be finitely continuous information of cardinality at most n. For each positive integer k, define an $(n+1)$-dimensional subspace G_k of D by

$$G_k = \text{span}\{ g_k, g_{k+1}, \ldots, g_{k+n} \}.$$

Since $\{ g_k, g_{k+1}, \ldots, g_{k+n} \}$ is an orthonormal basis for G_k, the mapping $H: \mathbb{R}^{n+1} \to G_k$ defined by

$$H(\gamma) = \sum_{i=0}^{n} \gamma_i g_{i+k} \qquad \forall \gamma \in \mathbb{R}^{n+1}$$

is an isometry of $\mathbb{R}^{n+1}$ onto G_k.

Let B denote the closed unit ball in Euclidean $(n+1)$-space $\mathbb{R}^{n+1}$. Since H is an isometry, it is easy to check that $H(B) = G_k \cap F$. We define $N_k: B \to \mathbb{R}^n$ by

$$N_k(\gamma) = N(H(\gamma)) \qquad \forall \gamma \in B.$$

Using the finite continuity of N, we see that N_k is continuous. By the Borsuk-Ulam theorem, there exists γ^* on the boundary of B such that $N_k(\gamma^*) = N_k(-\gamma^*)$.

Let
$$f_k = H(\gamma^*) = \sum_{j=0}^{n} \gamma_j^* g_{j+k} \,,$$

so that
$$N(f_k) = N_k(\gamma^*) = N_k(-\gamma^*) = N(-f_k).$$

Since H is an isometry and γ^* is on the boundary of B, we have $\|\pm f_k\| = 1$; since $\pm f_k \in D$, this implies that $\pm f_k \in F$.

Now let ϕ be any algorithm using the information N. From §4.2 , we have

$$e(\phi, N) \geq r(N) \geq \tfrac{1}{2}d(N) \geq \sup_{\substack{f, \tilde{f} \in F \\ Nf = N\tilde{f}}} \|Sf - S\tilde{f}\| \geq \tfrac{1}{2}\|Sf_k - S(-f_k)\| = \|Sf_k\| \,.$$

Since the eigenvectors $\{e_i\}_{i=1}^{\infty}$ of $L^* L$ are orthonormal and correspond to decreasing eigenvalues $\{\lambda_i\}_{i=1}^{\infty}$ and $\gamma^* \in \mathbb{R}^{n+1}$ has unit norm, we see that

$$\|Sf_k\|^2 = \sum_{j=0}^{n} \frac{\gamma_j^{*2}}{\lambda_{j+k}^2} \geq \frac{1}{\lambda_k^2} \sum_{j=0}^{n} \gamma_j^{*2} = \frac{1}{\lambda_k^2} \,.$$

Hence
$$e(\phi, N) \geq \|Sf_k\| \geq \frac{1}{\lambda_k} \,.$$

Since k is arbitrary and $\lim_{k \to 0} \lambda_k = 0$, we have $e(\phi, N) = \infty$. □

Hence we see that we cannot solve the Fredholm problem of the first kind in the Hilbert case with finite error using finitely continuous information. Although this result is of independent interest, we can also use this result to show that the Fredholm problem of the first kind cannot be solved using arbitrary linear information of finite cardinality:

THEOREM 6.4.2.4. *For the Fredholm problem of the first kind in the Hilbert case, there exists no algorithm using (nonadaptive or adaptive) arbitrary linear adaptive information of finite cardinality having finite error.*

PROOF. Since the problem is linear, it suffices to prove the result for non-adaptive information. But linear nonadaptive information is finitely continuous, since any linear functional with finite-dimensional domain is continuous. Hence the radius of arbitrary nonadaptive linear information is infinite by the previous theorem. □

Hence in the Hilbert case, it is impossible to solve the Fredholm problem of the first kind with finite error with an algorithm using finitely continuous information or an algorithm using linear (and possibly adaptive) information.

Finally, we may use the results of this section to determine the ε-complexity of our problem:

THEOREM 6.4.2.5. *If either*

    (i) *the solution operator is unbounded and continuous linear information is permissible,*

   (ii) *the problem is a Fredholm problem of the first kind and finitely continuous information is permissible, or*

  (iii) *the problem is a Fredholm problem of the first kind and arbitrary linear information is permissible,*

then

$$\mathrm{comp}(\varepsilon, \Lambda) = \infty \qquad \forall \varepsilon > 0. \qquad\qquad \square$$

In short, ill-posed problems are *strongly noncomputable*: we cannot find a finite-error approximation to an ill-posed problem with finite cost, if we want to use the worst case setting and the absolute error criterion.

Notes and Remarks

NR 6.4.2:1 The results in this subsection are from Werschulz (1987*a, c*).

NR 6.4.2:2 Pour-El and Richards (1983) investigated the computation of linear transformations of Banach spaces from the viewpoint of classical computability theory. They proved that the only linear transformations that map computable input data onto computable solutions are bounded linear transformations. Hence their result is that only bounded operators preserve computability.

NR 6.4.2:3 We have only considered nonlinear information that is finitely continuous. This is because some restriction must be made on the class of permissible information. If we allow *arbitrary* nonlinear information, all the problems that we study in this book turn out to be trivial. That is, we find that the there exist algorithms using information of cardinality 1 (i.e. using a single information evaluation) whose error is arbitrarily small. For further details, see pp. 153–157 of GTOA.

6.4.3   The residual error criterion

We now discuss the solution of ill-posed problems in the worst case setting under the residual error criterion. As described in §4.5, we are given a bounded compact injective problem operator $L\colon G \to F_1$ with range D dense in F_1, along with a subset $F = T^{-1}(BX)$, where $T\colon F_1 \to X$ is a bounded linear surjective restriction operator. Our solution operator $S\colon F \cap D \to G$ is given by

$$u = Sf \quad \text{iff} \quad Lu = f \qquad \forall f \in F \cap D.$$

We are also given an embedding $E\colon F_1 \to W$. Then for any algorithm ϕ using information N, the residual of ϕ is

$$e^{\mathrm{resid}}(\phi, N) = \sup_{f \in F \cap D} \left\| E\big(L\phi(Nf) - f\big) \right\|.$$

Here, the spaces F_1, G, X, and W are all Banach spaces. We choose $\Lambda = \Lambda^*$, i.e. continuous linear information is permissible.

Clearly, we recover the formulation in the previous subsection by choosing $X = F_1$ and $T = I$. We use this more general formulation to allow both normed and seminormed cases in applications in which F_1 is a Sobolev space.

Note that in our development of the results for the residual error criterion, we never required any further properties of the operator L. In particular, the results in §4.5 all hold for compact L. This means that we can use these results to solve ill-posed problems arising as Fredholm problems of the first kind in the worst case setting under the residual error criterion.

The reader should review §4.5 for the general results. Here, one may find formulas for minimal radii and diameter, criteria for the existence of almost optimal error linear algorithms, and complexity bounds. In particular, note that the problem is convergent iff E is compact.

Rather than restate all these results for a general Fredholm problem of the first kind, we will immediately analyze an application described in the previous subsection, namely, the inversion of the finite Laplace transform on the unit interval $I = (0, 1)$. For $r \geq 0$, we let $F_1 = H^r(I)$ and $G = L_2(I)$. We define our problem operator $L: G \to F_1$ by

$$(Lu)(s) = \int_0^1 e^{-st} u(t)\, dt \qquad \forall\, u \in G.$$

As we have shown in §6.4.1, L is a self-adjoint compact injection with dense range, as required.

Problem definition. To complete the description of our problem, we must define the space W in which the residual is measured and the set F. First, we choose $W = L_2(I)$, so that $E: F_1 \to W$ is the usual embedding of $H^r(I)$ into $L_2(I)$. Next, we distinguish between two different classes F of problem elements. In the *normed case*, we let $F = BH^r(I)$ be the unit ball of F_1, whereas in the *seminormed case*, we let $F = \mathcal{B}H^r(I)$ be the semiunit ball of F_1. Note that for the normed case, we have

$$X = H^r(I) \qquad \text{and} \qquad Tf = f \quad \forall f \in H^r(I),$$

while for the seminormed case, we have

$$X = L_2(I) \qquad \text{and} \qquad Tf = f^{(r)} \quad \forall f \in H^r(I).$$

Radius of information. We are dealing with a Hilbert case with $F_1 = H^r(I)$ and $W = L_2(I)$. From §4.5.2 and §4.5.3, we see that adaptive information is no stronger than nonadaptive information. Hence, let N

be nonadaptive information. Using Corollary 4.5.3.2, we find that in the normed case,

$$r^{\text{resid}}(N) = \sup_{Nh=0} \frac{\|h\|_{L_2(I)}}{\|h\|_{H^r(I)}} = \sup_{\substack{Nh=0 \\ \|h\|_{H^r(I)} \leq 1}} \|h\|_{L_2(I)},$$

whereas

$$r^{\text{resid}}(N) = \sup_{Nh=0} \frac{\|h\|_{L_2(I)}}{|h|_{H^r(I)}} = \sup_{\substack{Nh=0 \\ \|h^{(r)}\|_{L_2(I)} \leq 1}} \|h\|_{L_2(I)}$$

in the seminormed case.

Minimal radius and optimal information. We claim that for any index $n \in \mathbb{N}$,

$$r^{\text{resid}}(n) = \Theta(n^{-r}) \qquad \text{as } n \to \infty$$

in both the normed and seminormed cases. Indeed, from Theorem 4.5.2.1, we see that

$$r^{\text{resid}}(n) = d^n\left(F, L_2(I)\right).$$

Let $S = E$ in the proofs of Theorems 5.4.1 and 5.6.1 (for the normed and seminormed cases, respectively). We then see that

$$d^n\left(F, L_2(I)\right) = \Theta(n^{-r}) \qquad \text{as } n \to \infty.$$

Hence

$$r^{\text{resid}}(n) = \Theta(n^{-r}) \qquad \text{as } n \to \infty,$$

as claimed.

We now derive nth (almost) optimal information. First, we consider the normed case. Given $r \geq 0$, we let $S_{n,k}$ be an n-dimensional subspace of $L_2(I)$, consisting of piecewise polynomials of degree k, for some $k \geq 0$. We require that the sequence $\{\mathcal{T}_n\}_{n=1}^{\infty}$ of triangulations be quasi-uniform. Let $\{s_1, \ldots, s_n\}$ be a basis for $S_{n,k}$. Define information $\hat{N}_n$ by

$$\hat{N}_n f = \begin{bmatrix} \langle f, s_1 \rangle_{L_2(I)} \\ \vdots \\ \langle f, s_n \rangle_{L_2(I)} \end{bmatrix} \qquad \forall f \in H^r(I).$$

From the formula for the radius of information and Theorem 5.4.3 (with $S = E$), we see that

$$r^{\text{resid}}(\hat{N}_n) = \sup_{\substack{\hat{N}_n h=0 \\ \|h\|_{H^r(I)} \leq 1}} \|h\|_{L_2(I)} = \Theta(n^{-r}) = \Theta\left(r^{\text{resid}}(n)\right) \qquad \text{as } n \to \infty,$$

and so $\hat{N}_n$ is almost optimal information.

Using the results in §4.5, we can describe nth optimal information for the normed case with $r > 0$. Since our problem is convergent (that is, $\lim_{n \to \infty} r^{\text{resid}}(n) = 0$) iff $r > 0$, this is the most important case in practice.

Since $T = I$, we see that $\ker T = 0$, and so $n^* = \dim \ker T = 0$ (see §4.5.3 and §4.5.4). Integrating by parts, it is easy to check that the eigenfunctions y_j and the eigenvalues κ_j^2 of $K = E^*E$ are the nonzero solutions y and κ^2 of the eigenproblem

$$\sum_{i=0}^{r} (-1)^i y^{(2i)}(s) = \kappa^{-2} y(s) \qquad \forall s \in [0, 1],$$

$$\sum_{i=0}^{j} (-1)^i y^{(r-j+2i)}(0) = \sum_{i=0}^{j} (-1)^i y^{(r-j+2i)}(1) = 0 \qquad (0 \le j \le r - 1).$$

Then information N_n defined by

$$N_n f = \begin{bmatrix} \langle f, y_1 \rangle_{H^r(I)} \\ \vdots \\ \langle f, y_n \rangle_{H^r(I)} \end{bmatrix} \qquad \forall f \in H^r(I)$$

is nth optimal information, and the nth minimal radius is

$$r^{\text{resid}}(n) = \kappa_{n+1}.$$

If $r = 1$, we can explicitly solve this eigenproblem. We find that

$$y_j(s) = \begin{cases} 1 & \text{if } j = 1 \\ \sqrt{2} \cos \left((j - 1)\pi s \right) & \text{if } j \ge 2 \end{cases}$$

and that

$$\kappa_j = \frac{1}{\sqrt{1 + \pi^2 (j - 1)^2}}.$$

Hence if $r = 1$, then

$$r^{\text{resid}}(n) = \frac{1}{\sqrt{1 + \pi^2 n^2}} \sim \frac{1}{\pi n} \qquad \text{as } n \to \infty.$$

Unfortunately, the exact eigenvalues and eigenfunctions are unknown for arbitrary r.

We now consider the seminormed case. Since $n^* = \dim \ker T = r$, we see that if $n < r$, then $r^{\text{resid}}(N) = \infty$. Suppose now that $n \ge r$. We again let $\mathcal{S}_{n,k}$ be an n-dimensional subspace of piecewise polynomials of degree k, with $\{\mathcal{T}_n\}_{n=1}^{\infty}$ quasi-uniform, and let $\hat{N}_n$ be information of cardinality n

consisting of $L_2(I)$-inner products with the basis functions of $S_{n,k}$. From Theorem 5.6.3 (with $S = E$), we find

$$r^{\text{resid}}(\hat{N}_n) = \Theta(n^{-r}) = \Theta\big(r^{\text{resid}}(n)\big) \qquad \text{as } n \to \infty.$$

Thus $\hat{N}_n$ is almost optimal information of cardinality n.

As with the normed case, we can find optimal information for the semi-normed case in terms of the solutions to an eigenproblem. For $n \geq r$, we find that nth optimal information N_n is given by

$$N_n f = \begin{bmatrix} \langle f, f_1^* \rangle_{L_2(I)} \\ \vdots \\ \langle f, f_r^* \rangle_{L_2(I)} \\ \langle f, y_1 \rangle_{L_2(I)} \\ \vdots \\ \langle f, y_{n-r} \rangle_{L_2(I)} \end{bmatrix} \qquad \forall f \in F_1$$

where

$$f_j^*(x) = \sqrt{\frac{2j-1}{2}} \, P_{j-1}(2x-1) \qquad (0 \leq x \leq 1)$$

(with P_i denoting the ith Legendre polynomial on $[-1, 1]$) and where y_j and κ_j^2 are the nonzero solutions y and κ^2 of the eigenproblem

$$(-1)^r y^{(2r)}(s) = \kappa^{-2} y(s) \qquad \forall s \in [0, 1],$$
$$y^{(i)}(0) = y^{(i)}(1) = 0 \qquad (r \leq i \leq 2r - 1);$$

see Tikhomirov (1976) for further discussion. The nth minimal radius is given by

$$r^{\text{resid}}(n) = \kappa_{n-r+1}.$$

If $r = 1$, we can find an explicit solution to this eigenproblem, namely, that

$$y_j(s) = \sqrt{2} \cos(j\pi s)$$

and

$$\kappa_j = \frac{1}{\pi j}.$$

Hence,

$$r^{\text{resid}}(n) = \kappa_n = \frac{1}{\pi n}.$$

As with the normed case, the explicit eigenfunctions and eigenvalues for arbitrary r are unknown.

Note that the optimal information, the minimal radius, and the problem complexity are all independent of the problem operator L. They only depend on the class F and the space W in which the residual is measured. That is, suppose that $L: L_2(I) \to H^r(I)$ is *any* compact injection with dense range. If we measure residual in the $L_2(I)$-norm and if F is given by either the unit or semiunit ball, then nth optimal information, nth minimal radius, and ε-complexity are as given above. The only explicit dependence on the operator L will occur when we try to find an optimal or almost optimal error algorithm, as we shall see presently.

Optimality of linear algorithms. We now ask whether linear algorithms are almost optimal for our problem. First, consider the normed case. Since the restriction operator T is the identity, we see that $\ker T = \{0\} \subseteq D$. Hence by Theorem 4.5.3.1, finite residual linear algorithms exist in the normed case. Furthermore, there always exist linear optimal residual algorithms in the normed case (modulo a factor of $1 + \delta$ for arbitrary $\delta > 0$).

We now turn to the seminormed case, with $r \geq 1$. The restriction operator T is now the rth derivative operator. We claim that $\ker T \not\subseteq D$, so that there exists no linear algorithm whose residual is finite (even though the problem is convergent).

Indeed, consider the constant function $f(s) \equiv 1$. Clearly $f \in \ker T$. Suppose that $f \in D = L\big(L_2(I)\big)$. Then there would exist $u \in L_2(I)$ such that $Lu = f$. Using the inversion integral in Debnath and Thomas (1976), we find that

$$u(t) = \lim_{R \to \infty} \frac{1}{2\pi i} \int_{-iR}^{iR} e^{st} f(s) \, ds = \lim_{R \to \infty} \frac{1}{2\pi i} \int_{-iR}^{iR} e^{st} \, ds = \lim_{R \to \infty} \frac{1}{\pi} \frac{\sin Rt}{t} .$$

But this limit does not belong to $L_2(I)$. Thus $f \notin D$, showing that $\ker T \not\subseteq D$, as required.

Minimal residual algorithms. We now seek the explicit form of an nth almost minimal residual algorithm, i.e. for $\delta > 0$, we seek an algorithm $\phi_{n,\delta}$ using information N_n of cardinality at most n for which

$$e^{\text{resid}}(\phi_{n,\delta}, N_n) \leq (1 + \delta) r^{\text{resid}}(n).$$

Since there exists no finite-residual linear algorithm for the seminormed case of inverting the finite Laplace transform, we consider only the normed case. Recall that $y_1, \ldots, y_n$ are the first n orthonormal eigenfunctions of the operator $E^* E$, where E is the injection of $H^r(I)$ into $L_2(I)$. The information N_n defined by

$$N_n f = \begin{bmatrix} \langle f, y_1 \rangle_{L_2(I)} \\ \vdots \\ \langle f, y_n \rangle_{L_2(I)} \end{bmatrix} \qquad \forall f \in H^r(I)$$

is nth optimal information, with

$$r^{\text{resid}}(N_n) = r^{\text{resid}}(n) = \kappa_{n+1},$$

the $(n+1)$st singular value of E. Thus,

$$r^{\text{resid}}(N_n) = r^{\text{resid}}(n) = \Theta(n^{-r}) \qquad \text{as } n \to \infty.$$

Since the range of L is dense in $L_2(I)$, there exist $u_1, \ldots, u_n \in L_2(I)$ such that

$$\|Lu_j - y_j\|_{L_2(I)} \le \frac{\delta \, r^{\text{resid}}(n)}{n}.$$

Defining the algorithm

$$\phi_{n,\delta}(N_n f) = \sum_{j=1}^{n} \langle f, y_j \rangle_{L_2(I)} u_j \qquad \forall f \in D \cap H^r(I),$$

we have

$$r^{\text{resid}}(n) \le e^{\text{resid}}(\phi_{n,\delta}, N_n) \le (1+\delta) r^{\text{resid}}(n)$$

by Theorem 4.5.3.2, and so $\phi_{n,\delta}$ is within at most δ of having the minimal residual among all algorithms using information of cardinality at most n. To be specific, let us consider the case $r = 1$. We found that

$$r^{\text{resid}}(n) = \frac{1}{\sqrt{1 + \pi^2 n^2}}$$

and the eigenfunctions are given by

$$y_j(s) = \begin{cases} 1 & \text{if } j = 1, \\ \sqrt{2}\cos\left((j-1)\pi s\right) & \text{if } j \ge 2. \end{cases}$$

Let $\delta > 0$ be given. Suppose we can find $u_{1,\delta}, \ldots, u_{n,\delta} \in L_2(I)$ such that

$$\|Lu_{j,\delta} - y_j\|_{L_2(I)} \le \frac{\delta \, r^{\text{resid}}(n)}{n} \qquad (1 \le j \le n).$$

From Theorem 4.5.3.2, we then see that the algorithm $\phi_{n,\delta}$ using N_n given by

$$\phi_{n,\delta}(N_n f) = \sum_{j=1}^{n} \langle f, y_j \rangle_{L_2(I)} u_{j,\delta} \qquad \forall f \in D \cap H^1(I)$$

satisfies

$$e^{\text{resid}}(\phi_{n,\delta}, N_n) \le (1+\delta) r^{\text{resid}}(N_n) = (1+\delta) r^{\text{resid}}(n).$$

Hence, to within δ, the algorithm $\phi_{n,\delta}$ has minimal residual among all algorithms using information of cardinality n for inverting the finite Laplace transform for problem elements in the unit ball of $H^1(I)$.

To find our functions $u_{j,\delta}$ for $1 \le j \le n$, we first choose an integer m such that

$$\frac{\sqrt{2}(n\pi)^{2m+2}}{(2m+2)!} < \frac{\delta\, r^{\text{resid}}(n)}{2n}. \tag{1}$$

Next, we choose

$$R > \frac{\sqrt{2}\, n}{\delta\, r^{\text{resid}}(n)} \sum_{l=0}^{m} \frac{(j\pi)^{2l}}{(2l)!}. \tag{2}$$

For any $k \in \mathbb{N}$, define a function $v_{k,R} : I \to \mathbb{R}$ by

$$v_{k,R} = \begin{cases} (-1)^i \binom{k}{i} R^{k+1} & \text{if } t \in [i/R, (i+1)/R) \text{ for some } i \in \{0, \dots, k\}, \\ 0 & \text{if } t \in [(k+1)/R, 1]. \end{cases}$$

Then for $j = 1, \dots, n$, we define our functions $u_{j,\delta} \in L_2(I)$ by

$$u_{j,\delta}(t) = \gamma_j \sum_{l=0}^{m} \frac{(-1)^l (j\pi)^{2l}}{(2l)!} v_{2l,R}(t),$$

where $\gamma_1 = 1$ and $\gamma_j = \sqrt{2}$ for $j \ge 2$. We need to show that

$$\|Lu_{j,\delta} - y_j\|_{L_2(I)} < \frac{\delta\, r^{\text{resid}}(n)}{n} \qquad (1 \le j \le n). \tag{3}$$

Define $p_k(s) \equiv s^k$ for $k \in \mathbb{N}$. We claim that

$$\|Lv_{k,R} - p_k\|_{L_2(I)} < \frac{1}{2R}. \tag{4}$$

Indeed,

$$(Lv_{k,R})(s) = R^{k+1} \sum_{i=0}^{k} (-1)^i \binom{k}{i} \int_{j/R}^{(j+1)/R} e^{-st}\, dt$$

$$= R^{k+1} \sum_{i=0}^{k} (-1)^i \binom{k}{i} [e^{-js/R} - e^{-(j+1)s/R}]$$

$$= \frac{R^{k+1}}{s} (1 - e^{-s/R}) \sum_{i=0}^{k} (-1)^i \binom{k}{i} e^{-js/R}$$

$$= \frac{R^{k+1}}{s} (1 - e^{-s/R})^{k+1}$$

$$= s^k \left[\frac{R(1 - e^{-s/R})}{s} \right]^{k+1}.$$

Since a simple calculation yields

$$(Lv_{0,R})(s) = \frac{R(1 - e^{-s/R})}{s},$$

this implies that

$$Lv_{k,R} = p_k \cdot Lv_{0,R}.$$

Now the usual remainder theorem for alternating series yields that

$$\left| (1 - e^{-s/R}) - \frac{s}{R} \right| < \frac{s^2}{2R^2},$$

and so

$$|(Lv_{0,R})(s) - 1| = \frac{R}{s} \left| (1 - e^{-s/R}) - \frac{s}{R} \right| < \frac{R}{s} \cdot \frac{s^2}{2R^2} < \frac{1}{2R}.$$

Hence for any $s \in I$,

$$|(Lv_{k,R})(s) - s^k| = |s|^k |(Lv_{0,R})(s) - 1| < |s|^k \cdot \frac{1}{2R} < \frac{1}{2R}.$$

Since the sup-norm dominates the L_2-norm on I, the desired inequality (4) holds, as claimed.

From (1), we find that

$$\left\| \gamma_j \sum_{l=0}^{m} \frac{(-1)^l (j\pi)^{2l}}{(2l)!} p_{2l} - y_j \right\|_{L_2(I)} < \frac{\delta \, r^{\mathrm{resid}}(n)}{2n}.$$

Moreover, from (2) and (4), we find

$$\gamma_j \left\| \sum_{l=0}^{m} \frac{(-1)^l (j\pi)^{2l}}{(2l)!} (Lv_{2l,R} - p_{2l}) \right\|_{L_2(I)} < \gamma_j \left[\sum_{l=0}^{m} \frac{(j\pi)^{2l}}{(2l)!} \right] \cdot \frac{1}{2R}$$

$$< \frac{\delta r^{\mathrm{resid}}(n)}{2n}.$$

Hence

$$\|Lu_{j,\delta} - y_j\|_{L_2(I)} \le \gamma_j \left\| \sum_{l=0}^{m} \frac{(-1)^l (j\pi)^{2l}}{(2l)!} (Lv_{2l,R} - p_{2l}) \right\|_{L_2(I)}$$

$$+ \left\| \gamma_j \sum_{l=0}^{m} \frac{(-1)^l (j\pi)^{2l}}{(2l)!} p_{2l} - y_j \right\|_{L_2(I)}$$

$$< \frac{\delta r^{\mathrm{resid}}(n)}{n},$$

proving (3), as required.

Complexity. We now analyze the ε-complexity of our problem in the normed case. Recall that we are trying to invert the finite Laplace transform for problem elements whose Sobolev r-norm is bounded by unity. If $r = 0$, then the problem is not convergent, and so $\text{comp}(\varepsilon) = \infty$ for sufficiently small ε. Suppose now that $r \geq 1$. We find that the ε-cardinality number is

$$m^{\text{resid}}(\varepsilon) = \Theta(\varepsilon^{-1/r}) \qquad \text{as } n \to \infty.$$

Let $n = m^{\text{resid}}(\varepsilon)+1$. Let N_n be nth optimal or almost optimal information. Choose δ satisfying

$$0 < \delta < \frac{r^{\text{resid}}(n-1)}{r^{\text{resid}}(n)} - 1.$$

We then find that $\phi_{n,\delta}$ is an almost optimal complexity algorithm. Furthermore, we find that

$$\text{comp}(\varepsilon) = \Theta(\varepsilon^{-1/r}) \qquad \text{as } \varepsilon \to 0,$$

giving the ε-complexity of the problem.

In short, we may summarize our results in the following

THEOREM 6.4.3.1. *For the problem of inverting the finite Laplace transform under the residual error criterion in the normed case $F = BH^r(I)$, the following hold if continuous linear information is permissible:*

(i) *The nth minimal radius is given by*

$$r^{\text{resid}}(n) = \Theta(n^{-r}) \qquad \text{as } n \to \infty.$$

(ii) *Let E be the embedding of $H^r(I)$ into $L_2(I)$. Then information N_n consisting of inner products with the first n orthonormal eigenfunctions of E^*E is nth optimal information.*

(iii) *For any $\delta > 0$, the linear algorithm $\phi_{n,\delta}$ using N_n satisfies*

$$r^{\text{resid}}(n) \leq e^{\text{resid}}(\phi_{n,\delta}, N_n) \leq (1+\delta)r^{\text{resid}}(n),$$

and so $\phi_{n,\delta}$ is, to within a factor of $1 + \delta$, an nth minimal error algorithm.

(iv) *The ε-complexity of the problem is*

$$\text{comp}(\varepsilon) = \Theta(\varepsilon^{-1/r}) \qquad \text{as } \varepsilon \to 0.$$

(v) *For $\varepsilon > 0$, let $n = m^{\text{resid}}(\varepsilon) + 1$ and let δ satisfy*

$$0 < \delta < \frac{r^{\text{resid}}(n-1)}{r^{\text{resid}}(n)} - 1.$$

Then

$$\text{cost}(\phi_{n,\delta}, N_n) = \Theta(\varepsilon^{-1/r}) \qquad \text{as } n \to \infty,$$

and so $\phi_{n,\delta}$ is an almost optimal complexity algorithm. $\square$

Finite element methods.   As with other problems we have studied, the optimal algorithms produced by the general techniques of information-based complexity require spectral information that is often hard to obtain, except for special cases. Since finite elements have been successful for elliptic PDE, elliptic systems, and integral equations of the second kind, it is natural to consider using them to solve our problem of inverting the finite Laplace transform in the worst case setting under the residual error criterion.

We consider only the case $r > 0$, since the problem is not convergent for $r = 0$. Since we know that linear algorithms have infinite error in the seminormed case with positive r, we will only consider the normed case. We will allow either $\Lambda = \Lambda^*$ or $\Lambda = \Lambda^{\text{std}}$, i.e. we will consider both continuous linear information and standard information. We will find that continuous linear information is not stronger than standard information, and we will show that the method of degree k is optimal iff $k \geq r - 1$.

There *will* be one important difference between the FEMs we construct here and those studied previously. The FEMs for the previous problems worked by finding a best approximation to u in the trial space of piecewise polynomials. Our FEM will work by finding a best approximation to $Lu = f$ in the test space of piecewise polynomials, and then use these coefficients to construct an approximation to u in a space whose image under L is "almost" the test space.

Let $\mathcal{S}_{n,k}$ be as in §6.3 of this chapter. That is, $\mathcal{S}_{n,k}$ consists of piecewise polynomials of degree k, with no inter-element continuity required. As always, we require that the sequence $\{\mathcal{T}_n\}_{n=1}^{\infty}$ of triangulations of the unit interval I be quasi-uniform. Moreover, as in Bramble and Nitsche (1973), we require that there is a constant C_1, independent of n, such that $\mathcal{S}_{n,k}$ has a basis $\{s_1, \ldots, s_n\}$ such that

$$\sum_{j=1}^{n} \alpha_j^2 \leq C_1 n \sum_{i,j=1}^{n} \alpha_i \alpha_j \langle s_j, s_i \rangle_{L_2(I)} \qquad \forall \alpha_1, \ldots, \alpha_n \in \mathbb{R}.$$

The usual bases having small support satisfy this property.

From Lemma 5.4.3, we know that there exists a positive constant C_2, independent of n, such that

$$\min_{s \in \mathcal{S}_{n,k}} \|f - s\|_{L_2(I)} \leq C_2 n^{-\mu} \|f\|_{H^r(I)} \qquad \forall f \in H^r(I),$$

with

$$\mu = \min\{k + 1, r\}.$$

Moreover, if we replace S in Theorem 5.4.2 by the injection E of $H^r(I)$ into $L_2(I)$, we see that there exists a positive constant C_3 and a function $f^* \in H^r(I)$ such that

$$\min_{v \in \mathcal{S}_{n,k}} \|f^* - v\|_{L_2(I)} \geq C_3 n^{-\mu} \tag{5}$$

and

$$\|f^*\|_{H^r(I)} = 1. \tag{6}$$

Our test space will be $\mathcal{S}_{n,k}$, and we define our finite element information (FEI) $N_{n,k}^{\mathrm{FE}}$ by

$$N_{n,k}^{\mathrm{FE}} f = \begin{bmatrix} \langle f, s_1 \rangle_{L_2(I)} \\ \vdots \\ \langle f, s_n \rangle_{L_2(I)} \end{bmatrix} \qquad \forall f \in F.$$

Next, we choose a space $\mathcal{W}_{n,k}$ such that $L\mathcal{W}_{n,k}$ is "almost" $\mathcal{S}_{n,k}$. Since the range of L is dense in $L_2(I)$, there exist functions $w_1, \ldots, w_n \in L_2(I)$ such that

$$\|Lw_j - s_j\|_{L_2(I)} \leq C_4 n^{-(\mu+1)}, \tag{7}$$

where

$$C_4 < \tfrac{1}{2} C_1^{-1/2} C_3.$$

Set $\mathcal{W}_{n,k} = \mathrm{span}\{w_1, \ldots, w_n\}$. We do not claim that $\{w_1, \ldots, w_n\}$ is a basis for $\mathcal{W}_{n,k}$, since $w_1, \ldots, w_n$ need not be linearly independent.

We are now ready to define our finite element method. For $n \in \mathbb{N}$, let

$$G = [g_{ij}]_{1 \leq i,j \leq n}$$

be the Gram matrix with entries

$$g_{ij} = \langle s_j, s_i \rangle_{L_2(I)} \qquad (1 \leq i, j \leq n).$$

Since $s_1, \ldots, s_n$ are linearly independent, G is nonsingular. For $f \in F$, let

$$b = N_{n,k}^{\mathrm{FE}} f.$$

We then let

$$a = \begin{bmatrix} \alpha_1 \\ \vdots \\ \alpha_n \end{bmatrix}$$

be the unique solution to

$$Ga = b.$$

Then we set

$$u_{n,k} = \sum_{j=1}^{n} \alpha_j w_j.$$

Since $u_{n,k}$ only depends on f through the information $N_{n,k}^{\mathrm{FE}} f$, we write

$$u_{n,k} = \phi_{n,k}^{\mathrm{FE}}(N_{n,k}^{\mathrm{FE}} f),$$

and say that $\phi_{n,k}^{\mathrm{FE}}$ is the *finite element method* (FEM) using $N_{n,k}^{\mathrm{FE}}$.

Since the basis functions have small supports, we once again find that the Gram matrix G is a banded matrix, whose bandwidth is independent of n. (Moreover, if there is no inter-element continuity in the space $\mathcal{S}_{n,k}$, we see that G is a block diagonal matrix, with blocksize independent of n.) Hence, we can compute $\alpha_1, \dots, \alpha_n$ in $\Theta(n)$ arithmetic operations.

We are now ready to give tight bounds on the residual of the FEM.

THEOREM 6.4.3.2. *Let*

$$\mu = \min\{k+1, r\}.$$

Then

$$e^{\mathrm{resid}}(\phi_{n,k}^{\mathrm{FE}}, N_{n,k}^{\mathrm{FE}}) = \Theta(n^{-\mu}) \qquad \text{as } n \to \infty.$$

PROOF. We first show the upper bound

$$e^{\mathrm{resid}}(\phi_{n,k}^{\mathrm{FE}}, N_{n,k}^{\mathrm{FE}}) = O(n^{-\mu}) \qquad \text{as } n \to \infty.$$

Let $f \in F \cap D$ and $n \in \mathbb{P}$. Set

$$f_{n,k} = \sum_{j=1}^{n} \alpha_j s_j.$$

It is easy to check that

$$\|f - f_{n,k}\|_{L_2(I)} = \min_{s \in \mathcal{S}_{n,k}} \|f - s\|_{L_2(I)},$$

and so

$$\|f - f_{n,k}\|_{L_2(I)} \le C_2 n^{-\mu} \|f\|_{H^r(I)} \le C_2 n^{-\mu}$$

by Lemma 5.4.3. Now

$$\|Lu_{n,k} - f_{n,k}\|_{L_2(I)} = \left\| \sum_{j=1}^{n} \alpha_j (Lw_j - s_j) \right\|_{L_2(I)}$$

$$\le \left[\sum_{j=1}^{n} \alpha_j^2 \right]^{1/2} \left[\sum_{j=1}^{n} \|Lw_j - s_j\|_{L_2(I)}^2 \right]^{1/2}$$

$$\le \left[C_1 n \|f_{n,k}\|_{L_2(I)}^2 \right]^{1/2} \left[n \cdot C_4^2 n^{-2(\mu+1)} \right]^{1/2}$$

$$= C_4 C_1^{1/2} n^{-\mu} \|f_{n,k}\|_{L_2(I)}.$$

Since $0 \in \mathcal{S}_{n,k}$, we have

$$\|f - f_{n,k}\|_{L_2(I)} = \inf_{s \in \mathcal{S}_{n,k}} \|f - s\|_{L_2(I)} \le \|f - 0\|_{L_2(I)} = \|f\|_{L_2(I)},$$

and so

$$\|f_{n,k}\|_{L_2(I)} \le \|f\|_{L_2(I)} + \|f - f_{n,k}\|_{L_2(I)} \le 2\|f\|_{L_2(I)} \le 2.$$

Hence

$$\|Lu_{n,k} - f_{n,k}\|_{L_2(I)} \le 2C_4 C_1^{1/2} n^{-\mu}.$$

Combining these results, we find

$$\|Lu_{n,k} - f\|_{L_2(I)} \le \|f - f_{n,k}\|_{L_2(I)} + \|Lu_{n,k} - f_{n,k}\|_{L_2(I)}$$
$$\le (C_2 + 2C_4 C_1^{1/2}) n^{-\mu}.$$

Since f is an arbitrary element of $F \cap D$, we have

$$e^{\text{resid}}(\phi_{n,k}^{\text{FE}}, N_{n,k}^{\text{FE}}) = \sup_{f \in F \cap D} \|Lu_{n,k} - f\|_{L_2(I)} \le (C_2 + 2C_4 C_1^{1/2}) n^{-\mu},$$

which completes the proof of the upper bound.

We next prove the lower bound

$$e^{\text{resid}}(\phi_{n,k}^{\text{FE}}, N_{n,k}^{\text{FE}}) = \Omega(n^{-\mu}) \qquad \text{as } n \to \infty.$$

Choose f^* satisfying (5) and (6), and let $u_{n,k}^* = \phi_{n,k}^{\text{FE}}(N_{n,k}^{\text{FE}} f^*)$. Since $\phi_{n,k}^{\text{FE}}$ is a linear algorithm using continuous information N, we see that

$$e^{\text{resid}}(\phi_{n,k}^{\text{FE}}, N_{n,k}^{\text{FE}}) = \sup_{f \in F \cap D} \|L\phi_{n,k}^{\text{FE}}(Nf) - f\|_{L_2(I)}$$
$$= \sup_{f \in F} \|L\phi_{n,k}^{\text{FE}}(Nf) - f\|_{L_2(I)} \ge \|Lu_{n,k}^* - f^*\|_{L_2(I)}.$$

Let $f_{n,k}^*$ be the best approximation to f^*, i.e.

$$\|f^* - f_{n,k}^*\|_{L_2(I)} = \min_{s \in \mathcal{S}_{n,k}} \|f^* - s\|_{L_2(I)},$$

Using the inequalities proven above, we have

$$e^{\text{resid}}(\phi_{n,k}^{\text{FE}}, N_{n,k}^{\text{FE}}) \ge \|Lu_{n,k}^* - f^*\|_{L_2(I)}$$
$$\ge \|f^* - f_{n,k}^*\|_{L_2(I)} - \|Lu_{n,k}^* - f_{n,k}^*\|_{L_2(I)}$$
$$\ge (C_3 - 2C_4 C_1^{1/2}) n^{-\mu}.$$

Since $C_3 > 2C_4 C_1^{1/2}$, the lower bound follows immediately. $\square$

Since $r^{\text{resid}}(n) = \Theta(n^{-r})$, we see that the FEM of degree k is an almost minimal error algorithm iff $k \ge r - 1$. Suppose that $k < r - 1$. As with problems studied previously, the FEM of too low a degree is not an almost minimal error algorithm because it does not use its information well. Any almost optimal error algorithm using FEI will always have error $\Theta(n^{-r})$, regardless of the value of k, as we see in

THEOREM 6.4.3.3. *Let ϕ_n be an almost optimal error algorithm using FEI. Then*

$$e^{\text{resid}}(\phi_n, N_n) = \Theta\big(r^{\text{resid}}(N_n)\big) = \Theta\big(r^{\text{resid}}(n)\big) = \Theta(n^{-r}) \qquad \text{as } n \to \infty.$$

PROOF. From the formula for the radius of information under the residual error criterion, we have

$$r^{\text{resid}}(N) = \sup_{\substack{Nh=0 \\ \|h\|_{H^r(I)} \leq 1}} \|h\|_{L_2(I)}.$$

From Theorem 5.4.3, the latter quantity is $\Theta(n^{-r})$ as $n \to \infty$. $\square$

We now discuss the complexity of using the FEM to solve our problem under the residual error criterion. Let

$$\text{cost}_k^{\text{FE}}(\varepsilon) = \inf\{\, \text{cost}(\phi_{n,k}^{\text{FE}}, N_{n,k}^{\text{FE}}) : e(\phi_{n,k}^{\text{FE}}, N_{n,k}^{\text{FE}}) \leq \varepsilon \,\}$$

denote the minimal cost of using the FEM of degree k to compute an ε-approximation. Note that we can determine the functions w_j that define the space $\mathcal{W}_{n,k}$ in advance, independently of any problem element f. Hence under our model of computation, we can ignore the cost of finding these functions. It then follows that

$$\text{cost}(\phi_{n,k}^{\text{FE}}, N_{n,k}^{\text{FE}}) = \Theta(n)$$

as $n \to \infty$. Using Theorem 6.4.3.1, Theorem 6.4.3.2, and Corollary 4.5.4.1, we then have the following

THEOREM 6.4.3.4.

(i) *The ε-complexity using continuous linear information is given by*

$$\text{comp}(\varepsilon) = \Theta(\varepsilon^{-1/r}) \qquad \text{as } \varepsilon \to 0.$$

(ii) *Let $\mu = \min\{k+1, r\}$. Then*

$$\text{cost}_k^{\text{FE}}(\varepsilon) = \Theta(\varepsilon^{-1/\mu}) \qquad \text{as } \varepsilon \to 0.$$

Hence, if $k \geq r - 1$, then the FEM of degree k is an almost optimal complexity algorithm, i.e.

$$\text{cost}_k^{\text{FE}}(\varepsilon) = \Theta\big(\text{comp}(\varepsilon)\big) = \Theta(\varepsilon^{-1/r}) \qquad \text{as } \varepsilon \to 0. \qquad \square$$

As with other problems we have studied, we pay a large asymptotic penalty for using an FEM of too low a degree. That is, for $k < r - 1$, we have

$$\frac{\mathrm{cost}_k^{\mathrm{FE}}(\varepsilon)}{\mathrm{comp}(\varepsilon)} = \Theta(\varepsilon^{-\zeta d}) \qquad \text{as } \varepsilon \to 0,$$

where

$$\zeta = \frac{1}{k+1} - \frac{1}{r} > 0.$$

Thus,

$$\lim_{\varepsilon \to 0} \frac{\mathrm{cost}_k^{\mathrm{FE}}(\varepsilon)}{\mathrm{comp}(\varepsilon)} = \infty.$$

Finite element methods with quadrature.   Next, we consider the finite element method with quadrature (FEMQ). We do this because the standard FEM uses finite element information (FEI), whereas the FEMQ uses standard information, which is usually more readily available than FEI with quadrature This FEMQ will use standard information, and will also have error $\Theta(n^{-\mu})$.

We describe the FEMQ, using the usual notation seen previously. As always, $\hat{K}$ denotes our reference element. Choosing weights $\hat{\omega}_1, \ldots, \hat{\omega}_J \in \mathbb{R}$ and nodes $\hat{b}_1 \ldots , \hat{b}_J \in \hat{K}$, we define a reference quadrature rule

$$\hat{I}\hat{v} = \sum_{j=1}^{J} \hat{\omega}_j \hat{v}(\hat{b}_j) \qquad \forall \hat{v} \in C\left(\overline{\hat{K}}\right).$$

Next, we move from the reference element to an element appearing in a triangulation. For $\tilde{n} \in \mathbb{P}$, we define weights and nodes over the element $K \in \mathcal{T}_{\tilde{n}}$ by

$$\omega_{j,K} = \det B_K \cdot \hat{\omega}_j \quad \text{and} \quad b_{j,K} = F_K(\hat{b}_j) \qquad (1 \le j \le J),$$

where $F_K \hat{x} \equiv B_K \hat{x} + b_K$ is the affine bijection of $\hat{K}$ onto K. Then the reference quadrature rule $\hat{I}$ over $\hat{K}$ induces a local quadrature rule I_K over K, which is defined by

$$I_K v = \sum_{j=1}^{J} \omega_{j,K} v(b_{j,K}).$$

Let

$$\mathcal{N}_{\tilde{n}} = \bigcup_{K \in \mathcal{T}_{\tilde{n}}} \bigcup_{j=1}^{J} \{b_{j,K}\}$$

denote the set of all nodes in all the elements belonging to $\mathcal{T}_{\tilde{n}}$. We define a new bilinear form $B_{\tilde{n}}$ on $\mathcal{S}_{\tilde{n},k}$ by

$$B_{\tilde{n}}(v,w) = \sum_{K \in \mathcal{T}_{\tilde{n}}} I_K(vw) \qquad \forall\, v, w \in \mathcal{S}_{\tilde{n},k}.$$

For $f \in F$, we define a linear functional $f_{\tilde{n}}$ on $\mathcal{S}_{\tilde{n},k}$ by

$$f_{\tilde{n}}(v) = \sum_{K \in \mathcal{T}_n} I_K(fv) \qquad \forall\, v \in \mathcal{S}_{\tilde{n},k}.$$

Of course, the bilinear form $B_{\tilde{n}}$ approximates $\langle \cdot, \cdot \rangle_{L_2(I)}$ over $\mathcal{S}_{\tilde{n},k}$, while the linear functional $f_{\tilde{n}}$ approximates $\langle f, \cdot \rangle_{L_2(I)}$ over $\mathcal{S}_{\tilde{n},k}$.

As with the FEM, we will require a space $\mathcal{W}_{\tilde{n},k}$ such that $L\mathcal{W}_{\tilde{n},k}$ is "almost" $\mathcal{S}_{\tilde{n},k}$. To define this space, we choose $w_1, \ldots, w_{\tilde{n}} \in L_2(I)$ such that

$$\|Lw_j - s_j\|_{L_2(I)} \le Cn^{-(\mu+1)}.$$

Then we set $\mathcal{W}_{\tilde{n},k} = \operatorname{span}\{w_1, \ldots, w_{\tilde{n}}\}$.

We are now ready to define our finite element method with quadrature. Let $n \in \mathbb{P}$. Choose $\tilde{n}$ to be the largest integer such that

$$\operatorname{card} \mathcal{N}_{\tilde{n}} \le n.$$

Note that $\tilde{n} = \Theta(n)$ as $n \to \infty$. Letting $\mathcal{N}_{\tilde{n}} = \{x_1, \ldots, x_{\tilde{n}}\}$, we choose $x_{\tilde{n}+1}, \ldots, x_n \notin \mathcal{N}_{\tilde{n}}$. For $f \in F$, define standard information $N^{\mathrm{Q}}_{n,k}$ of cardinality n by

$$N^{\mathrm{Q}}_{n,k}f = \begin{bmatrix} f(x_1) \\ \vdots \\ f(x_n) \end{bmatrix} \qquad \forall f \in F.$$

Define

$$\tilde{G} = [\tilde{g}_{i,j}]_{1 \le i,j \le \tilde{n}} \qquad \text{and} \qquad \tilde{b} = \begin{bmatrix} f_{\tilde{n}}(s_1) \\ \vdots \\ f_{\tilde{n}}(s_{\tilde{n}}) \end{bmatrix},$$

where

$$\tilde{g}_{i,j} = B_{\tilde{n}}(s_j, s_i) \qquad (1 \le i, j \le \tilde{n}).$$

Let

$$\tilde{a} = \begin{bmatrix} \tilde{\alpha}_1 \\ \vdots \\ \tilde{\alpha}_{\tilde{n}} \end{bmatrix}$$

be the solution to the problem

$$\tilde{G}\tilde{a} = \tilde{b}.$$

Then we let

$$\phi^Q_{n,k}(N^Q_{n,k}f) = \tilde{u}_{n,k} = \sum_{j=1}^{\tilde{n}} \tilde{a}_j w_j.$$

denote the *finite element method with quadrature* (FEMQ).

We let

$$\nu = \min\{k+1, r\},$$

and assume that

(i) $r \geq 1$, and

(ii) $\hat{I}$ is exact of degree $2k + \nu - 1$ over the reference element $\hat{K}$.

Condition (i) guarantees that standard information consisting of function values to be continuous, whereas condition (ii) means our quadrature scheme must be accurate enough to ensure that

$$B_{\tilde{n}}(v, w) = \langle v, w \rangle_{L_2(I)} \qquad \forall\, v, w \in \mathcal{S}_{\tilde{n},k}.$$

THEOREM 6.4.3.5. *The FEMQ is well defined for all indices n, and*

$$e^{\mathrm{resid}}(\phi^Q_{n,k}, N^Q_{n,k}) = \Theta(n^{-\mu}) \qquad \text{as } n \to \infty.$$

PROOF. From the exactness condition, we see that $\tilde{G}$ is the original Gram matrix G for the pure FEM using $\mathcal{S}_{\tilde{n},k}$. Hence $\tilde{G}$ is invertible for all $n \in \mathbb{P}$, and the FEMQ is well defined.

We next prove the lower bound

$$e^{\mathrm{resid}}(\phi^Q_{n,k}, N^Q_{n,k}) = \Omega(n^{-\mu}) \qquad \text{as } n \to \infty.$$

Suppose first that $k + 1 \geq r$. Then $\mu = r$. Since $r \geq 1$, we have $\Lambda^{\mathrm{std}} \subset \Lambda^*$. Using Theorem 6.4.3.1, we thus have

$$e^{\mathrm{resid}}(\phi^Q_{n,k}, N^Q_{n,k}) \geq r^{\mathrm{resid}}(n, \Lambda^{\mathrm{std}}) \geq r^{\mathrm{resid}}(n, \Lambda^*) = \Theta(n^{-r}).$$

Now suppose that $k + 1 < r$, so that $\mu = k + 1$. Defining f^* as in Theorem 6.4.3.2, we find that

$$e^{\mathrm{resid}}(\phi^Q_{n,k}, N^Q_{n,k}) \geq \inf_{s \in \mathcal{S}_{\tilde{n},k}} \|Lw - f^*\|_{L_2(I)} = \Theta(n^{-(k+1)}).$$

The lower bound follows immediately.

We only need to prove the upper bound

$$e^{\mathrm{resid}}(\phi^Q_{n,k}, N^Q_{n,k}) = O(n^{-\mu}) \qquad \text{as } n \to \infty.$$

Let $f \in F \cap D$. Define

$$\tilde{f}_{n,k} = \sum_{j=1}^{\tilde{n}} \tilde{\alpha}_j s_j.$$

Since $B_{\tilde{n}}(v, w) = \langle v, w \rangle_{L_2(I)}$ for $v, w \in \mathcal{S}_{n,k}$, we may use Strang's Lemma (Lemma A.3.2) to see that

$$\|f - \tilde{f}_{n,k}\|_{L_2(I)} \leq C \left[\inf_{v \in \mathcal{S}_{n,k}} \left(\|f - v\|_{L_2(I)} \right. \right.$$

$$\left. \left. + \sup_{w \in \mathcal{S}_{n,k}} \frac{|\langle f, w \rangle_{L_2(I)} - f_{\tilde{n}}(w)|}{\|w\|_{L_2(I)}} \right) \right].$$

From Lemma 5.4.3, there exists $v \in \mathcal{S}_{\tilde{n},k}$ such that

$$\|f - v\|_{L_2(I)} \leq C n^{-\mu} \|f\|_{H^r(I)},$$

since $\tilde{n} = \Theta(n)$. Moreover, from the proof of Theorem 5.5.2, we have

$$|\langle f, w \rangle_{L_2(I)} - f_{\tilde{n}}(w)| \leq C n^{-\mu} \|f\|_{H^r(I)} \|w\|_{L_2(I)} \qquad \forall\, w \in \mathcal{S}_{\tilde{n},k}.$$

Combining these results, we find that

$$\|f - \tilde{f}_{n,k}\|_{L_2(I)} \leq C n^{-\mu} \|f\|_{H^r(I)} \leq C n^{-\mu}.$$

As in the proof of Theorem 6.4.3.2, we find that

$$\|L\tilde{u}_{n,k} - \tilde{f}_{n,k}\|_{L_2(I)} \leq C n^{-\mu}.$$

Combining these results, we find

$$\|Lu_{n,k} - f\|_{L_2(I)} \leq \|f - f_{n,k}\|_{L_2(I)} + \|Lu_{n,k} - f_{n,k}\|_{L_2(I)} \leq C n^{-\mu}.$$

Since f is an arbitrary element of $F \cap D$, we have

$$e^{\mathrm{resid}}(\phi_{n,k}^{\mathrm{FE}}, N_{n,k}^{\mathrm{FE}}) = \sup_{f \in F \cap D} \|Lu_{n,k} - f\|_{L_2(I)} \leq C n^{-\mu},$$

which completes the proof of the upper bound, and (hence) the theorem. □

Of course, the difficult part of all this is finding the space $\mathcal{W}_{n,k}$ (for the pure FEM) or $\mathcal{W}_{\tilde{n},k}$ (for the FEMQ). We illustrate one possible construction in the following

EXAMPLE. Let $\mathcal{S}_{n,k}$ be a finite element space of degree k and dimension n, defined over a uniform partition of I. We suppose that $\mathcal{S}_{n,k} \subset H^{p+1}(I)$ for some $p \in \mathbb{N}$, and hence $k \geq p$. Let $\{s_1, \ldots, s_n\}$ be a standard basis for $\mathcal{S}_{n,k}$, having small supports. Our task is to find functions $\{w_1, \ldots, w_n\}$ such that (7) holds.

We do this as follows. First, we approximate s_j with a Legendre series

$$s_{j,m} = \sum_{j=0}^{m} \alpha_{j,i} G_i,$$

where $G_i(x) = P_i(2x - 1)$, with P_i denoting the usual Legendre polynomial of degree i. Then

$$\alpha_{j,i} = (2n + 1) \int_0^1 s_j G_i.$$

Writing

$$G_i(x) = \sum_{l=0}^{i} \beta_{i,l} x^l,$$

we let

$$w_{j,m,R} = \sum_{i=0}^{m} \alpha_{j,i} \sum_{l=0}^{i} \beta_{i,l} v_{l,R},$$

where

$$v_{l,R} = \begin{cases} (-1)^i \binom{l}{i} R^{l+1} & \text{if } t \in [i/R, (i+1)/R) \text{ for some } i \in \{0, \ldots l\}, \\ 0 & \text{if } t \in [(l+1)/R, 1] \end{cases}$$

as before. Letting $w_j = w_{j,m,R}$, it only remains to choose j and R so that (7) holds.

Since the support of s_j has length $\Theta(n^{-1})$, we see that

$$\|s_j\|_{L_2(I)} \leq An^{-1} \tag{8}$$

for some positive constant A. From Theorem 1.3.2 of Gui (1988), we have

$$\|s_j - s_{j,m}\|_{L_2(I)} \leq \left(\frac{e}{4m}\right)^{p+1} \|s_j^{(p+1)}\|_{L_2(I)}. \tag{9}$$

Using the inverse inequality (Lemma A.2.3.4), we see that

$$\|s_j^{(p+1)}\|_{L_2(I)} \leq Cn^{p+1} \|s_j\|_{L_2(I)} \leq Bn^p \tag{10}$$

for some $B > 0$.

We claim that

$$\|Lw_{j,m,R}\|_{L_2(I)} \leq \frac{A}{2R\sqrt{3}} \cdot \frac{1}{n} + B \left(\frac{e}{4m}\right)^{p+1} n^p.$$

Indeed, let $p_l(x) \equiv x^l$. Since $Lv_{l,R} = p_l \cdot Lv_{0,R}$, we have

$$Lw_{j,m,R} = \sum_{i=0}^{m} \alpha_{j,i} \sum_{l=0}^{i} \beta_{i,l} p_l \cdot Lv_{l,R}$$

$$= \left(\sum_{i=0}^{m} \alpha_{j,i} G_i\right) Lv_{0,R} = s_{j,m} \cdot Lv_{0,R},$$

and so

$$\|Lw_{j,m,R} - s_j\|_{L_2(I)} = \|s_{j,m} \cdot Lv_{0,R} - s_j\|_{L_2(I)}$$
$$\leq \|s_{j,m}(Lv_{0,R} - p_0)\|_{L_2(I)} + \|s_j - s_{j,m}\|_{L_2(I)}. \tag{11}$$

Using Bessel's inequality, we have

$$\|s_{j,m}(Lv_{0,R} - p_0)\|_{L_2(I)} \leq \|s_{j,m}\|_{L_2(I)} \|Lv_{0,R} - p_0\|_{L_2(I)}$$
$$\leq \|s_j\|_{L_2(I)} \|Lv_{0,R} - p_0\|_{L_2(I)}. \tag{12}$$

Since

$$\|Lv_{0,R} - p_0\|^2_{L_2(I)} \leq \int_0^1 \left(\frac{s}{2R}\right)^2 ds = \frac{1}{3 \cdot (2R)^2},$$

the desired inequality follows from (8)–(12).

Finally, choose

$$R \geq \frac{A}{\sqrt{3}C_4} n^\mu \quad \text{and} \quad m \geq \frac{1}{4}\left(\frac{2B}{C_4}\right)^{1/(p+1)} n^{(\mu+p+1)/(p+1)}.$$

Then (7) holds, as required.

Note that the larger we choose p (and thus k), the smaller we need to pick m, the degree of the Legendre series approximating s_j. This means that we have a tradeoff. On the one hand, we need only choose $k = r - 1$ for the FEMQ to be almost optimal. On the other hand, if we choose k this small, then the prescription given above for choosing m yields that $m = \Theta(n^2)$, which will be prohibitively expensive for large n (unless we ignore the cost of computing w_j, considering this merely as a precomputation). □

We now discuss the complexity of using the FEMQ to solve our problem under the residual error criterion. Let

$$\text{cost}_k^Q(\varepsilon) = \inf\{\,\text{cost}(\phi_{n,k}^Q, N_{n,k}^Q) : e(\phi_{n,k}^Q, N_{n,k}^Q) \leq \varepsilon\,\},$$

denote the minimal cost of using the FEMQ of degree k to compute an ε-approximation As above, we may precompute $w_1, \ldots, w_{\tilde{n}}$, and so

$$\text{cost}(\phi_{n,k}^Q, N_{n,k}^Q) = \Theta(n)$$

as $n \to \infty$. From Theorem 6.4.3.1, Theorem 6.4.3.5, and Corollary 4.5.4.1, we have the following

THEOREM 6.4.3.6.

(i) *The ε-complexity using standard information is given by*

$$\mathrm{comp}(\varepsilon, \Lambda^{\mathrm{std}}) = \Theta(\varepsilon^{-1/r}) \qquad \text{as } \varepsilon \to 0.$$

(ii) *Let $\mu = \min\{k + 1, r\}$. Then*

$$\mathrm{cost}_k^Q(\varepsilon) = \Theta(\varepsilon^{-1/\mu}) \qquad \text{as } \varepsilon \to 0.$$

Hence, if $k \geq r - 1$, then the FEMQ of degree k is an almost optimal complexity algorithm, i.e.

$$\mathrm{cost}_k^Q(\varepsilon) = \Theta\big(\mathrm{comp}(\varepsilon, \Lambda^{\mathrm{std}})\big) = \Theta(\varepsilon^{-1/r}) \qquad \text{as } \varepsilon \to 0. \qquad \square$$

Not surprisingly, we pay a large asymptotic penalty for using an FEMQ of too low a degree. That is, for $k < r - 1$, we have

$$\frac{\mathrm{cost}_k^Q(\varepsilon)}{\mathrm{comp}(\varepsilon, \Lambda^{\mathrm{std}})} = \Theta(\varepsilon^{-\zeta d}) \qquad \text{as } \varepsilon \to 0,$$

where

$$\zeta = \frac{1}{k+1} - \frac{1}{r} > 0.$$

Thus,

$$\lim_{\varepsilon \to 0} \frac{\mathrm{cost}_k^Q(\varepsilon)}{\mathrm{comp}(\varepsilon, \Lambda^{\mathrm{std}})} = \infty.$$

Finally, we compare the strength of continuous linear information with standard information. From Theorem 6.4.3.4 and Theorem 6.4.3.6, we see that

$$\mathrm{comp}(\varepsilon, \Lambda^{\mathrm{std}}) = \Theta\big(\mathrm{comp}(\varepsilon, \Lambda^*)\big) \qquad \text{as } \varepsilon \to 0.$$

Hence for this problem, standard information is as powerful as arbitrary continuous linear information.

Notes and Remarks

NR 6.4.3:1 Much of the material in this subsection is based on Werschulz (1990). The material on finite element techniques is new.

6.5 Ordinary differential equations

We now discuss the complexity of initial value problems for ordinary differential equations (ODE) in the worst case setting. Let D be a bounded open convex subset of $\mathbb{R}^{d+1}$. We let F_1 denote the space of all $f: \mathbb{R}^{d+1} \to \mathbb{R}^d$ with support in D, with $D^\alpha f$ continuous for all $|\alpha| \le r$, which is a Banach space under the norm

$$\|f\|_r = \sum_{|\alpha| \le r} \|D^\alpha f\|_{\mathrm{sup}}.$$

Let F be the unit ball of F_1, and let $G = C([0, 1])$ under the sup norm. Our solution operator $S: F \to G$ is then defined by $Sf = z$, where z is the solution to the problem

$$\begin{aligned} z'(x) &= f\big(x, z(x)\big) \qquad 0 < x < 1, \\ z(0) &= \eta. \end{aligned} \tag{1}$$

We consider the complexity of the problem (1) for different classes of information.

6.5.1   Standard information

Suppose that $\Lambda = \Lambda^{\mathrm{std}}$, i.e. we can evaluate $D^\alpha f(p)$ for any multi-index α and any $p \in \mathbb{R}^{d+1}$. One of the most well-known methods for approximating the solution of (1) is the Taylor series method. This algorithm starts by partitioning $[0, 1]$ as

$$0 = x_0 < x_1 < \cdots < x_m = 1,$$

where $x_j = j/m$. For $0 \le j \le m - 1$, it then computes an approximation l_j to z on $[x_j, x_{j+1}]$ by

$$\begin{aligned} l_j(x) &= z_j + \sum_{j=0}^{r-1} \frac{(x - x_j)^{j+1}}{(j+1)!} \left(\frac{d}{dx}\right)^j f(x, y(x))\bigg|_{\substack{x=x_j \\ y(x)=z_j}}, \\ z_{j+1} &= l_j(x_{j+1}). \end{aligned}$$

Here, $z_0 = \eta$.

Let $h = 1/m$ denote the *stepsize* of the partition. Our computation requires information $N_{\mathrm{T},r,h} f$ consisting of evaluations of $D^\alpha f(p_j)$ for $|\alpha| \le r$ and $0 \le j \le m - 1$; the cardinality of $N_{\mathrm{T},r,h}$ is proportional to h^{-1}. Let $\phi_{\mathrm{T},r,h}$ denote a continuous function on $[0, 1]$ that coincides with l_j on $[x_j, x_{j+1}]$ for $0 \le j \le m - 1$. We say that $\phi_{\mathrm{T},r,h}$ is the *Taylor series algorithm* using *Taylor information* $N_{\mathrm{T},r,h}$.

We then have the following result of Kacewicz (1984):

THEOREM 6.5.1.1. *Suppose that only standard information is permissible, i.e. $\Lambda = \Lambda^{\text{std}}$.*

(i) *The complexity of the ODE (1) is*

$$\text{comp}(\varepsilon, \Lambda^{\text{std}}) = \Theta(\varepsilon^{-1/r}) \qquad \text{as } \varepsilon \to 0.$$

(ii) *The Taylor series algorithm $\phi_{\text{T},r,h}$ using stepsize $h = \Theta(\varepsilon^{-1/r})$ as $\varepsilon \to 0$ satisfies*

$$e(\phi_{\text{T},r,h}, N_{\text{T},r,h}) \leq \varepsilon$$

and

$$\text{cost}(\phi_{\text{T},r,h}, N_{\text{T},r,h}) = \Theta(\varepsilon^{-1/r}).$$

Hence the Taylor series algorithm is a nearly optimal complexity algorithm. □

REMARK. Recall that the problems we have studied so far have been intractable, in the sense that the complexity has depended exponentially on the dimension of the region. However, ODE are not intractable in this sense. □

Of course, most texts on the numerical solution of ODE dismiss the use of Taylor series methods. This is because Taylor series information is difficult to obtain if f is a complicated function. It will be more useful to look at algorithms using restricted standard information, consisting of only values of f, with no derivatives allowed. Examples of such algorithms include more standard methods, such as Runge-Kutta methods and multistep methods.

In particular, consider an rth-order Runge-Kutta method, used on a grid over $[0,1]$ having stepsize h. This method computes approximations z_j to $z(x_j)$ for $0 \leq j \leq m$ by

$$z_{j+1} = z_j + h \sum_{l=1}^{p} \gamma_l f_{jl},$$

where

$$f_{jl} = f\left(x_j + \beta_j h, z_j + h \sum_{i=0}^{l-1} \alpha_{li} f_{ji}\right) \qquad (1 \leq l \leq s, 0 \leq j \leq m-1).$$

Here, the coefficients α_{li}, β_l, and γ_l are chosen so that the method is of order r, i.e. exact for all $z \in P_r\big((x_j, x_{j+1})\big)$.

Note that this computation requires information $N_{\text{RK},r,h}$ consisting of evaluations of f at sm points, i.e. the cardinality of $N_{\text{RK},r,h}$ is proportional to h^{-1}. Let $\phi_{\text{RK},r,h}$ denote the function whose restriction to $[x_{ir}, x_{(i+1)r}]$ (for $0 \leq i \leq m/r - 1$) is the Lagrange polynomial of degree r interpolating the data $(x_{ir}, z_{ir}), \ldots, (x_{(i+1)r}, z_{(i+1)r})$. We say that $\phi_{\text{RK},r,h}$ is the rth-order *Runge-Kutta algorithm* using *Runge-Kutta information* $N_{\text{RK},r,h}$.

THEOREM 6.5.1.2.

(i) *There exists a constant C_r, independent of h, such that*

$$e(\phi_{\mathrm{RK},r,h}, N_{\mathrm{RK},r,h}) \le C_r h^r.$$

(ii) *The rth-order Runge-Kutta algorithm $\phi_{\mathrm{RK},r,h}$ using stepsize $h = \lceil (C_r/\varepsilon)^{1/r} \rceil$ satisfies*

$$\mathrm{cost}(\phi_{\mathrm{RK},r,h}, N_{\mathrm{RK},r,h}) = O(\varepsilon^{-1/r}) = \Theta\big(\mathrm{comp}(\varepsilon, \Lambda^{\mathrm{std}})\big),$$

and is thus a nearly optimal complexity algorithm for solving the ODE (1).

PROOF. It is well-known that if $f \in F_1$, then

$$|z(x_j) - z_j| \le C h^r \|z^{(r+1)}\|_{\sup} \qquad (0 \le j \le m),$$

with C independent of f, z, and h. Hence we can view $\phi_{\mathrm{RK},r,h}(N_{\mathrm{RK},r,h}f)$ as an approximation to $z = Sf$ using noisy information at the points $x_0, \dots, x_m$. From pp. 368–369 of Lee *et al.* (1987), we then see that

$$\|Sf - \phi_{\mathrm{RK},r,h}(N_{\mathrm{RK},r,h}f)\|_{\sup} \le C_r h^r \|f\|_r,$$

for some r independent of f and h, which establishes (i). Now using the definition of $h^*(\varepsilon)$, we find that $e(\phi_{\mathrm{RK},r,h^*(\varepsilon)}, N_{\mathrm{RK},r,h^*(\varepsilon)}) \le \varepsilon$. Since $\mathrm{cost}(\phi_{\mathrm{RK},r,h}, N_{\mathrm{RK},r,h}) = O(h^{-1})$, the remainder of the theorem holds immediately. □

6.5.2 Other information

Recall that when we looked at the complexity of elliptic PDE, there was a penalty for using standard information rather than continuous linear information (i.e. finite element inner products). Alternatively, there is an improvement in the complexity if we use continuous linear information instead of standard information. Kacewicz (1984) shows that this also happens for the ODE (1) above.

For $0 \le j \le m - 1$, recall the definition of l_j in the description of the Taylor series method $\phi_{\mathrm{T},r,h}$. Define a new function $\tilde{l}_j$ on $[x_j, x_{j+1}]$ by

$$\tilde{l}_j(x) = l_j(x) + v_j(x) \qquad x \in [x_j, x_{j+1}],$$

where

$$I_j = \int_{x_j}^{x_{j+1}} f\big(t, l_j(t)\big)\, dt,$$

$$v_j(x) = \frac{x - x_j}{h} \big(y_j + I_j - l_j(x_{j+1})\big),$$

$$y_{j+1} = y_j + I_j.$$

Here, $y_0 = \eta$.

Note that this computation requires information $N_{\mathrm{TI},r,h}$, consisting of $N_{\mathrm{T},r,h}$ along with the the integrals $I_0, \ldots, I_{m-1}$. Clearly, the cardinality of $N_{\mathrm{TI},r,h}$ is proportional to h^{-1}. Let $\phi_{\mathrm{TI},r,h}$ denote a continuous function on $[0,1]$ that coincides with $\tilde{l}_j$ on $[x_j, x_{j+1}]$ for $0 \le j \le m-1$. We say that $\phi_{\mathrm{TI},r,h}$ is the *Taylor-integral algorithm* using *Taylor-integral information* $N_{\mathrm{TI},r,h}$.

We then have the following result of Kacewicz (1984):

THEOREM 6.5.2.1. *Suppose that continuous linear information is permissible, i.e. $\Lambda = \Lambda^*$.*

(i) *The complexity of the ODE (1) is*

$$\mathrm{comp}(\varepsilon, \Lambda^*) = \Theta(\varepsilon^{-1/(r+1)}) \qquad \text{as } \varepsilon \to 0.$$

(ii) *The Taylor-integral algorithm $\phi_{\mathrm{TI},r,h}$ using stepsize $h = \Theta(\varepsilon^{-1/r})$ as $\varepsilon \to 0$ satisfies*

$$e(\phi_{\mathrm{TI},r,h}, N_{\mathrm{TI},r,h}) \le \varepsilon$$

  and

$$\mathrm{cost}(\phi_{\mathrm{TI},r,h}, N_{\mathrm{TI},r,h}) = \Theta(\varepsilon^{-1/(r+1)}).$$

  Hence the Taylor-integral algorithm is a nearly optimal complexity algorithm. □

Finally, we can consider the use of finitely continuous information Λ^{fc} for the ODE (1). Recall (as in our study of ill-posed problems in §6.4), that (possibly nonlinear) information is finitely continuous if its restriction to a finite-dimensional subspace is continuous. Using the Borsuk-Ulam theorem (as we did in §6.4), Kacewicz showed that finitely continuous information is no stronger than continuous linear information, i.e. that

$$\mathrm{comp}(\varepsilon, \Lambda^{\mathrm{fc}}) = \Theta(\varepsilon^{-1/(r+1)}).$$

Moreover, since any continuous linear information is finitely continuous, the Taylor-integral algorithm is nearly optimal among all algorithms using finitely continuous information.

Notes and Remarks

NR 6.5.2:1 Most of the recent work on the complexity of ODE has been done by B. Z. Kacewicz, although earlier work may be found in Werschulz (1980), which is based on the Ph.D. dissertation of the author. Most of the results in this section are found on pp. 186–188 of IBC, although the result on optimality of Runge-Kutta methods appears to be new. For further details, the reader should consult Kacewicz (1984).

NR 6.5.2:2 Note that in our discussion of the optimality of Runge-Kutta methods, we have been rather cavalier about the dependence of the combinatory cost on the regularity r. The question of minimizing the combinatory cost is closely related to the question of finding $s_{\min}(r)$, the minimal number of stages s required for a Runge-Kutta method of given order r. The best result known to date is that

$$r + b(r) \leq s_{\min}(r) \leq \begin{cases} \frac{1}{8}(3r^2 - 10r + 24) & \text{for } r \text{ even,} \\ \frac{1}{8}(3r^2 - 4r + 9) & \text{for } r \text{ odd.} \end{cases}$$

Here, b is a function for which $b(r) = \Omega(r)$ as $r \to \infty$. Hence the best lower bound known for $s_{\min}(r)$ is linear in r, while the best upper bound is quadratic. For further details, see pp. 191–194 of Butcher (1987).

NR 6.5.2:3 Additional material on the approximate solution of ODE may be found in many standard books on numerical analysis. Moreover, there are many books devoted specifically to this topic. The text Butcher (1987) mentioned above is fairly recent. Moreover, Henrici (1962) remains as a good readable "classical" introduction to the field.

7

The average case setting

7.1 Introduction

In this chapter, we change our point of view, considering our problems in an average case setting, rather than a worst case setting.

Why do we wish to go to a new setting? One could always appeal to sheer mathematical interest. However, there is a simple, pragmatic justification for leaving the worst case setting, namely, that many of the problems we want to solve are too hard under the worst case setting. We found that elliptic PDE, elliptic systems, and the Fredholm problem of the second kind were intractable, in the sense that their complexity was exponential in the dimension of the domain. Moreover, those problems are nonconvergent if the problem elements are only smooth enough that existence and uniqueness of solutions hold. Things were even worse for ill-posed problems, which we found to be strongly noncomputable in the sense that the cost of finite error approximation is infinite, no matter how large we allow the error to be.

Note that these negative results are problem invariants, independent of any algorithm used to solve the problem. Hence, if we want to overcome these difficulties, we have to abandon the worst case setting. With this in mind, we now turn to an average case setting. Here, we assume that our space of problem elements is equipped with a Gaussian measure; we then measure error and cost on the average, rather than at a worst problem element.

We briefly outline the contents of this chapter. First, we recall some basic definitions of measure theory. Next, we describe general results from IBC about the average case setting.

After that, we go on to specific applications. First, we deal with "shift-invariant" problems, i.e. problems for which a "shift theorem" holds. This class of problems includes elliptic PDE, elliptic systems, and Fredholm problems of the second kind. One of our main results in this chapter is that going from a worst case setting has the effect of increasing the problem regularity. If the measure is sufficiently well-behaved, we find that these shift-invariant problems are no longer intractable. Moreover, recall that we found finite element methods (both the FEM and FEMQ) of sufficiently high degree to be almost optimal algorithms for these problems. We extend this result from the worst case setting to the average case setting, again finding conditions that are necessary and sufficient for an FEM to be almost optimal on the average.

Next, we discuss ill-posed problems. Recall that in the worst case setting, we found that if the solution operator was unbounded, then the radius of information (and hence, the problem complexity) was infinite. However, the story for the average case setting with a Gaussian measure is completely different, since ill-posed problems are always solvable in the average case setting. We show that the error can be made arbitrarily small (i.e. the problem is convergent), and we find optimal information and algorithms for computing ε-approximations.

Finally, we briefly discuss the probabilistic setting, which is related to the average case setting. Here we use a worst case error criterion, except that now we have a probability measure on the set of problem elements, and we ignore possible bad behavior on a set of small measure. Hence our goal is to compute an (ε, δ)-approximation, i.e. to guarantee that the error be at most ε outside a set of measure δ. For shift-invariant problems with δ fixed, the results for the probabilistic setting are essentially the same as those for the average case setting; in particular, we prove that the finite element method of sufficiently high degree is arbitrarily close to being almost optimal. We also show that ill-posed problems are solvable in the probabilistic setting. That is, we prove that the probabilistic (ε, δ)-complexity is finite, and we exhibit optimal information and algorithms.

7.2   Some basic measure theory

We briefly recall some basic definitions and results of measure theory. For further details on basic measure theory, the reader is invited to peruse any of the standard texts that cover measure and integration, such as Dunford and Schwartz (1963), Riesz and Sz.-Nagy (1955), or Shilov and Gurevich (1966). Since we will be especially interested in probability measures defined over infinite-dimensional spaces, we additionally recommend Gihman and Skorohod (1974), Kuo (1975), Parthasarathy (1967), Skorohod (1974), and Vakhania (1981). In what follows, we will let X be a separable normed linear space.

The *Borel σ-field* $\mathfrak{B}(X)$ is the smallest class of subsets of X containing all open subsets of X that is closed under complements and countable unions. A subset of X is said to be *(Borel) measurable* if it belongs to $\mathfrak{B}(X)$.

A *measure* on X is a countably additive function $\mu: \mathfrak{B}(X) \to [0, +\infty]$ for which $\mu(\emptyset) = 0$. A measure μ is *complete* if every subset of a set of zero measure is itself a set of zero measure. A measure μ is a *probability measure* on X if $\mu(X) = 1$.

A function $f \colon X \to Y$ (with Y a normed linear space) is said to be *measurable* if $f^{-1}(A) \in \mathfrak{B}(X)$ for any $A \in \mathfrak{B}(Y)$. In particular, a function $f \colon X \to [0, +\infty)$ is measurable iff $f^{-1}((-\infty, a])$ is measurable for all $a \in \mathbb{R}$.

We next recall the definition of integral with respect to a measure. If $f \colon X \to [0, +\infty)$ is a nonnegative measurable function and E is a measurable subset of X, then we define

$$\int_E f(x)\, \mu(dx) = \sup_{\{E_j\}} \sum_{j=1}^{\infty} \mu(E_j) \inf_{x \in E_j} f(x),$$

the supremum being with respect to all measurable and countable partitions of E. If $f \colon X \to \mathbb{R}$ is an arbitrary measurable function, we define

$$\int_E f(x)\, \mu(dx) = \int_E f_+(x)\, \mu(dx) - \int_E f_-(x)\, \mu(dx),$$

where

$$f_+(x) = \max\{0, f(x)\} \qquad \text{and} \qquad f_-(x) = \max\{0, -f(x)\}.$$

EXAMPLE: *Lebesgue measure.* Let $X = \mathbb{R}^n$. The classical *Lebesgue measure* on X is defined by

$$\mu\left(\prod_{j=1}^{n} I_j\right) = \prod_{j=1}^{n} \text{length}(I_j),$$

for (open, closed, or half-open) intervals $I_1, \ldots, I_n$. If $B \subset \mathbb{R}^n$ has finite nonzero Lebesgue measure, the measure μ_B defined by

$$\mu_B(A) = \frac{\mu(A \cap B)}{\mu(B)}$$

is a probability measure on B. $\qquad\qquad\square$

EXAMPLE: *Normal distribution (Gaussian measure).* Let $X = \mathbb{R}^n$. For an $n \times n$ symmetric, positive definite matrix M and a vector $m \in \mathbb{R}^n$, the *Gaussian measure* (or *normal distribution*) with *mean element* m and *correlation operator (matrix)* M is defined by

$$\mu(A) = \frac{1}{(2\pi \det M)^{n/2}} \int_A \exp\left[-\tfrac{1}{2}\left(M^{-1}(x - m)\right)^T (x - m)\right] dx$$

for any Borel set $A \in \mathfrak{B}(X)$. The mean element m and correlation operator M are respectively characterized by the conditions

$$m^T x = \int_{\mathbb{R}^n} f^T x\, \mu(df) \qquad \forall\, x \in X$$

and

$$(Mx)^T y = \int_{\mathbb{R}^n} \left((f - m)^T x\right)\left((f - m)^T y\right) \mu(df) \qquad \forall\, x, y \in X. \qquad \square$$

EXAMPLE: *Wiener measure.* Let $I = [0,1]$, and let $X = \{f \in C(I) : f(0) = 0\}$ under the sup-norm. Let $u_0 = 0$. Then *Wiener measure w* on X is characterized by

$$w\left(\{f \in X : (f(t_1), \ldots, f(t_n)) \in B\}\right)$$
$$= \prod_{j=1}^{n} \frac{1}{\sqrt{2\pi(t_j - t_{j-1})}} \int_B \exp\left(-\tfrac{1}{2} \sum_{j=1}^{n} \frac{(u_j - u_{j-1})^2}{t_j - t_{j-1}}\right) du_1 \ldots du_n,$$

for all $n \in \mathbb{P}$, $B \in \mathfrak{B}(\mathbb{R}^n)$, and partitions $0 = t_0 < t_1 < \cdots < t_n \leq 1$. □

We now describe Gaussian measures, which extend the familiar normal distribution to spaces that need not be finite-dimensional. They are our standard measures in infinite-dimensional spaces.

Recall that the *characteristic functional* $\psi_\mu \colon X^* \to \mathbb{C}$ of a measure μ on X is defined to be

$$\psi_\mu(\lambda) = \int_X \exp\left(i\lambda(x)\right) \mu(dx) \qquad \forall \lambda \in X^*,$$

where $i = \sqrt{-1}$. It is important to note that the characteristic functional of a measure uniquely determines that measure.

A probability measure μ on a separable Banach space X is said to be *Gaussian* if there exists $C_\mu \colon X^* \to X$ and $m_\mu \in X$ such that the characteristic functional ψ_μ of μ has the form

$$\psi_\mu(\lambda) = \exp\left[i\lambda(m_\mu) - \tfrac{1}{2}\lambda(C_\mu \lambda)\right] \qquad \forall \lambda \in X^*,$$

where $i = \sqrt{-1}$. For such a Gaussian measure, m_μ is the *mean element* of μ, i.e.

$$\lambda(m_\mu) = \int_X \lambda(x) \mu(dx) \qquad \forall \lambda \in X^*,$$

and C_μ is the *correlation operator* of μ, i.e.

$$\lambda_1(C_\mu \lambda_2) = \int_X \lambda_1(x - m_\mu)\lambda_2(x - m_\mu) \mu(dx) \qquad \forall \lambda_1, \lambda_2 \in X^*.$$

Equivalently, μ is Gaussian with mean element m_μ and correlation operator C_μ iff for any $\lambda \in X^*$ and $d \in \mathbb{R}$,

$$\mu(\{x \in X : \lambda(x) \leq d\}) = \frac{1}{\sqrt{2\pi\lambda(C_\mu \lambda)}} \int_{-\infty}^{d} \exp\left(\frac{-(t - \lambda(m_\mu))^2}{2\lambda(C_\mu \lambda)}\right) dt.$$

This gives a relation between Gaussian measures on normed linear spaces and the familiar Gaussian distribution on the real numbers.

EXAMPLE: *Wiener measure (continued)*. Wiener measure is a Gaussian measure. Its mean is zero, and its correlation operator C_w is $\lambda_{x_2}(C_w \lambda_{x_1}) = \min\{x_1, x_2\}$ for $\lambda_{x_j} = f(x_j)$. ☐

Notes and Remarks

NR 7.2:1 This review is based on the material in the Appendix to IBC, which contains a more leisurely review of basic measure theory than that presented here.

NR 7.2:2 The characterization of correlation operators for Gaussian measures on Banach spaces is an open problem. However, if X is a separable Hilbert space, this problem has been solved. Recall that the *trace* of a linear operator V on a Hilbert space X is

$$\text{trace}\, V = \sum_{j=1}^{\infty} \langle V x_j, x_j \rangle_X,$$

where $\{x_j\}_{j=1}^{\infty}$ is an orthonormal system in X. (See Gel'fand and Vilenkin (1964) for further details.) Then V is the correlation operator of a Gaussian measure iff V is symmetric, nonnegative define, and trace V is finite, in which case we have trace $V = \int_X \|x\|^2 \mu(dx)$. For further details see pp. 349–350 of Gihman and Skorohod (1974).

7.3 General results for the average case setting

As always, we are given a solution operator $S \colon F \to G$ defined on a set F of problem elements and a class Λ of permissible information operations on F. We wish to approximate Sf for $f \in F$ using permissible information. Hence, our approximations take the form $\phi(Nf)$, where N is permissible information, and $\phi \colon N(F) \to G$ is an algorithm using N.

We now wish to find (almost) optimal algorithms, measuring cost on the average. Let μ be a measure on F. If ϕ is an algorithm using information N, then its (average case) error and cost were respectively defined (in Chapter 3) to be

$$e^{\text{avg}}(\phi, N) = \left[\int_F \|Sf - \phi(Nf)\|^2 \, \mu(df) \right]^{1/2}$$

and

$$\text{cost}^{\text{avg}}(\phi, N) = \int_F \text{cost}(\phi, N, f) \, \mu(df).$$

For $\varepsilon \geq 0$, the (average case) ε-complexity is then given by

$$\text{comp}^{\text{avg}}(\varepsilon, \Lambda) = \inf\{\, \text{cost}^{\text{avg}}(\phi, N) : N \text{ using } \Lambda \text{ and } e^{\text{avg}}(\phi, N) \leq \varepsilon \,\}.$$

If $\Lambda = \Lambda^*$, i.e. any continuous linear functional is permissible, then we write $\text{comp}^{\text{avg}}(\varepsilon)$ for $\text{comp}^{\text{avg}}(\varepsilon, \Lambda)$. An algorithm ϕ_ε using information N_ε (consisting of evaluations from Λ) is said to be *almost optimal* if

$$e^{\text{avg}}(\phi_\varepsilon, N_\varepsilon) \leq \varepsilon$$

and

$$\text{cost}^{\text{avg}}(\phi_\varepsilon, N_\varepsilon) = \Theta\big(\text{comp}^{\text{avg}}(\varepsilon, \Lambda)\big) \quad \text{as } \varepsilon \to 0,$$

the Θ-constant being independent of ε.

In what follows, we now specialize to the following classes of problems, which is general enough to include most of the applications that we will study:

(i) The solution operator S is (the restriction of) a continuous linear operator $S \colon F_1 \to G$, where F_1 is a real separable Banach space and G is a real separable Hilbert space.

(ii) The class F of problem elements is either the whole space F_1 or a ball of radius q, i.e. $F = qB(F_1)$ for some $q \in (0, +\infty]$.

(iii) $\Lambda \subseteq \Lambda^*$, i.e. only continuous linear functionals are permissible.

(iv) F_1 is equipped with a Gaussian measure μ with mean zero and correlation operator $C_\mu \colon F_1^* \to F_1$. If F_1 is a Hilbert space, we shall identify F_1^* with F_1 when this is convenient, so that C_μ is an operator on F_1.

(v) If F is a ball of finite radius q, then we will use the truncated Gaussian measure μ_q on F, defined by

$$\mu_q(A) = \frac{\mu(A \cap F)}{\mu(F)} \qquad \forall A \in \mathfrak{B}(F_1),$$

as our measure on F.

Hence we are studying a continuous linear problem, under the assumption that the class of problem elements is equipped with a probability measure. We will remove the restriction that the solution operator be continuous in §7.5, where we study ill-posed problems in the average case setting.

Note that our class of problem elements has changed somewhat. In the worst case setting, our problem elements were generally taken to be the unit (or semi-unit) ball BW of a Sobolev space W. Knowing the worst case complexity with problem elements BW, it is easy to determine the complexity with problem elements a q-multiple qBW of the unit ball. Indeed, the nth minimal worst case radius gets multiplied by a factor of q. Since for the problems we studied, the nth minimal radius (over BW) was $\Theta(n^{-p})$ for some $p > 0$, it follows that the nth minimal error and ε-complexity over qBW are $\Theta(q \cdot n^{-p})$ and $\Theta\big((q/\varepsilon)^{1/p}\big)$, respectively. Hence we can determine the worst case complexity over a ball of arbitrary radius from that over the unit ball. In particular, the complexity over the whole space is infinite.

Since we will eventually wish to compare the worst and average cases, we must choose the same class of problem elements in each case. So, we choose a ball of the space as our class of problem elements. For technical reasons, we must start off with the case $q = \infty$, i.e. we assume that the class of problem elements is the whole space. Since the worst case complexity for this class of problem elements is infinite, we then wish to analyze the case of finite q. If q is not too small (relative to trace C_μ), then the truncated measure μ_q preserves the essential properties of the Gaussian measure μ. It then turns out that we can determine the average case complexity for finite q from that for $q = \infty$. Hence, our plan of attack will be to describe results for $q = \infty$, and then use those results to find the complexity for $q < \infty$.

So, assume that $q = \infty$, i.e. $F = F_1$. We follow the same general ideas for determining ε-complexity in the average case as we did in the worst case. Hence, we must determine the minimal average error achievable by algorithms using information of cardinality at most n.

For information N, we let

$$r^{\text{avg}}(N) = \inf_{\phi \text{ using } N} e^{\text{avg}}(\phi, N)$$

denote the *radius of information*, which is the minimal average error achievable by algorithms using information N. We wish to compute $r^{\text{avg}}(N)$ and to find an optimal error algorithm using N. Suppose that

$$Nf = \begin{bmatrix} \lambda_1(f) \\ \vdots \\ \lambda_n(f) \end{bmatrix} \qquad \forall f \in F_1,$$

i.e. N is linear nonadaptive information of cardinality n. Without loss of generality, we may assume that

$$\lambda_i(C_\mu \lambda_j) = \delta_{i,j} \qquad (1 \leq i, j \leq n).$$

For any $y \in \mathbb{R}^n$, the *μ-spline* interpolating y is given by

$$\sigma_N(y) = \sum_{j=1}^{n} y_j C_\mu \lambda_j.$$

Define the *μ-spline algorithm* ϕ^{s} by

$$\phi^{\text{s}}(Nf) = S\sigma_N(Nf) = \sum_{j=1}^{n} \lambda_j(f) SC_\mu \lambda_j \qquad \forall f \in F_1.$$

We then have

THEOREM 7.3.1. *The μ-spline algorithm is an optimal error algorithm using N, and*

$$e^{\mathrm{avg}}(\phi^{\mathrm{s}}, N) = r^{\mathrm{avg}}(N) = \left[\int_{F_1} \|Sf\|^2 \, \mu(df) - \sum_{j=1}^{n} \|SC_\mu \lambda_j\|^2 \right]^{1/2}. \qquad \square$$

Now that we know how to find an algorithm using given information whose average case error is minimal among all algorithms using that information, we now look for the information of given cardinality whose radius is minimal. We first consider nonadaptive information. Let Λ_n denote the set of all nonadaptive information operators using at most n information operations from Λ. Then the nth *minimal radius of information*

$$r^{\mathrm{avg}}(n, \Lambda) = \inf_{N \in \Lambda_n} r^{\mathrm{avg}}(N)$$

is the minimal average error among all algorithms using nonadaptive information of average cardinality at most n. Information $N_n^* \in \Lambda_n$ is said to be nth *optimal information* if $r^{\mathrm{avg}}(N_n^*) = r^{\mathrm{avg}}(n, \Lambda)$. As in the worst case, we write $r^{\mathrm{avg}}(n)$ instead of $r^{\mathrm{avg}}(n, \Lambda^*)$ whenever $\Lambda = \Lambda^*$, i.e. when any continuous linear functional is permissible. We say that an algorithm ϕ_n^* using nonadaptive information $N_n^* \in \Lambda_n$ is an nth *minimal error algorithm* if

$$e^{\mathrm{avg}}(\phi_n^*, N_n^*) = r^{\mathrm{avg}}(n, \Lambda),$$

i.e. ϕ_n^* has minimal average error among all algorithms using n nonadaptive information evaluations from Λ. Note that the μ-spline algorithm using nth optimal information is an nth minimal error algorithm.

We now describe nth optimal nonadaptive information for the case $\Lambda = \Lambda^*$, i.e. where any continuous linear functional is a permissible information operation, as well as an nth minimal error algorithm. Since G is a Hilbert space, we may identify the normed dual G^* with G. Then $C_\nu = SC_\mu S^*$ is the correlation operator of the a priori Gaussian measure $\nu = \mu S^{-1}$ on the solution elements. The operator C_ν is self-adjoint, nonnegative definite, and has finite trace. Let $\{u_j\}_{j=1}^\infty$ be a complete orthonormal basis for G consisting of eigenvectors for C_ν, with

$$C_\nu u_j = \gamma_j u_j \qquad (j = 1, 2, \dots)$$

and

$$\gamma_1 \geq \gamma_2 \geq \dots \geq 0 \qquad \text{and} \qquad \lim_{j \to \infty} \gamma_j = 0.$$

For any index j, we let

$$\lambda_j^*(f) = \gamma_j^{-1/2} \langle Sf, u_j \rangle,$$

where we have assumed that $\gamma_j > 0$ for simplicity. Define information N_n^* of cardinality n by

$$N_n^* f = \begin{bmatrix} \lambda_1^*(f) \\ \vdots \\ \lambda_n^*(f) \end{bmatrix} \qquad \forall f \in F_1,$$

and let ϕ_n^* be the spline algorithm using N_n^*, so that

$$\phi_n^*(N_n^* f) = \sum_{j=1}^{n} \lambda_j^*(f) S C_\mu \lambda_j^* = \sum_{j=1}^{n} \langle Sf, u_j \rangle u_j \qquad \forall f \in F_1.$$

We then have

THEOREM 7.3.2. *Let $\Lambda = \Lambda^*$. The algorithm ϕ_n^* is an nth minimal error algorithm and N_n^* is nth optimal information, i.e.*

$$e^{\mathrm{avg}}(\phi_n^*, N_n^*) = r^{\mathrm{avg}}(N_n^*) = r^{\mathrm{avg}}(n) = \left(\sum_{j=n+1}^{\infty} \gamma_j \right)^{1/2}. \qquad \square$$

It is also possible to derive general formulas for $r^{\mathrm{avg}}(n, \Lambda)$ when $\Lambda \subseteq \Lambda^*$. See p. 233 of IBC for details.

Next, we consider adaptive information, our main goal being to determine whether adaption is significantly stronger than nonadaption. Since we allow information of varying cardinality, we let

$$\mathrm{card}^{\mathrm{avg}} N = \int_F n(f) \, \mu(df)$$

denote the *average cardinality* of N, where $n(f)$ is the number of information operations used in calculating Nf (see Chapter 3). Since $\mathrm{card}^{\mathrm{avg}} N$ is given by an average, it need not be an integer.

We now state a result giving sufficient conditions for nonadaptive information to be almost as strong as adaptive information of varying cardinality. Let us say that a sequence $\alpha_1, \alpha_2, \ldots$ is *convex* if

$$\alpha_n \leq \tfrac{1}{2}(\alpha_{n-1} + \alpha_{n+1}) \qquad \forall n \geq 2.$$

Important examples of convex sequences include
  (i) the sequence $\{r^{\mathrm{avg}}(n)^2\}_{n=1}^{\infty}$ of the squares of nth minimal average radii for the case $\Lambda = \Lambda^*$, and
  (ii) the sequence $\{n^{-p}\}_{n=1}^{\infty}$, where $p \geq 0$.
We also say that the sequence $\beta_1, \beta_2, \ldots$ is *semiconvex* if there exist positive numbers a and b and a convex sequence $\alpha_1, \alpha_2, \ldots$ such that

$$a\,\alpha_n \leq \beta_n \leq b\,\alpha_n \qquad \forall n \geq 1.$$

We then have

THEOREM 7.3.3. *Let N^{a} be adaptive information of finite average cardinality* $\mathrm{card}^{\mathrm{avg}}\, N^{\mathrm{a}}$. *Let N_n^* be nth optimal nonadaptive information from the class Λ_n, with $n = \lceil \mathrm{card}^{\mathrm{avg}}\, N^{\mathrm{a}} \rceil$.*

(i) *If $\{r^{\mathrm{avg}}(n,\Lambda)^2\}_{n=1}^{\infty}$ is convex, then*

$$r^{\mathrm{avg}}(N_n^*) \le r^{\mathrm{avg}}(N^{\mathrm{a}}),$$

i.e. nonadaption is as powerful as adaption.

(ii) *If $\{r^{\mathrm{avg}}(n,\Lambda)^2\}_{n=1}^{\infty}$ is semiconvex (with constants a and b), then*

$$r^{\mathrm{avg}}(N_n^*) \le \sqrt{\frac{b}{a}} r^{\mathrm{avg}}(N^{\mathrm{a}}),$$

i.e. nonadaption is almost as powerful as adaption. □

In particular, we see that nonadaption is as powerful as adaption if $\Lambda = \Lambda^*$.

Having investigated minimal error algorithms, we now consider problem complexity in the average case setting. For $\varepsilon \ge 0$, we let

$$m^{\mathrm{avg}}(\varepsilon,\Lambda) = \inf\{\mathrm{card}^{\mathrm{avg}}\, N : N \text{ using } \Lambda \text{ such that } r^{\mathrm{avg}}(N) \le \varepsilon\}$$

denote the average ε-*cardinality number* of our problem. We also let

$$n^{\mathrm{avg}}(\varepsilon,\Lambda) = \min\{n \in \mathbb{N} : r^{\mathrm{avg}}(n,\Lambda) \le \varepsilon\}$$

denote the minimal number of *nonadaptive* information evaluations that are needed to compute an ε-approximation. We then find

THEOREM 7.3.4. *Let Λ denote the class of permissible information operations.*

(i) *For any $\varepsilon \ge 0$, the average problem complexity satisfies*

$$c\, m^{\mathrm{avg}}(\varepsilon,\Lambda) \le \mathrm{comp}^{\mathrm{avg}}(\varepsilon,\Lambda) \le (c+2)\, m^{\mathrm{avg}}(\varepsilon,\Lambda).$$

(ii) *Suppose that $\{r^{\mathrm{avg}}(n,\Lambda)^2\}_{n=1}^{\infty}$ is convex, and let $\varepsilon \ge 0$. Then*

$$\lceil m^{\mathrm{avg}}(\varepsilon,\Lambda) \rceil = n^{\mathrm{avg}}(\varepsilon,\Lambda).$$

Let $n = n^{\mathrm{avg}}(\varepsilon,\Lambda)$. If N_n^* is nth optimal nonadaptive information from the class Λ and ϕ_n^* is the μ-spline algorithm using N_n^*, then

$$e^{\mathrm{avg}}(\phi_n^*,N_n^*) \le \varepsilon$$

and
$$c\,n \le \mathrm{cost}^{\mathrm{avg}}(\phi_n^*,N_n^*) \le (c+2)n - 1,$$

and so ϕ_n^* is an almost optimal complexity algorithm.

(iii) *Suppose that* $\{r^{\mathrm{avg}}(n,\Lambda)^2\}_{n=1}^\infty$ *is semiconvex. Then*

$$m^{\mathrm{avg}}(\varepsilon,\Lambda) = n^{\mathrm{avg}}\big(\Theta(\varepsilon),\Lambda\big) \qquad \text{as } \varepsilon \to 0.$$

For $\varepsilon \geq 0$, *let* $n = n^{\mathrm{avg}}(\varepsilon,\Lambda)$. *If* N_n^* *is nth optimal nonadaptive information from the class* Λ *and* ϕ_n^* *is the μ-spline algorithm using* N_n^*, *then*

$$e^{\mathrm{avg}}(\phi_n^*, N_n^*) \leq \varepsilon$$

and for $c \gg 1$,

$$\mathrm{cost}^{\mathrm{avg}}(\phi_n^*, N_n^*) = c\, m^{\mathrm{avg}}\big(\Theta(\varepsilon),\Lambda\big) \qquad \text{as } \varepsilon \to 0.$$

(iv) *Suppose that* $\{r^{\mathrm{avg}}(n,\Lambda)^2\}_{n=1}^\infty$ *is semiconvex and that*

$$\mathrm{comp}^{\mathrm{avg}}\big(\varepsilon(1+\eta),\Lambda\big) = \mathrm{comp}^{\mathrm{avg}}(\varepsilon,\Lambda)\big(1+O(\eta)\big) \qquad \text{as } \eta \to 0.$$

Then

$$\mathrm{cost}^{\mathrm{avg}}(\phi_n^*, N_n^*) = \Theta\big(c\, m^{\mathrm{avg}}(\varepsilon,\Lambda)\big) = \Theta\big(\mathrm{comp}^{\mathrm{avg}}(\varepsilon,\Lambda)\big) \qquad \text{as } \varepsilon \to 0,$$

i.e. ϕ_n^* *is an almost optimal complexity algorithm.* $\square$

So far, we have dealt only with the case $F = F_1$. What happens when $F = B_q \equiv qB(F_1)$ is a ball of finite radius q? Clearly as q increases, the complexity for finite q ought to approach the complexity for $q = \infty$, i.e. the complexity for $F = F_1$. In what follows, we assume that q is chosen large enough that

$$\mu(B_q) \geq \tfrac{1}{2}(\sqrt{21} - 3) \simeq 0.8. \tag{1}$$

This is not a restrictive condition, since $\mu(B_q)$ goes to one exponentially as q goes to infinity (see **NR 7.3:4**).

THEOREM 7.3.5. *Let* $\mathrm{comp}^{\mathrm{avg}}(\varepsilon,\Lambda,q)$ *denote the average case ε-complexity, with* Λ *denoting the class of permissible information functionals and with problem elements* B_q.

(i) *Suppose that q is chosen large enough that (1) holds. Let* $x = 1 - \mu(B_q)$. *Then*

$$\frac{c}{c+2}\,\frac{1-x-\sqrt{3x}}{1-x}\,\mathrm{comp}^{\mathrm{avg}}\left(\varepsilon\sqrt{\frac{1-x}{1-x-\sqrt{3x}}},\Lambda\right)$$
$$\leq \mathrm{comp}^{\mathrm{avg}}(\varepsilon,\Lambda,q) \leq \frac{1}{1-x}\,\mathrm{comp}^{\mathrm{avg}}(\varepsilon\sqrt{1-x},\Lambda).$$

(ii) *Let*

$$a^* = (2\sup\{\,\lambda(C_\mu\lambda) : \lambda \in F_1^*, \|\lambda\| = 1\,\})^{-1}.$$

If

$$\operatorname{comp}^{\operatorname{avg}}\big(\varepsilon(1+\eta),\Lambda\big) = \operatorname{comp}^{\operatorname{avg}}(\varepsilon,\Lambda)\big(1+O(\eta)\big) \qquad \textit{as } \eta \to 0,$$

then for any $a < a^*$ we have

$$\operatorname{comp}^{\operatorname{avg}}(\varepsilon,\Lambda,q) = \operatorname{comp}^{\operatorname{avg}}(\varepsilon,\Lambda)\big[1 + o\big(\exp(-\tfrac{1}{2}q^2 a)\big)\big]\big(1+O(c^{-1})\big)$$

as $q \to \infty$ and $c \to \infty$.   □

Hence, $\operatorname{comp}^{\operatorname{avg}}(\varepsilon,\Lambda,q) \simeq \operatorname{comp}^{\operatorname{avg}}(\varepsilon,\Lambda)$ for reasonably large q and c.

Notes and Remarks

NR 7.3:1 The material in this section is taken from Chapter 6 of IBC.

NR 7.3:2 Note that we are using an absolute error criterion in our average case setting. We will discuss what happens if we use other error criteria to solve a particular problem in that section of this chapter discussing said problem.

NR 7.3:3 In the definition of average error, it is implicitly assumed that the function $f \mapsto \|\phi(Nf)\|^2$ is integrable. It is possible to extend the definition of average error to ϕ for which this need not hold. See pp. 205–206 of IBC for details.

NR 7.3:4 We have the following results showing how quickly $\mu(B_q) \to 1$ as $q \to \infty$. (See pp. 258–259 of IBC for further details.)

(i) Define a^* as in Theorem 7.3.5. Then

$$\mu(B_q) = 1 - \exp\big[-q^2 a^*(1+o(1))\big] \qquad \textit{as } q \to \infty.$$

See Borell (1975, 1976) for details.

(ii) Using Fernique's theorem (see p. 141 of Araujo and Giné 1980), we have

$$0 \le 1 - \mu(B_q) \le \exp(-q^2 a)\int_{F_1} \exp(a\|f\|^2)\,\mu(df)$$

for any $a < a^*$.

(iii) Suppose that F_1 is a separable Hilbert space. Then

$$0 \le 1 - \mu(B_q) \le \frac{\exp(-q^2 a)}{\sqrt{1 - 2a\operatorname{trace}C_\mu}}$$

for any $a < a^*$. In particular, suppose that the eigenvalues γ_j of C_μ go to zero as j^{-p} for some $p > 1$. Setting $a = (p-1)/(4p)$, we find that

$$0 \le 1 - \mu(B_q) \le \sqrt{2}\exp\left(\frac{-q^2(p-1)}{4p}\right).$$

7.4   Complexity of shift-invariant problems

In this section, we discuss the average case complexity of a wide class of problems, which we call shift-invariant problems. These problems are those that admit a shift theorem, including elliptic PDEs, elliptic systems, and the Fredholm problem of the second kind. Our main result is that for a large class of measures, the finite element method of sufficiently high degree is an almost optimal algorithm for shift-invariant problems. Furthermore, if the eigenvalues of the correlation operator of the measure go to zero reasonably quickly, we no longer find that the complexity is exponential in the dimension, i.e. the problem is no longer intractable.

Notes and Remarks

NR 7.4:1 Much of the material in this section is from Werschulz (1989), which was mainly an investigation of the optimality of the FEM for solving elliptic PDEs in the average case setting. New material in this section includes the analysis of other shift-invariant problems, the average case upper bound for the FEM of arbitrary degree, a sharper estimate in Theorem 7.4.2.4 (along with a cleaner proof), and the analysis of standard information and the FEMQ.

7.4.1   Problem definition

In what follows, we use Sobolev spaces that may be spaces of either scalar- or vector-valued functions. Let Ω be a C^∞-region in $\mathbb{R}^d$, and let $m, s \geq 0$. Let $H_E^m(\Omega)$ be a closed subspace of $H^m(\Omega)$ that contains $H_0^m(\Omega)$. A (possibly unbounded) linear operator L on $L_2(\Omega)$ is said to be (m, s)-*shift-invariant* if there exists $\tilde{m} \geq m$ such that for any $a \in [m, \tilde{m}]$, there is a constant $\theta > 0$ such that

$$\theta^{-1}\|Lu\|_{H^{a-s}(\Omega)} \leq \|u\|_{H^a(\Omega)} \leq \theta\|Lu\|_{H^{a-s}(\Omega)} \qquad \forall u \in H^a(\Omega) \cap H_E^m(\Omega).$$

That is, L is shift-invariant iff it admits a "shift theorem."

EXAMPLES.

(i)   The $2m$th-order elliptic partial differential equations that were studied in Chapter 5 are $(m, 2m)$-shift-invariant, with $\tilde{m} = \infty$. The space $H_E^m(\Omega)$ is the space of all elements of $H^m(\Omega)$ satisfying the essential boundary conditions.

(ii)   Elliptic systems, as studied in §6.2, are (t, t)-shift-invariant, with $H_E^t(\Omega) = H^t(\partial)$ and $\tilde{m} = \infty$.

(iii)   The Fredholm problem of the second kind studied in §6.3 is $(0, 0)$-shift-invariant in the Hilbert case, with with $H_E^0(\Omega) = L_2(\Omega)$. Here, $\tilde{m}$ is the largest integer such that the function $x \mapsto \partial_x^\alpha k(x, y)$ is continuous for all multi-indices α such that $|\alpha| = \tilde{m}$.   □

REMARK. In the remainder of this section, we shall simplify the exposition a great deal by assuming that $\tilde{m} = \infty$. This will cause no problems with elliptic PDEs and systems. For the Fredholm problem, we have $\tilde{m} = \infty$ if the kernel k is C^∞ in its first variable. If not, we will have to keep track of an upper limit on the smoothness of problem elements that can be shifted. While not impossible, this is a bit tedious. We prefer to simplify matters by not keeping track of this upper limit. $\square$

Suppose that L and L^* are (m, s)-shift-invariant. Let $r \geq -m$. Then for any $f \in H^r(\Omega)$, we claim that there exists a unique $u \in H_E^m(\Omega) \cap H^{r+s}(\Omega)$ such that $Lu = f$. That is, we claim that $L \colon H_E^m(\Omega) \cap H^{r+s}(\Omega) \to H^r(\Omega)$ is a bijection. Indeed, since L is shift-invariant, we see that L is an injection with closed range, while the shift-invariance of L^* implies that $\ker L^* = \{0\}$. Hence,

$$\text{range } L = \overline{\text{range } L} = (\ker L^*)^\perp = \{0\}^\perp = H^r(\Omega),$$

and so L is surjective. Hence, L is bijective, as claimed.

To define our problem, we need to define our solution operator, the class of problem elements, permissible information, and a measure on the problem elements.

First, we define our solution operator. Let L and L^* be (m, s)-shift-invariant, and let $l \in [0, \min\{m, s\}]$. Write $H_E^l(\Omega)$ for the closure of $H_E^m(\Omega)$ under the $H^l(\Omega)$-norm. Let $r \geq -m$. Our solution operator is then $S \colon H^r(\Omega) \to H_E^l(\Omega)$, where for any $f \in H^r(\Omega)$, we let $Sf = u$ be the unique element of $H_E^m(\Omega) \cap H^{r+s}(\Omega)$ such that $Lu = f$.

Our class of problem elements will be a subset of $H^r(\Omega)$. We will have to do some background work to define this subset. Let A and A^* be (m', s')-shift-invariant for some $m', s' \geq 0$. Then A is a bijection of $H_*^{s'}(\Omega)$ onto $L_2(\Omega)$, and A^* is a bijection of $H_*^{2s'}(\Omega)$ onto $H_*^{s'}(\Omega)$. Here, $H_*^{js'}(\Omega)$ is a subspace of $H^{js'}(\Omega)$ that contains $H_0^{js'}(\Omega)$, for $j = 1$ and $j = 2$. So, A^*A is a self-adjoint bijection of $H_*^{2s'}(\Omega)$ onto $L_2(\Omega)$. Let $\hat{A} = (A^*A)^{1/2}$.

LEMMA 7.4.1.1. *There exist functions $z_1, z_2, \cdots \in C^\infty(\Omega)$ forming an orthonormal basis of $L_2(\Omega)$, as well as positive scalars $\alpha_1 \leq \alpha_2 \leq \ldots$, such that*

$$\hat{A} z_j = \alpha_j z_j \qquad \forall j \in \mathbb{P},$$

with

$$\alpha_j = \Theta(j^{s'/d}) \qquad \text{as } j \to \infty.$$

PROOF. Let $T = \hat{A}^{-2}$, as a linear transformation of $L_2(\Omega)$, and let $\lambda_j(T)$ denote the jth eigenvalue of T. Since $T L_2(\Omega) = H_*^{2s'}(\Omega)$ and $s' > 0$, we know that T is a compact operator by the Rellich-Kondrasov theorem

(Lemma A.2.2.3). Hence T has pure point spectrum, and there exist functions $z_1, z_2, \cdots \in C^\infty(\Omega)$ forming an orthonormal basis of $L_2(\Omega)$ such that

$$T z_j = \lambda_j(T) z_j \qquad \forall j \in \mathbb{P}.$$

From general results on Gelfand widths on Hilbert spaces (see, for example, Pinkus 1985), we know that

$$d^j\left(B T L_2(\Omega), L_2(\Omega)\right) = \lambda_{j+1}(T) \qquad \forall j \in \mathbb{N},$$

and that

$$d^j\left(B T L_2(\Omega), L_2(\Omega)\right) = \Theta(j^{-2s'/d}) \qquad \text{as } j \to \infty.$$

Letting $\alpha_j = \lambda_j^{-1/2}(T)$, the result follows. $\qquad\qquad\qquad\square$

Let $t \in \mathbb{R}$. We define an operator $\hat{A}^{t/s'}$ on $L_2(\Omega)$ by

$$\hat{A}^{t/s'} v = \sum_{j=1}^\infty \alpha_j^{t/s'} \langle v, z_j \rangle_{L_2(\Omega)} z_j.$$

Note that since $s' > 0$, the operator $\hat{A}^{t/s'}$ is unbounded. We also define an inner product by

$$\langle v, w \rangle_t = \langle \hat{A}^{t/s'} v, \hat{A}^{t/s'} w \rangle_{L_2(\Omega)} = \sum_{j=1}^\infty \alpha_j^{2t/s'} \langle v, z_j \rangle_{L_2(\Omega)} \langle w, z_j \rangle_{L_2(\Omega)},$$

with corresponding norm

$$\|v\|_t = \langle v, v \rangle_t^{1/2} = \|\hat{A}^{t/s'} v\|_{L_2(\Omega)} = \left[\sum_{j=1}^\infty \alpha_j^{2t/s'} \langle v, z_j \rangle_{L_2(\Omega)}^2 \right]^{1/2}.$$

Set

$$H_*^t(\Omega) = \{ v \in L_2(\Omega) : \|v\|_t < \infty \},$$

i.e. $H_*^t(\Omega)$ is the domain of $\hat{A}^{t/s'}$. We then have

LEMMA 7.4.1.2.

(i) $H_*^t(\Omega)$ is a Hilbert space under the norm $\|\cdot\|_t$.

(ii) $\|\cdot\|_t$ is equivalent to the usual Sobolev norm $\|\cdot\|_{H^t(\Omega)}$ on $H_*^t(\Omega)$, i.e. there exists a constant $\sigma \geq 1$ such that

$$\sigma^{-1}\|v\|_t \leq \|v\|_{H^t(\Omega)} \leq \sigma\|v\|_t \qquad \forall v \in H_*^t(\Omega).$$

(iii) $\{H_*^t(\Omega)\}_{t \in \mathbb{R}}$ is a decreasing family of Hilbert spaces satisfying the interpolation property.

(iv) $H_*^t(\Omega)$ is a closed subspace of $H^t(\Omega)$ that contains $H_0^t(\Omega)$.

(v) $H_*^0(\Omega) = L_2(\Omega)$ and $\|\cdot\|_0 = \|\cdot\|_{L_2(\Omega)}$.

PROOF. For (i)–(iii), see Berezanskii (1968). The other items follow immediately. ☐

Now we are ready to define our problem elements. For $q \in (0, \infty]$, our set F of problem elements will be $F = qBH_*^r(\Omega)$. That is, if q is finite, then F will be a q-multiple of the unit ball in $H_*^r(\Omega)$, whereas if $q = \infty$, then F will be the whole space $H_*^r(\Omega)$.

We study two classes of permissible information operations. First, we will consider the case $\Lambda = \Lambda^*$, i.e. we will allow any continuous linear functional as information. Later, we will restrict our attention to the case $\Lambda = \Lambda^{\text{std}}$ of standard information, i.e. we can only evaluate a problem element of one of its derivatives at a point in Ω.

Finally, we describe our measure. For $r \geq -m$, let

$$y_j = \alpha_j^{-r/s'} z_j \qquad (j = 1, 2, \dots).$$

Then $\{y_j\}_{j=1}^\infty$ is an orthonormal basis for $H_*^r(\Omega)$. Let $\beta_1 \geq \beta_2 \geq \cdots > 0$ be chosen satisfying

$$\sum_{j=1}^\infty \beta_j < \infty,$$

a typical choice being $\beta_j = \Theta(j^{-p})$ as $j \to \infty$, for some $p > 1$. Our measure μ on the space $H_*^r(\Omega)$ is now chosen to be a Gaussian measure with zero mean and positive definite correlation operator C_μ defined by

$$C_\mu f = \sum_{j=1}^\infty \beta_j \langle f, y_j \rangle_r y_j = \sum_{j=1}^\infty \beta_j \langle f, z_j \rangle_0 z_j \qquad \forall f \in H_*^r(\Omega).$$

If $F = H_*^r(\Omega)$, we use the Gaussian measure μ as our measure on F, while if $F = qBH_*^r(\Omega)$, we use the truncated Gaussian measure μ_q as our measure on F.

We now give several examples to show that this construction includes a rich set of possible classes of problem elements and measures.

EXAMPLE: *Wiener measure.* Let w denote the classical Wiener measure on $C^*(I)$. We define a measure μ on

$$H_*^r(I) = \{ v \in H^r(I) : (D^j v)(0) = 0 \text{ for } 0 \leq j \leq r - 1 \}$$

by

$$\mu(B) = w(D^r B \cap C_*(I)) \qquad \forall \text{ Borel sets } B \subseteq H_*^r(I).$$

Following Papageorgiou and Wasilkowski (1986), let $m' = s' = 2r + 2$, and consider the eigenproblem

$$(-1)^{r+1} D^{2r+2} z = \alpha z \qquad \text{in } I,$$

$$(D^j z)(0) = (D^{r+1+j} z)(1) = 0 \qquad (0 \leq j \leq r).$$

This problem has an $L_2(I)$-orthonormal basis $\{z_j\}_{j=1}^{\infty}$ of eigenvectors, with the corresponding eigenvalues $\{\alpha_j\}_{j=1}^{\infty}$ satisfying

$$\alpha_j \sim (j\pi)^{2r+2} \qquad \text{as } j \to \infty.$$

The inner product on $H_*^t(\Omega)$ is then given by

$$\langle v, w \rangle_t = \sum_{j=1}^{\infty} \alpha_j^{t/(r+1)} \langle v, z_j \rangle_{L_2(I)} \langle w, z_j \rangle_{L_2(I)}.$$

Since $C_\mu z_j = \alpha_j^{-1} z_j$, we have

$$C_\mu f = \sum_{j=1}^{\infty} \beta_j \langle f, y_j \rangle_r y_j \qquad \forall f \in H_*^r(I)$$

with

$$y_j = \alpha_j^{-1/(r+1)} z_j \qquad \text{and} \qquad \beta_j = \alpha_j^{-1/(r+1)} \sim (j\pi)^{-2}. \qquad \square$$

EXAMPLE: *A problem-dependent measure.*   One might think that important directions for the original problem should be the important directions for our measure. If $s > 0$, we may choose $m' = m$, $s' = s$, and $A = L$. We choose our measure μ on $H_*^r(I)$ as a Gaussian measure with mean zero and correlation operator

$$C_\mu f = \sum_{j=1}^{\infty} \beta_j \langle f, y_j \rangle_r y_j \qquad \forall f \in H_*^r(I)$$

with $\beta_1 \geq \beta_2 \geq \cdots > 0$ any sequence such that

$$\sum_{j=1}^{\infty} \beta_j < \infty. \qquad \square$$

EXAMPLE: *A measure on the whole space $H^r(\Omega)$.*   We now give a measure with $H_*^r(\Omega) = H^r(\Omega)$. Choose $m' = s' = 2r$, and take

$$Av = \sum_{|\alpha|, |\beta| \leq r} (-1)^{|\alpha|} D^{\alpha+\beta} v$$

with boundary conditions

$$\partial_n^{j+r} v = 0 \quad \text{on } \partial\Omega \qquad (0 \leq j \leq r-1).$$

It is easy to check that $\| \cdot \|_r = \| \cdot \|_{H^r(\Omega)}$ and $H_*^r(\Omega) = H^r(\Omega)$. The measure μ is then chosen as for the problem-dependent measure. $\qquad \square$

EXAMPLE: *Periodic functions in one dimension.*　Let $I = (0,1)$, $m' = s' = 2$, and choose

$$Av = -v'' \quad \text{in } I$$

with boundary conditions

$$u(0) = u(1) = 0.$$

The eigenvalues and $L_2(I)$-orthonormal eigenvalues are given by

$$z_j(x) = \frac{1}{\sqrt{2}} \sin j\pi x,$$
$$\alpha_j = j^2 \pi^2.$$

Our measure μ is then chosen as for the problem-dependent measure.　　□

Notes and Remarks

NR 7.4.1:1 As mentioned earlier, we are mainly dealing with the absolute error criterion in this chapter. What can we say about the complexity of shift-invariant problems under other error criteria?

We first note that for a (m,s)-shift-invariant problem, all Θ-results for the absolute error criterion in the $\| \cdot \|_{H^l(\Omega)}$ norm also hold under the residual error criterion in the $\| \cdot \|_{H^{l-s}(\Omega)}$-norm. Thus, if we use the average case residual defined by

$$e^{\text{resid}}(\phi, N) = \left[\int_F \|f - L(\phi(Nf))\|^2_{H^{l-s}(\Omega)} \, \mu(df) \right]^{1/2},$$

then minimal error and complexity will be the same as for the absolute error criterion in the average case setting.

In what follows, we use the following result, which we prove later: for any shift-invariant problem, there is a positive constant ω such that

$$\text{comp}^{\text{avg}}(\varepsilon, \Lambda) = \Theta(\varepsilon^{-1/\omega}) \quad \text{as } \varepsilon \to 0.$$

It then follows that

$$\text{comp}^{\text{avg}}(b\varepsilon, \Lambda) = \Theta\big(\text{comp}^{\text{avg}}(\varepsilon, \Lambda)\big) \quad \text{as } \varepsilon \to 0$$

for any positive b.

Under the normalized error criterion, we let

$$e^{\text{nor}}(\phi, N) = \left[\int_F \frac{\|Sf - \phi(Nf)\|^2_{H^l(\Omega)}}{\|f\|^2_{H^r(\Omega)}} \right]^{1/2} \mu(df).$$

From Theorem 6.2.1 in Chapter 6 of IBC, we then have

$$\text{comp}^{\text{nor}}(\varepsilon, \Lambda) = \Theta\big(\text{comp}^{\text{avg}}(\varepsilon, \Lambda)\big) \quad \text{as } \varepsilon \to 0.$$

We may also use the relative error criterion, in which

$$e^{\mathrm{rel}}(\phi, N) = \left[\int_F \frac{\|Sf - \phi(Nf)\|^2_{H^l(\Omega)}}{\|Sf\|^2_{H^l(\Omega)}} \right]^{1/2} \mu(df).$$

Using Theorem 6.1.2 in Chapter 6 of IBC, we find that

$$\mathrm{comp}^{\mathrm{rel}}(\varepsilon, \Lambda) = \Theta\big(\mathrm{comp}^{\mathrm{avg}}(\varepsilon, \Lambda)\big) \qquad \text{as } \varepsilon \to 0.$$

In summary, we find that once we know the average case complexity of a shift-invariant problem under the absolute error criterion, we can determine the complexity under residual, normalized, and relative error criteria.

NR 7.4.1:2 The construction of our family $\{H^t_*(\Omega)\}_{t \in \mathbb{R}}$, requires that $s' > 0$. If $s' = 0$, this construction does not work, because all the spaces in the scale turn out to be $L_2(\Omega)$. In particular, this technique cannot be used to define a problem-dependent scale (and measure) for the Fredholm problem of the second kind. However, if the kernel k of the compact operator K has only limited smoothness, the eigenfunctions of K^*K can be used to define a problem-dependent scale and measure for the Fredholm problem of the second kind.

7.4.2 Continuous linear information

In this subsection, we assume that arbitrary continuous linear information is permissible, i.e. $\Lambda = \Lambda^*$. For future reference, we recall the following result for the worst case setting:

THEOREM 7.4.2.1. *Let $\Lambda = \Lambda^*$. In the worst case setting, the nth minimal radius is given by*

$$r^{\mathrm{wor}}(n) = \Theta(q \cdot n^{-(r+s-l)/d}) \qquad \text{as } n \to \infty$$

and the ε-complexity is given by

$$\mathrm{comp}^{\mathrm{wor}}(\varepsilon) = \Theta\left(\left(\frac{q}{\varepsilon}\right)^{d/(r+s-l)} \right) \qquad \text{as } \varepsilon \to 0.$$

PROOF. The proof of the estimate for $r^{\mathrm{wor}}(n)$ is almost the same as in Theorem 6.2.2.1. Since the problem is a linear problem in a Hilbert space setting, the estimate of $\mathrm{comp}^{\mathrm{wor}}(\varepsilon)$ follows from the results in §4.4.5. ☐

Again, we note that complexity is exponential in d for shift-invariant problems. Moreover, the problem is non-convergent (i.e. the ε-complexity is infinite for sufficiently small ε) whenever $r + s = l$. What does this mean? Suppose we decide to measure error in the most "natural" norm for our problem and that we only have enough smoothness in our problem elements to guarantee existence and uniqueness of solutions. Then the problem is nonconvergent, since $r + s = l$. More precisely, we have the following:

   (i) For the 2mth-order elliptic boundary value problem, the natural norm to measure error is the energy norm, which is equivalent to $\|\cdot\|_{H^m_E(\Omega)}$. For $s = 2m$ and $l = m$, we thus find that the problem is non-convergent if $r = -m$.

  (ii) For elliptic systems with parameter t, the natural norm (corresponding to the energy norm for elliptic boundary value problems) is the $L_2(\Omega)$-norm. For $s = t$ and $l = 0$, we see that the problem is non-convergent if $r = -t$.

 (iii) For the Fredholm problem of the second kind, the natural norm is the $L_2(\Omega)$-norm. For $s = l = 0$, we see that the problem is non-convergent if $r = 0$.

Hence we see that if we measure error in the natural norm for these problems, *the complexity is infinite if the problem elements are only smooth enough to guarantee existence and uniqueness of solutions.*

Average case complexity. Now that we have defined our measure, we can (in principle) apply the general techniques of the previous section to find optimal algorithms, as well as determine the minimal error and complexity. However, this is difficult to do, except in the simplest cases. For an example, see **NR 7.4.2:1**, where we show that for the problem-dependent measure, the nth minimal error algorithm is S applied to the truncated standard Fourier series of the problem element.

Since the complexity for finite q is roughly the same as that for $q = \infty$, we need only consider the case $q = \infty$ in what follows.

We first compute the nth minimal average case radius:

THEOREM 7.4.2.2. *The nth minimal average radius is given by*

$$r^{\mathrm{avg}}(n) = \Theta\left(\left(\sum_{j=n+1}^{\infty} \beta_j \alpha_j^{-2(r+s-l)/s'}\right)^{1/2}\right) \qquad \text{as } n \to \infty.$$

PROOF. Let $E: H^r_*(\Omega) \to H^{l-s}(\Omega)$ be the identity injection $Ef \equiv f$. Write

$$e^{\mathrm{avg}}(\phi, N, E) = \left[\int_{H^r_*(\Omega)} \|f - \phi(Nf)\|^2_{H^{l-s}(\Omega)} \mu(df)\right]^{1/2},$$
$$r^{\mathrm{avg}}(N, E) = \inf_{\phi \text{ using } N} e^{\mathrm{avg}}(\phi, N, E),$$
$$r^{\mathrm{avg}}(n, E) = \inf_{N: \operatorname{card} N \leq n} r^{\mathrm{avg}}(N, E).$$

That is, we wish to approximate elements of $H^r_*(\Omega)$ in the $H^{l-s}(\Omega)$-norm on the average.

We claim that

$$r^{\text{avg}}(n) = \Theta\big(r^{\text{avg}}(n, E)\big) \qquad \text{as } n \to \infty.$$

Indeed, let N be information of cardinality n and let σ_N be the μ-spline for N. Then $S\sigma_N$ is an optimal error algorithm for our variational boundary-value problem, i.e.

$$r^{\text{avg}}(N) = e^{\text{avg}}(S\sigma_N, N) = \left[\int_{H_*^r(\Omega)} \|Sf - S(\sigma_N(Nf))\|_{H^l(\Omega)}^2 \, \mu(df) \right]^{1/2}$$

and σ_N is an optimal error algorithm for the auxiliary approximation problem mentioned in the previous paragraph, i.e.

$$r^{\text{avg}}(N, E) = e^{\text{avg}}(\sigma_N, N, E) = \left[\int_{H_*^r(\Omega)} \|f - \sigma_N(Nf)\|_{H^{l-s}(\Omega)}^2 \, \mu(df) \right]^{1/2}.$$

Since the problem is shift-invariant, there exists a constant $\theta \geq 1$ such that

$$\theta^{-1}\|g\|_{H^{l-s}(\Omega)} \leq \|Sg\|_{H^l(\Omega)} \leq \theta\,\|g\|_{H^{l-s}(\Omega)} \qquad \forall\, g \in H^{l-s}(\Omega).$$

Combining these last three formulas, we see that

$$\theta^{-1} r^{\text{avg}}(N, E) \leq r^{\text{avg}}(N) \leq \theta\, r^{\text{avg}}(N, E).$$

Taking the infimum over all information N of cardinality at most n, we find

$$\theta^{-1} r^{\text{avg}}(n, E) \leq r^{\text{avg}}(n) \leq \theta\, r^{\text{avg}}(n, E),$$

as required.

Hence to complete the proof of this theorem, we need only show that

$$r^{\text{avg}}(n, E) = \left(\sum_{j=n+1}^{\infty} \beta_j \alpha_j^{-2(r+s-l)/s'} \right)^{1/2}.$$

From Theorem 7.3.2, we see that

$$r^{\text{avg}}(n, E) = \left(\sum_{j=n+1}^{\infty} \kappa_j \right)^{1/2},$$

where κ_j is the jth eigenvalue of the operator $EC_\mu E^*$ on the space $H^{l-s}(\Omega)$. Since $\{\alpha_j^{(s-l)/s'} z_j\}_{j=1}^{\infty}$ is an orthonormal basis for $H^{l-s}(\Omega)$, our expression for $r^{\text{avg}}(n, E)$ follows once we show that

$$EC_\mu E^* z_j = \beta_j \alpha_j^{-2(r+s-l)/s'} z_j.$$

Since E^* is the adjoint of E, we have

$$\langle E^* z_j, \alpha_i^{-r/s'} z_i \rangle_r = \langle z_j, \alpha_i^{-r/s'} z_i \rangle_{l-s},$$

and so

$$
\begin{aligned}
E^* z_j &= \sum_{i=1}^{\infty} \langle E^* z_j, \alpha_i^{-r/s'} z_i \rangle_r \alpha_i^{-r/s'} z_i \\
&= \sum_{i=1}^{\infty} \langle z_j, z_i \rangle_{l-s} \alpha_i^{-2r/s'} z_i \\
&= \sum_{i=1}^{\infty} \sum_{q=1}^{\infty} \alpha_q^{2(l-s)/s'} \langle z_j, z_q \rangle_0 \langle z_i, z_q \rangle_0 \alpha_i^{-2r/s'} z_i \\
&= \alpha_j^{-2(r+s-l)/s'} z_j.
\end{aligned}
$$

Hence

$$E C_\mu E^* z_j = \alpha_j^{-2(r+s-l)/s'} C_\mu z_j = \beta_j \alpha_j^{-2(r+s-l)/s'} z_j,$$

as required.   □

We now specialize this result to the case of $\beta_j = \Theta(j^{-p})$ as $j \to \infty$ for some $p > 1$. This case includes Wiener measure.

COROLLARY 7.4.2.1. *Suppose that*

$$\beta_j = \Theta(j^{-p}) \qquad as\ j \to \infty$$

for some $p > 1$. Let

$$\omega^* = \frac{r+s-l}{d} + \frac{p-1}{2}.$$

Then

$$r^{\mathrm{avg}}(n) = \Theta(n^{-\omega^*}) \qquad as\ n \to \infty$$

and

$$\mathrm{comp}^{\mathrm{avg}}(\varepsilon) = \Theta(\varepsilon^{-1/\omega^*}) \qquad as\ \varepsilon \to 0.$$

Furthermore, if $n = m^{\mathrm{avg}}(\varepsilon) = \Theta(\varepsilon^{-1/\omega^*})$, then the algorithm ϕ_n^* using information N_n^* is an almost optimal complexity algorithm.

PROOF. This follows immediately from the previous theorem, the estimate $\alpha_j = \Theta(j^{s'/d})$ as $j \to \infty$, and the relation between minimal error and optimal complexity algorithms mentioned in the previous section.   □

Hence, if the jth eigenvalue of the correlation operator of our measure goes to zero like $\Theta(j^{-p})$, then the nth minimal error improves by a factor of $\Theta(n^{-(p-1)/2})$ when going from the worst case to the average case. The problem complexity improves analogously when changing settings. In other words, using the measure μ in the average case setting is equivalent to increasing the smoothness of the problem elements by $\frac{1}{2}d(p-1)$ in the worst case setting.

This means that if p is only mildly dependent on d, then our problem is no longer intractable. For instance, suppose that p is independent of d. Then $r^{\mathrm{avg}}(n) = O(n^{-(p-1)/2})$ as $n \to \infty$, and so $\mathrm{comp}^{\mathrm{avg}}(\varepsilon) = O(\varepsilon^{-2/(p-1)})$ as $\varepsilon \to 0$. So, the complexity is no longer exponential in d, showing that the problem is tractable in the average case setting. Of course, these O-estimates are gross overestimates, but they *do* serve to show that the problem complexity is not inherently exponential.

Note that we have only determined minimal error and problem complexity in the average case for a special class of measures. Although this class includes many measures of common interest, the next step will be to extend these results to include measures that cannot be obtained via the techniques of this section.

Finite element methods. Of course, the general prescription for finding an average case optimal algorithm is useless except for the simplest problems. Once again guided by the success of the finite element method for shift-invariant problems in the worst case setting, we now investigate the FEM in the average case setting.

Recall that we are trying to solve a shift-invariant problem, as defined above. Suppose that $\{\mathcal{T}_n\}_{n=1}^{\infty}$ is a quasi-uniform family of partitions of Ω. For any $n \in \mathbb{P}$, we let $\mathcal{S}_{n,k}$ be an n-dimensional subspace of $H_E^m(\Omega)$ consisting of $\mathcal{T}_n$-piecewise polynomials of degree k. Let $N_{n,k}^{\mathrm{FE}}$ denote *finite element information* (FEI) of cardinality n consisting of $L_2(\Omega)$-inner products with a basis for $\mathcal{S}_{n,k}$. We say that an algorithm $\phi_{n,k}^{\mathrm{FE}}$ using $N_{n,k}^{\mathrm{FE}}$ is a *finite element method* (FEM) of degree k if the following hold:

(i) For any $f \in F$, we have $\phi_{n,k}^{\mathrm{FE}}(N_{n,k}^{\mathrm{FE}}f) \in \mathcal{S}_{n,k}$.

(ii) We have the following standard error estimate: if any $a \geq -m$ and $f \in H^a(\Omega)$, then

$$\|Sf - \phi_{n,k}^{\mathrm{FE}}(N_{n,k}^{\mathrm{FE}}f)\|_{H^l(\Omega)} \leq Cn^{-(\min\{k+1,a+s\}-l)/d}\|f\|_{H^a(\Omega)},$$

for some positive constant C, independent of f and n.

(iii) We have the following standard lower bound: if $K \in \mathcal{T}_n$, with $\partial K \cap \partial\Omega = \emptyset$, and $u \in P_{k+1}(K)$, then

$$\inf_{s \in P_k(K)} |u - s|_{H^l(K)} \geq Cn^{-(k+1-l)/d}|u|_{H^{k+1}(K)},$$

for some positive constant C, independent of u, K, and n.

Of course, the standard FEMs (or LSFEMs) described in Chapter 5 and §§6.2 and 6.3 are all FEMs under this definition. Using these conditions, it is easy to see that we have the following result in the worst case setting:

THEOREM 7.4.2.3.   *In the worst case setting, let*

$$\eta_{\mathrm{FE}} = \frac{\min\{k+1, r+s\} - l}{d}.$$

(i)   *The worst case error of the FEM of degree k is*

$$e^{\mathrm{wor}}(\phi_{n,k}^{\mathrm{FE}}, N_{n,k}^{\mathrm{FE}}) = \Theta(q \cdot n^{-\eta_{\mathrm{FE}}}) \qquad \text{as } n \to \infty,$$

   and so the FEM of degree k is an almost minimal error algorithm iff $k \geq r + s - 1$.

(ii)   *Let*

$$\mathrm{FEM}_k^{\mathrm{wor}}(\varepsilon) = \inf\{\,\mathrm{cost}^{\mathrm{wor}}(\phi_{n,k}^{\mathrm{FE}}, N_{n,k}^{\mathrm{FE}}) : e^{\mathrm{wor}}(\phi_{n,k}^{\mathrm{FE}}, N_{n,k}^{\mathrm{FE}}) \leq \varepsilon\,\}$$

   denote the cost of using the FEM of degree k to compute an ε-approximation. Then

$$\mathrm{FEM}_k^{\mathrm{wor}}(\varepsilon) = \Theta\left(\left(\frac{q}{\varepsilon}\right)^{1/\eta_{\mathrm{FE}}}\right) \qquad \text{as } \varepsilon \to 0.$$

   Hence the FEM of degree k is an almost optimal complexity algorithm in the worst case setting iff

$$k \geq r + s - 1. \qquad \qquad \qquad \square$$

Hence, an FEM of sufficiently high degree is almost optimal in the worst case setting. We now show that this is also true in the average case setting. In the remainder of this section, we will assume that

$$\beta_j = \Theta(j^{-p}) \qquad \text{as } j \to \infty$$

for some $p > 1$. Let

$$\omega_{\mathrm{FE}} = \frac{\min\{k+1, r+s+\frac{1}{2}d(p-1)\} - l}{d}.$$

THEOREM 7.4.2.4.   *The average case error of the FEM of degree k satisfies*

$$e^{\mathrm{avg}}(\phi_{n,k}^{\mathrm{FE}}, N_{n,k}^{\mathrm{FE}}) = \Theta(n^{-\omega_{\mathrm{FE}}}) \qquad \text{as } n \to \infty.$$

PROOF.   We first show that

$$e^{\mathrm{avg}}(\phi_{n,k}^{\mathrm{FE}}, N_{n,k}^{\mathrm{FE}}) = O(n^{-\omega_{\mathrm{FE}}}) \qquad \text{as } n \to \infty \qquad (1)$$

The proof for the case $\omega_{\text{FE}} = 0$ being trivial, we assume without loss of generality that $\omega_{\text{FE}} > 0$.

We write $P_n f = \phi_{n,k}^{\text{FE}}(N_{n,k}^{\text{FE}} f)$ for $f \in F$. Let $\sigma_2(\cdot)$ denote the Hilbert-Schmidt norm, i.e. for an operator $K \colon H_*^r(\Omega) \to H^l(\Omega)$ and *any* orthonormal sequence $e_1, e_2 \ldots$ in $H_*^r(\Omega)$,

$$\sigma_2(K) = \left(\sum_{j=1}^{\infty} \| K e_j \|_{H^l(\Omega)}^2 \right)^{1/2},$$

see Chapter 1 of Kuo (1975). It is easy to check that

$$\left[\int_{H_*^r(\Omega)} \| T f \|_{H^l(\Omega)}^2 \, \mu(df) \right]^{1/2} = \sigma_2(T C_\mu^{1/2})$$

for any bounded linear operator $T \colon H_*^r(\Omega) \to H_E^l(\Omega)$, and so

$$e^{\text{avg}}(\phi_{n,k}^{\text{FE}}, N_{n,k}^{\text{FE}}) = \sigma_2\big((S - P_n) C_\mu^{1/2}\big).$$

In what follows, we assume without loss of generality that $n = 2^q$ for some integer q.

We claim that

$$\sigma_2\big((S - P_n) C_\mu^{1/2}\big) = \sigma_2\left(\sum_{j=0}^{\infty} (P_{2^{q+j+1}} - P_{2^{q+j}}) C_\mu^{1/2} \right).$$

Indeed, let $\{e_i\}_{i=1}^{\infty}$ be an orthonormal basis for $H_*^r(\Omega)$, so that

$$\left(\sigma_2\big((S - P_n) C_\mu^{1/2}\big) \right)^2 = \sum_{i=1}^{\infty} \| (S - P_n) C_\mu^{1/2} e_j \|_{H^l(\Omega)}^2$$

and

$$\left(\sigma_2\left(\sum_{j=0}^{\infty} (P_{2^{q+j+1}} - P_{2^{q+j}}) S C_\mu^{1/2} \right) \right)^2$$

$$= \sum_{i=1}^{\infty} \left\| \sum_{j=0}^{\infty} (P_{2^{q+j+1}} - P_{2^{q+j}}) C_\mu^{1/2} e_i \right\|_{H^l(\Omega)}^2.$$

Let

$$\rho_a = \| (S - P_{2^{a+1}}) C_\mu^{1/2} \|,$$

where $\| \cdot \|$ denotes the operator norm from $H_*^r(\Omega)$ into $H_E^l(\Omega)$. Since $C_\mu^{1/2}$ is compact and $\lim_{a\to\infty} P_{2^{a+1}} f = Sf$ for all $f \in H_*^r(\Omega)$, we see that $\lim_{a\to\infty} \rho_a = 0$. Now for any index i, we have

$$
\left| \|(S - P_n)C_\mu^{1/2}e_i\|_{H^l(\Omega)} - \left\| \sum_{j=0}^{t}(P_{2^{q+j+1}} - P_{2^{q+j}})C_\mu^{1/2}e_i \right\|_{H^l(\Omega)} \right|
$$

$$
\leq \left\| (S - P_n)C_\mu^{1/2}e_i - \sum_{j=0}^{t}(P_{2^{q+j+1}} - P_{2^{q+j}})C_\mu^{1/2}e_i \right\|_{H^l(\Omega)}
$$

$$
= \|(S - P_{2^{q+t+1}})C_\mu^{1/2}e_i\|_{H^l(\Omega)} \leq \rho_{q+t},
$$

and hence

$$
\left| \|(S - P_n)C_\mu^{1/2}e_i\|_{H^l(\Omega)}^2 - \left\| \sum_{j=0}^{t}(P_{2^{q+j+1}} - P_{2^{q+j}})C_\mu^{1/2}e_i \right\|_{H^l(\Omega)}^2 \right|
$$

$$
\leq \left[\|(S - P_n)C_\mu^{1/2}e_i\|_{H^l(\Omega)} + \left\| \sum_{j=0}^{t}(P_{2^{q+j+1}} - P_{2^{q+j}})C_\mu^{1/2}e_i \right\|_{H^l(\Omega)} \right]
$$

$$
\times \left| \|(S - P_n)C_\mu^{1/2}e_i\|_{H^l(\Omega)} - \left\| \sum_{j=0}^{t}(P_{2^{q+j+1}} - P_{2^{q+j}})C_\mu^{1/2}e_i \right\|_{H^l(\Omega)} \right|
$$

$$
\leq \left(2\|(S - P_n)C_\mu^{1/2}e_i\|_{H^l(\Omega)} + \rho_{q+t} \right)\rho_{q+t}
$$

$$
\leq A\rho_{q+t},
$$

for

$$
A = 2\|(S - P_n)C_\mu^{1/2}\| + \sup_a \rho_a.
$$

Letting $t \to \infty$, we have $\rho_{q+t} \to 0$, and so

$$
\|(S - P_n)C_\mu^{1/2}e_i\|_{H^l(\Omega)}^2 = \left\| \sum_{j=0}^{\infty}(P_{2^{q+j+1}} - P_{2^{q+j}})C_\mu^{1/2}e_i \right\|_{H^l(\Omega)}^2.
$$

Summing over all indices i, the result follows as claimed.

Hence

$$
\sigma_2\big((S - P_n)C_\mu^{1/2}\big) = \sigma_2\left(\sum_{j=0}^{\infty}(P_{2^{q+j+1}} - P_{2^{q+j}})C_\mu^{1/2} \right)
$$

$$
\leq \sum_{j=0}^{\infty} \sigma_2\big((P_{2^{q+j+1}} - P_{2^{q+j}})C_\mu^{1/2}\big),
$$

since σ_2 is an operator norm. Moreover, we have

$$\sigma_2\big((P_{2^q+j+1} - P_{2^q+j})C_\mu^{1/2}\big) \le \sigma_2\big((S - P_{2^q+j})C_\mu^{1/2}\big) + \sigma_2\big((S - P_{2^q+j+1})C_\mu^{1/2}\big).$$

Now there exists a positive constant C, independent of q and j, such that

$$\|(S - P_{2^q+j+1})f\|_{H^l(\Omega)} \le C\|(S - P_{2^q+j})f\|_{H^l(\Omega)} \qquad \forall f \in H^r(\Omega),$$

from which it follows that

$$\sigma_2\big((S - P_{2^q+j+1})C_\mu^{1/2}\big) \le C\sigma_2\big((S - P_{2^q+j})C_\mu^{1/2}\big).$$

Hence

$$\sigma_2\big((S - P_n)C_\mu^{1/2}\big) \le C \sum_{j=0}^{\infty} \sigma_2\big((S - P_{2^q+j})C_\mu^{1/2}\big).$$

Recall that $y_1, y_2, \ldots$ is an orthonormal sequence in $H_*^r(\Omega)$, where $y_i = \alpha_i^{-r/s'} z_i$ for all indices i. Combining our results so far, we see that

$$e^{\text{avg}}(\phi_{n,k}^{\text{FE}}, N_{n,k}^{\text{FE}}) = \sigma_2\big((S - P_n)C_\mu^{1/2}\big)$$
$$\le C \sum_{j=0}^{\infty} \left[\sum_{i=1}^{\infty} \|(S - P_{2^q+j})C_\mu^{1/2} y_i\|_{H^l(\Omega)}^2 \right]^{1/2}. \qquad (2)$$

We claim that there is a positive constant C, independent of q and j, such that

$$\left[\sum_{i=1}^{\infty} \|(S - P_{2^q+j})C_\mu^{1/2} y_i\|_{H^l(\Omega)}^2 \right]^{1/2} \le C \cdot (2^{q+j})^{-\omega_{\text{FE}}}. \qquad (3)$$

Indeed, using the definition of ω_{FE} and the standard error estimate with $a = r + \frac{1}{2}d(p-1)$, we see that there is a positive constant C, independent of q, such that, for any indices i and j,

$$\|(S - P_{2^q+j})C_\mu^{1/2} y_i\|_{H^l(\Omega)} \le C(2^{q+j})^{-\omega_{\text{FE}}} \|C_\mu^{1/2} y_i\|_{H^{r+d(p-1)/2}(\Omega)}.$$

But

$$\|C_\mu^{1/2} y_i\|_{H^{r+d(p-1)/2}(\Omega)} = \beta_i^{1/2} \|y_i\|_{H^{r+d(p-1)/2}(\Omega)}$$
$$= \alpha_i^{-r/s'} \beta_i^{1/2} \|z_i\|_{H^{r+d(p-1)/2}(\Omega)}.$$

Since $z_i \in H_*^{r+d(p-1)/2}(\Omega)$, there is a constant C, independent of i, such that

$$\|z_i\|_{H^{r+d(p-1)/2}(\Omega)} \le C\|z_i\|_{r+d(p-1)/2}$$
$$= C\alpha_i^{(r+d(p-1)/2)/s'} \|z_i\|_{L_2(\Omega)} = C\alpha_i^{(r+d(p-1)/2)/s'}.$$

So,

$$\|C_\mu^{1/2} y_i\|_{H^{r+d(p-1)/2}(\Omega)} \le C\beta_i^{1/2} \alpha_i^{(d(p-1)/2)/s'} \le C i^{-(p+1)/2},$$

the latter since $\beta_i = \Theta(i^{-p})$ and $\alpha_i = \Theta(i^{s'/d})$ as $i \to \infty$. Hence

$$\left[\sum_{i=1}^{\infty} \|(S - P_{2^{q+j}})C_\mu^{1/2} y_i\|_{H^l(\Omega)}^2\right]^{1/2} \le C(2^{q+j})^{-\omega_{\mathrm{FE}}} \left[\sum_{i=1}^{\infty} i^{-(p+1)}\right]^{1/2}.$$

Since $p > 0$, the infinite series $\sum_{i=1}^{\infty} i^{-(p+1)}$ converges, and so (3) holds, as claimed.

Using (2) and (3), we have

$$e^{\mathrm{avg}}(\phi_{n,k}^{\mathrm{FE}}, N_{n,k}^{\mathrm{FE}}) \le C \sum_{j=0}^{\infty} (2^{q+j})^{-\omega_{\mathrm{FE}}} = C n^{-\omega_{\mathrm{FE}}} \sum_{j=0}^{\infty} (2^{\omega_{\mathrm{FE}}})^{-j}.$$

Since $\omega_{\mathrm{FE}} > 0$, the infinite series is convergent, and so

$$e^{\mathrm{avg}}(\phi_{n,k}^{\mathrm{FE}}, N_{n,k}^{\mathrm{FE}}) = O(n^{-\omega_{\mathrm{FE}}}) \qquad \text{as } n \to \infty,$$

as required, completing the proof of (1).

We next show that

$$e^{\mathrm{avg}}(\phi_{n,k}^{\mathrm{FE}}, N_{n,k}^{\mathrm{FE}}) = \Omega(n^{-\omega_{\mathrm{FE}}}) \qquad \text{as } n \to \infty \qquad (4)$$

Since the algorithm $\phi_{n,k}^{\mathrm{FE}}$ uses information $N_{n,k}^{\mathrm{FE}}$ of cardinality n, Corollary 7.4.2.1 implies that

$$e^{\mathrm{avg}}(\phi_{n,k}^{\mathrm{FE}}, N_{n,k}^{\mathrm{FE}}) \ge r^{\mathrm{avg}}(n) = \Theta(n^{-\omega^*}) \qquad \text{as } n \to \infty.$$

Hence, if we can show

$$e^{\mathrm{avg}}(\phi_{n,k}^{\mathrm{FE}}, N_{n,k}^{\mathrm{FE}}) = \Omega(n^{-(k+1-l)/d}) \qquad \text{as } n \to \infty, \qquad (5)$$

then the lower bound (4) will follow immediately.

We now prove (5). As in the proof of Theorem 5.4.2, let Ω^0 be the interior of a hypercube whose closure is contained in Ω, and let

$$\mathcal{T}_n^0 = \{\, K \in \mathcal{T}_n : K \subseteq \overline{\Omega^0} \,\}.$$

Fix $n_0 \in \mathbb{P}$, and let Ω^1 be an open region such that

$$\overline{\Omega^1} \subseteq \bigcup_{K \in \mathcal{T}_n^0} K \qquad \forall n \ge n_0.$$

Let

$$V = \left\{ v \in C_0^\infty(\Omega) : v|_{\Omega^0} \in P_{k+1}(\Omega^0) \setminus P_k(\Omega^0) \right\},$$

i.e. V consists of C_0^∞-functions, whose restriction to Ω^0 are polynomials of *exact* degree $k+1$.

Choose nonzero $v \in V$, and let $g = Lv$. Note that $g \in C_0^\infty(\Omega)$. Furthermore, g is nonzero since v is nonzero. We claim that there exists (nonzero) $f \in H_*^r(\Omega)$ such that $C_\mu^{1/2} f = g$. Indeed, write

$$g = \sum_{i=1}^\infty \langle g, z_i \rangle_{L_2(\Omega)} z_i.$$

Define

$$f = \sum_{i=1}^\infty \beta_i^{-1/2} \langle g, z_i \rangle_{L_2(\Omega)} z_i.$$

Of course, $C_\mu^{1/2} f = g$; it only remains to show that $f \in H_*^r(\Omega)$. But

$$\|f\|_r^2 = \sum_{i=1}^\infty \beta_i^{-1} \alpha_i^{2r/s'} \langle g, z_i \rangle_{L_2(\Omega)}^2.$$

Since

$$\beta_i^{-1} = \Theta(i^p) = \Theta(\alpha_i^{pd/s'}) \qquad \text{as } i \to \infty,$$

we have

$$\|f\|_r^2 \leq C \sum_{i=1}^\infty \alpha_i^{(2r+pd)/s'} \langle g, z_i \rangle_{L_2(\Omega)}^2 = C \|g\|_{r+pd/2}^2.$$

Since $g \in C_0^\infty(\Omega)$, we know that $\|g\|_{r+pd/2} < \infty$, and so $\|f\|_r < \infty$, i.e. $f \in H_*^r(\Omega)$.

Let $e_1 = f/\|f\|_{H^r(\Omega)}$, and let $e_2, e_3, \dots$ be an orthonormal basis for the orthogonal complement of $\mathrm{span}\{e_1\}$ in $H_*^r(\Omega)$. As before, we find that

$$e^{\mathrm{avg}}(\phi_{n,k}^{\mathrm{FE}}, N_{n,k}^{\mathrm{FE}}) = \sigma_2\big((S-P_n)C_\mu^{1/2}\big) = \left[\sum_{i=1}^\infty \|(S-P_n)C_\mu^{1/2} e_i\|_{H^l(\Omega)}^2 \right]^{1/2}$$

$$\geq \frac{\|(S-P_n)C_\mu^{1/2} f\|_{H^l(\Omega)}}{\|f\|_{H^r(\Omega)}} = C \|(S-P_n)C_\mu^{1/2} f\|_{H^l(\Omega)},$$

for some positive constant C, independent of n.

Since $v = SC_\mu^{1/2} f$ is a member of $\mathcal{V}$ and $P_n C_\mu^{1/2} f \in \mathcal{S}_{n,k}$, we may use the standard lower bound (as in the proof of Theorem 5.4.2) to find that

$$\|(S - P_n)C_\mu^{1/2} f\|_{H^l(\Omega)} \geq \inf_{s \in \mathcal{S}_{n,k}} |v - s|_{H^l(\Omega)}$$

$$\geq \left[\sum_{K \in \mathcal{T}_n^0} \inf_{s \in P_k(K)} |v - s|_{H^l(K)}^2 \right]^{1/2}$$

$$\geq Cn^{-(k+1-l)/d} \left[\sum_{K \in \mathcal{T}_n^0} |v|_{H^{k+1}(K)}^2 \right]^{1/2}$$

$$\geq Cn^{-(k+1-l)/d} |v|_{H^{k+1}(\Omega^1)} .$$

Combining this result with the previous inequality, we see that (5) follows immediately, completing the proof of (4) and of the theorem.   □

Comparing the previous two theorems, we immediately have

COROLLARY 7.4.2.2.   *The FEM of degree k is an almost minimal average error algorithm iff $k \geq s + r + \frac{1}{2}d(p-1) - 1$.*   □

For $\varepsilon > 0$, we now wish to compute

$$\mathrm{FEM}_k^{\mathrm{avg}}(\varepsilon) = \inf\{ \mathrm{cost}^{\mathrm{avg}}(\phi_{n,k}^{\mathrm{FE}}, N_{n,k}^{\mathrm{FE}}) : e^{\mathrm{avg}}(\phi_{n,k}^{\mathrm{FE}}, N_{n,k}^{\mathrm{FE}}) \leq \varepsilon \},$$

the cost of using the FEM to compute an ε-approximation in the average case setting. Of course, the FEM is a linear algorithm. Since the FEM of sufficiently high degree is an almost minimal error algorithm, we may use the previous two theorems and the results in §7.3 to prove the following

COROLLARY 7.4.2.3.   *The average case cost of computing an ε-approximation with the FEM of degree k is*

$$\mathrm{FEM}_k^{\mathrm{avg}}(\varepsilon) = \Theta(\varepsilon^{-1/\omega_{\mathrm{FE}}}) \qquad \text{as } \varepsilon \to 0.$$

Hence the FEM of degree k is an almost optimal complexity algorithm iff $k \geq s + r + \frac{1}{2}d(p-1) - 1$.   □

We now comment about the FEM in the average case setting. First, note that there is once again a large asymptotic penalty to be paid when we use a FEM of too low a degree. That is, for $k < s + r + \frac{1}{2}d(p-1) - 1$, we have

$$\frac{\mathrm{FEM}_k^{\mathrm{avg}}(\varepsilon)}{\mathrm{comp}^{\mathrm{avg}}(\varepsilon)} = \Theta(\varepsilon^{-d\delta}) \qquad \text{as } \varepsilon \to 0,$$

where

$$\delta = \frac{1}{k+1-l} - \frac{1}{r+s-l+\frac{1}{2}d(p-1)} > 0.$$

Thus,

$$\lim_{\varepsilon \to 0} \frac{\text{FEM}_k^{\text{avg}}(\varepsilon)}{\text{comp}^{\text{avg}}(\varepsilon)} = \infty.$$

Next, note that minimal error and complexity in the average case setting for the measure C_μ with eigenvalues $\beta_j = \Theta(j^{-p})$ and problem elements of smoothness r are the same as that for the worst case setting and problem elements of smoothness $r' = r + \frac{1}{2}d(p-1)$. Similarly, our upper bound for the average case error of the FEM is the same as that in the worst case setting, except that we replace r by r'. In other words, allowing the use of the measure μ effectively increases the smoothness of the problem by $\frac{1}{2}d(p-1)$.

Finally, note that we have only determined the average case complexity of using the FEM to compute an ε-approximation for a specific class of measures. It will be important to extend these results to include measures that are not members of this class.

Notes and Remarks

NR 7.4.2:1 Let $q = \infty$. We claim that for the problem-dependent measure

$$r^{\text{avg}}(n) = \left(\sum_{j=n+1}^{\infty} \beta_j \alpha_j^{-(r+s-l)/s} \right)^{1/2},$$

nth optimal information is

$$N_n^* f = \begin{bmatrix} \langle f, z_1 \rangle_{L_2(\Omega)} \\ \vdots \\ \langle f, z_n \rangle_{L_2(\Omega)} \end{bmatrix},$$

and

$$\phi_n^*(N_n^* f) = \sum_{j=1}^{n} \alpha_j^{-1} \langle f, z_j \rangle_{L_2(\Omega)} z_j = S\left(\sum_{j=1}^{n} \langle f, z_j \rangle_{L_2(\Omega)} z_j \right)$$

is an nth minimal error algorithm. (Note that this algorithm is independent of the eigenvalues $\{\beta_j\}_{j=1}^{\infty}$ defining the measure μ.)

Let

$$u_j = \alpha_j^{-l/s} z_j \qquad (j = 1, 2, \ldots),$$

so that $\{u_j\}_{j=1}^{\infty}$ is an orthonormal basis for $H_0^l(\Omega)$. We claim that for any index j,

$$C_\nu u_j = \beta_j \alpha_j^{-2(r+s-l)/s} u_j.$$

Indeed, first note that $L z_j = \alpha_j z_j$ implies that $S z_j = \alpha_j^{-1} z_j$. Since we must have $S^* z_j = \sigma_j z_j$ for some positive scalar σ_j, we compute

$$\alpha_j^{-1} \|z_j\|_l^2 = \langle \alpha_j^{-1} z_j, z_j \rangle_l = \langle S z_j, z_j \rangle_l = \langle z_j, S^* z_j \rangle_r$$
$$= \sigma_j \langle z_j, z_j \rangle_r = \sigma_j \|z_j\|_r^2,$$

and so

$$\sigma_j = \alpha_j^{-1} \frac{\|z_j\|_l^2}{\|z_j\|_r^2} = \alpha_j^{2(l-r)/s-1}.$$

Hence

$$C_\nu z_j = SC_\mu S^* z_j = \beta_j \alpha_j^{2(l-r)/s-2} z_j,$$

and so

$$C_\nu u_j = \alpha_j^{-l/s} C_\nu z_j = \beta_j \alpha_j^{-2(r+s-l)/s} u_j,$$

as claimed.

From the general theory (see §7.2), we see that the nth minimal radius is as claimed, that

$$\tilde{N}_n^* f = \begin{bmatrix} \beta_1 \alpha_1^{-2(r+s-l)/s} \langle Sf, u_1 \rangle_l \\ \vdots \\ \beta_n \alpha_n^{-2(r+s-l)/s} \langle Sf, u_n \rangle_l \end{bmatrix}$$

is nth optimal information, and that

$$\tilde{\phi}_n^*(\tilde{N}_n^* f) = \sum_{j=1}^n \langle Sf, u_j \rangle_l u_j$$

is an nth minimal error algorithm. We can simplify the formula for the algorithm by noting that

$$\langle Sf, z_j \rangle_l = \alpha_j^{2l/s} \langle Sf, z_j \rangle_0 = \alpha_j^{2l/s} \langle f, Sz_j \rangle_0 = \alpha_j^{2l/s-1} \langle f, z_j \rangle_0$$

and so

$$\langle Sf, u_j \rangle_l = \alpha_j^{-l/s} \langle Sf, z_j \rangle_l = \alpha_j^{l/s-1} \langle f, z_j \rangle_0.$$

Since

$$u_j = \alpha_j^{-l/s} z_j,$$

we thus have

$$\tilde{\phi}_n^*(\tilde{N}_n^* f) = \sum_{j=1}^n \alpha_j^{-1} \langle f, z_j \rangle_{L_2(\Omega)} z_j \equiv \phi_n^*(N_n^* f),$$

which shows that N_n^* and ϕ_n^* are nth optimal information and an nth minimal error algorithm, respectively.

7.4.3   Standard information

In the previous subsection, we studied the average case complexity of shift-invariant problems. We found conditions that were necessary and sufficient for the finite element method to be optimal. However, all this analysis applies only if finite element information (FEI) is permissible information. As we mentioned in Chapters 5 and 6, the permissibility of FEI is a stronger assumption than we are usually willing to allow in practice. Instead, we usually assume that evaluation of a problem element (or one of its derivatives) at a point in the domain Ω is the only permissible information. Hence, we now investigate the complexity of shift-invariant problems, assuming that only standard information is available. By *standard information*, we once again mean information for which $\Lambda = \Lambda^{\mathrm{std}}$, where

$$\lambda \in \Lambda^{\mathrm{std}} \quad \text{iff} \quad \exists \,\text{multi-index } \alpha \text{ and } x \in \overline{\Omega} : \lambda(f) = (D^\alpha f)(x).$$

The most important standard information is standard information using only function evaluations, in which the multi-index α is always chosen to be zero. Hence, standard information using only function evaluations has the form

$$Nf = \begin{bmatrix} f(x_1) \\ \vdots \\ f(x_{n(f)}) \end{bmatrix},$$

where $x_1, \ldots, x_{n(f)} \in \overline{\Omega}$. As usual, this information has *fixed cardinality* iff $n(f) \equiv n$, for some n independent of any problem element f, and has *varying cardinality* otherwise; it will be *nonadaptive* iff $x_1, \ldots, x_n$ are fixed, independent of any particular problem element f, and *adaptive* otherwise.

Of course, we will be especially interested in finite element methods with quadrature. Suppose that $\{\mathcal{T}_n\}_{n=1}^\infty$ is a quasi-uniform family of partitions of Ω. For any $\tilde{n} \in \mathbb{P}$, we let $\mathcal{S}_{\tilde{n},k}$ be an n-dimensional subspace of $H_E^m(\Omega)$ consisting of $\mathcal{T}_{\tilde{n}}$-piecewise polynomials of degree k. Let $N_{n,k}^Q$ denote standard information of the form

$$N_{n,k}^Q f = \begin{bmatrix} f(x_1) \\ \vdots \\ f(x_n) \end{bmatrix}.$$

We say that an algorithm $\phi_{n,k}^Q$ using $N_{n,k}^Q$ is a *finite element method with quadrature* (FEMQ) of degree k if the following hold:

(i) For any $f \in F$, we have $\phi_{n,k}^{\mathrm{FE}}(N_{n,k}^{\mathrm{FE}} f) \in \mathcal{S}_{\tilde{n},k}$, where $\tilde{n} = \Theta(n)$ as $n \to \infty$.

(ii) We have the following standard error estimate: for any $a > \frac{1}{2} d$ and $f \in H^a(\Omega)$,

$$\|Sf - \phi_{n,k}^Q(N_{n,k}^Q f)\|_{H^l(\Omega)} \le C n^{-(\min\{k+1-l,a\})/d} \|f\|_{H^a(\Omega)},$$

for some positive constant C, independent of f and n.

(iii) We have the following standard lower bound: if $K \in \mathcal{T}_{\tilde{n}}$, with $\partial K \cap \partial \Omega = \emptyset$, and $u \in P_{k+1}(K)$, then

$$\inf_{s \in P_k(K)} |u - s|_{H^l(K)} \ge C \tilde{n}^{-(k+1-l)/d} |u|_{H^{k+1}(K)},$$

for some positive constant C, independent of u, K, and $\tilde{n}$.

Once again, the standard FEMQs (or LSFEMQs) described in Chapter 5 and §§6.2 and 6.3 are all FEMQs under this definition.

For the purposes of comparison, we recall our results for the worst case setting:

THEOREM 7.4.3.1.   *Let $q \in (0, \infty]$, and let $H_*^r(\Omega)$ be a closed subspace of $H^r(\Omega)$ containing $H_0^r(\Omega)$, with $r > \frac{1}{2}d$. Suppose we are solving a shift-invariant problem with problem elements $F = qBH_*^r(\Omega)$. Let*

$$\eta_Q = \frac{\min\{k + 1 - l, r\}}{d}.$$

(i) *The nth minimal radius is*

$$r^{\mathrm{wor}}(n, \Lambda^{\mathrm{std}}) = \Theta(q \cdot n^{-r/d}) \qquad \text{as } n \to \infty.$$

(ii) *The worst case error of the FEM of degree k is*

$$e^{\mathrm{wor}}(\phi_{n,k}^Q, N_{n,k}^Q) = \Theta(n^{-\eta_Q}) \qquad \text{as } n \to \infty,$$

and so the FEMQ of degree k is an almost minimal error algorithm iff $k \geq r + l - 1$.

(iii) *The ε-complexity is*

$$\mathrm{comp}^{\mathrm{wor}}(\varepsilon, \Lambda^{\mathrm{std}}) = \Theta\left(\left(\frac{q}{\varepsilon}\right)^{d/r}\right) \qquad \text{as } \varepsilon \to 0.$$

(iv) *Let*

$$\mathrm{FEMQ}_k^{\mathrm{wor}}(\varepsilon) = \inf\{\,\mathrm{cost}^{\mathrm{wor}}(\phi_{n,k}^{\mathrm{FE}}, N_{n,k}^{\mathrm{FE}}) : e^{\mathrm{wor}}(\phi_{n,k}^{\mathrm{FE}}, N_{n,k}^{\mathrm{FE}}) \leq \varepsilon\,\}$$

denote the cost of using the FEMQ of degree k to compute an ε-approximation. Then

$$\mathrm{FEMQ}_k^{\mathrm{wor}}(\varepsilon) = \Theta\left(\left(\frac{q}{\varepsilon}\right)^{1/\eta_Q}\right) \qquad \text{as } \varepsilon \to 0.$$

Hence the FEMQ of degree k is an almost optimal complexity algorithm in the worst case setting iff

$$k \geq r + l - 1. \qquad\qquad \square$$

Hence, an FEMQ of sufficiently high degree is almost optimal in the worst case setting. We now investigate to what extent this is also true in the average case setting. Recall that we are assuming that the eigenvalues $\beta_1, \beta_2, \ldots$ of the correlation operator C_μ of the measure μ satisfy

$$\beta_j = \Theta(j^{-p}) \qquad \text{as } j \to \infty$$

for some $p > 1$. Let

$$\omega_{\text{std}} = \frac{r}{d} + \frac{p-1}{2},$$

and recall that

$$\omega^* = \frac{r+s-l}{d} + \frac{p-1}{2}.$$

We first prove a lower bound on the nth minimal average radius of standard information.

THEOREM 7.4.3.2. *The nth minimal average radius is bounded from below by*

$$r^{\text{avg}}(n, \Lambda^{\text{std}}) = \Omega(n^{-\min\{\omega_{\text{std}}+1/2, \omega^*\}}) \qquad \text{as } n \to \infty.$$

PROOF. Since $r^{\text{avg}}(n, \Lambda^{\text{std}}) \geq r^{\text{avg}}(n) = \Theta(n^{-\omega^*})$, we need only show that $r^{\text{avg}}(n, \Lambda^{\text{std}}) = \Omega(n^{-(\omega_{\text{std}}+1/2)})$ as $n \to \infty$.

To do this, we choose nonadaptive standard information N of cardinality n, and let $\sigma_N(y)$ be the μ-spline interpolating $y = Nf$. Then $S\sigma_N$ is an optimal error algorithm using N. Since $l \geq 0$, the shift-invariance of our problem implies that

$$r^{\text{avg}}(N)^2 \geq \int_{H_*^r(\Omega)} \|Sf - S\sigma_N(Nf)\|_{L_2(\Omega)}^2 \, \mu(df)$$

$$\geq \theta^{-2} \int_{H_*^r(\Omega)} \|f - \sigma_N(Nf)\|_{H^{-s}(\Omega)}^2 \, \mu(df)$$

$$= \theta^{-2} \sum_{i=1}^{\infty} \left\| (E - \sigma_N \circ N) C_\mu^{1/2} e_i \right\|_{H^{-s}(\Omega)}^2,$$

where $E \colon H_*^r(\Omega) \to H^{-s}(\Omega)$ is the identity injection and $\{e_i\}_{i=1}^{\infty}$ is any orthonormal basis for $H_*^r(\Omega)$.

Let Ω^0 be a C^∞ region whose closure is in $\text{int}\,\Omega$. As in Theorem 5.5.1, there exists $z \in C_0^\infty(\Omega)$ such that

$$z(x) = 0 \qquad \forall x \in \Omega \setminus \Omega^0,$$

$$Nz = 0,$$

$$\|z\|_{r+pd/2} = 1,$$

and

$$\int_{\Omega^0} z(x)\,dx \geq Cn^{-r/d+p/2}.$$

Since $z \in C_0^\infty(\Omega)$, and $C_0^\infty(\Omega) \subset H_*^{r+pd/2}(\Omega) = C_\mu^{1/2}(H_*^r(\Omega))$, there exists $w \in H_*^r(\Omega)$ such that $C_\mu^{1/2}w = z$. Moreover, since $\beta_i^{-1} = \Theta(i^{pd/s'})$ as $i \to \infty$, there is a positive constant C, independent of n, z, and w, such that

$$\|w\|_r^2 = \sum_{i=1}^\infty \beta_i^{-1}\alpha_i^{2r/s'}\langle z, z_i\rangle_{L_2(\Omega)}^2 \geq C\sum_{i=1}^\infty \alpha_i^{(2r+pd)/s'}\langle z, z_i\rangle_{L_2(\Omega)}^2$$
$$= C\|z\|_{r+pd/2} = C.$$

Let $\{e_i\}_{i=1}^\infty$ be any orthonormal basis for $H_*^r(\Omega)$ such that $e_1 = w/\|w\|_r$. Then

$$r^{\text{avg}}(N) \geq \theta^{-1}\left\|(E - \sigma_N \circ N)C_\mu^{1/2}e_1\right\|_{H^{-s}(\Omega)}$$
$$= \theta^{-1}\left\|C_\mu^{1/2}e_1\right\|_{H^{-s}(\Omega)} \geq \frac{\|z\|_{H^{-s}(\Omega)}}{C\theta}.$$

Choose $g^* \in H_0^s(\Omega)$ such that $g^*|_{\Omega^0} \equiv 1$ and $\|g^*\|_{H_0^s(\Omega)} \leq 2$. Since the support of z is contained in Ω^0, we find that there is a constant C, independent of z, g, and n, such that

$$\|z\|_{H^{-s}(\Omega)} = \sup_{g \in H_0^s(\Omega)} \frac{\int_\Omega g(x)z(x)\,dx}{\|g\|_{H^s(\Omega)}}$$
$$\geq \tfrac{1}{2}\int_\Omega g^*(x)z(x)\,dx = \tfrac{1}{2}\int_{\Omega^0} z(x)\,dx \geq Cn^{-r/d+p/2}.$$

Combining the last two inequalities, we see that $r^{\text{avg}}(N) = \Omega(n^{-(\omega_{\text{std}}+1/2)})$. Since N is arbitrary nonadaptive standard information of cardinality n, the result follows by taking the infimum over all such N. □

We now determine the error of the FEMQ in the average case setting. Let

$$\omega_Q = \frac{\min\{k+1-l, r+\tfrac{1}{2}d(p-1)\}}{d}.$$

THEOREM 7.4.3.3.   Let $r + \tfrac{1}{2}(p-2) > 0$. The error of the FEMQ of degree k is bounded from above by

$$e^{\text{avg}}(\phi_{n,k}^Q, N_{n,k}^Q) = O(n^{-\omega_Q}) \qquad \text{as } n \to \infty$$

and from below by

$$e^{\text{avg}}(\phi_{n,k}^Q, N_{n,k}^Q) = \Omega(n^{-\min\{\omega_Q, \omega_{\text{std}}+1/2\}}) \qquad \text{as } n \to \infty.$$

PROOF. First, we claim that

$$e^{\mathrm{avg}}(\phi_{n,k}^{\mathrm{Q}}, N_{n,k}^{\mathrm{Q}}) = O(n^{-\omega_{\mathrm{Q}}}) \qquad \text{as } n \to \infty.$$

Indeed, we assume without loss of generality that $n = 2^q$ and that $\omega_{\mathrm{Q}} > 0$. Let $y_i = \alpha_i^{-r/s'} z_i$ for all indices i. As in the proof of Theorem 7.4.2.4, we have

$$e^{\mathrm{avg}}(\phi_{n,k}^{\mathrm{FE}}, N_{n,k}^{\mathrm{FE}}) \le C \sum_{j=0}^{\infty} \left[\sum_{i=1}^{\infty} \|(S - Q_{2^{q+j}})C_\mu^{1/2} y_i\|_{H^l(\Omega)}^2 \right]^{1/2},$$

where $Q_{2^{q+j}} f = \phi_{2^{q+j},k}^{\mathrm{Q}}(N_{2^{q+j},k}^{\mathrm{Q}} f)$. From the definition of ω_{Q} and the standard error estimate for the FEMQ with $a = r + \frac{1}{2}d(p-1)$, we have

$$\|(S - Q_{2^{q+j}})C_\mu^{1/2} y_i\|_{H^l(\Omega)} \le C(2^{q+j})^{-\omega_{\mathrm{Q}}} \|C_\mu^{1/2} y_i\|_{H^{r+d(p-1)/2}(\Omega)}.$$

From the proof of Theorem 7.4.2.4, we have

$$\|C_\mu^{1/2} y_i\|_{H^{r+d(p-1)/2}(\Omega)} \le C \cdot i^{-(p+1)/2}.$$

Since the sum $\sum_{i=1}^{\infty} i^{-(p+1)}$ is convergent, this implies that

$$\left[\sum_{i=1}^{\infty} \|(S - Q_{2^{q+j}})C_\mu^{1/2} y_i\|_{H^l(\Omega)}^2 \right]^{1/2} \le C \cdot (2^{q+j})^{-\omega_{\mathrm{Q}}}.$$

Hence

$$e^{\mathrm{avg}}(\phi_{n,k}^{\mathrm{Q}}, N_{n,k}^{\mathrm{Q}}) \le C \sum_{j=0}^{\infty} (2^{q+j})^{-\omega_{\mathrm{Q}}} \le C n^{-\omega_{\mathrm{Q}}} \sum_{j=0}^{\infty} (2^{\omega_{\mathrm{Q}}})^{-j}.$$

Since $\omega_{\mathrm{Q}} > 0$, the sum converges, and so we have

$$e^{\mathrm{avg}}(\phi_{n,k}^{\mathrm{Q}}, N_{n,k}^{\mathrm{Q}}) = O(n^{-\omega_{\mathrm{Q}}}) \qquad \text{as } n \to \infty,$$

as claimed.

Next, we turn to the lower bound. Define f as in the proof of (5) in Theorem 7.4.2.4. Since $Q_n C_\mu^{1/2} f \in \mathcal{S}_{\tilde{n},k}$, we find that

$$e^{\mathrm{avg}}(\phi_{n,k}^{\mathrm{Q}}, N_{n,k}^{\mathrm{Q}}) \ge C \|(S - Q_n)C_\mu^{1/2} f\|_{H^l(\Omega)} \ge C \inf_{s \in \mathcal{S}_{\tilde{n},k}} |SC_\mu^{1/2} f - s|_{H^l(\Omega)}$$

$$\ge C n^{-(k+1-l)/d}.$$

On the other hand, we may use the previous theorem to see that

$$e^{\mathrm{avg}}(\phi^{\mathrm{Q}}_{n,k}, N^{\mathrm{Q}}_{n,k}) \geq r^{\mathrm{avg}}(n, \Lambda^{\mathrm{std}}) = \Omega(n^{-\min\{\omega_{\mathrm{std}}+1/2,\omega^*\}}) \qquad \text{as } n \to \infty.$$

Our desired lower bound follows from these last two inequalities.　□

Combining the previous two theorems, we find that there exist positive constants C_1, C_2, independent of n, such that

$$C_1 n^{-\min\{\omega_{\mathrm{std}}+1/2,\omega^*\}} \leq r^{\mathrm{avg}}(n, \Lambda^{\mathrm{std}}) \leq C_2 n^{-\omega_{\mathrm{std}}}$$

for n sufficiently large. Hence there is a gap between the lower and upper bounds on $r^{\mathrm{avg}}(n, \Lambda^{\mathrm{std}})$ iff $\omega_{\mathrm{std}} + \frac{1}{2} < \omega^*$, i.e. iff $s - l > \frac{1}{2}d$. The following are examples of such problems for which there is no gap:

(i)   the Fredholm problem of the second kind (since $s = l = 0$),
(ii) elliptic systems of order t, the error being measured in the $H^t(\Omega)$-norm,
(iii) first-order elliptic systems in $\mathbb{R}^d$ for $d \geq 2$ (since $s - l \leq 1 - 0 \leq \frac{1}{2}d$).

Note that if there is no gap, then the last two theorems imply that the FEMQ of sufficiently high degree is an almost minimal average error algorithm. In particular, the FEMQ is almost minimal error for the problems mentioned above.

In light of the evidence produced above, as well as our previous experience that finite element methods with quadrature of sufficiently high degree are optimal, it is reasonable to state the following

CONJECTURE 7.4.3.1. *Let $r + \frac{1}{2}(p - 2) > 0$. Then*

$$r^{\mathrm{avg}}(n, \Lambda^{\mathrm{std}}) = \Theta(n^{-\omega_{\mathrm{std}}}) \qquad \text{as } n \to \infty.$$ 　☒

That is, we believe that the upper bound is tight, and the lower bound needs to be improved. Of course, this conjecture holds trivially if $s - l \leq \frac{1}{2}d$. For further discussion of this conjecture, see **NR 7.4.3:1**.

In the remainder of this subsection, we study minimal error and optimal complexity algorithms for this problem, assuming that Conjecture 7.4.3.1 is satisfied. If this conjecture does not hold, then we can proceed as explained in **NR 7.4.3:2**.

Using Theorems 7.4.3.2 and 7.4.3.3, we easily find

COROLLARY 7.4.3.1.   *Suppose that Conjecture 7.4.3.1 holds. The nth minimal average radius of standard information is then given by*

$$r^{\mathrm{avg}}(n, \Lambda^{\mathrm{std}}) = \Theta(n^{-\omega_{\mathrm{std}}}) \qquad \text{as } n \to \infty,$$

and the error of the FEMQ is

$$e^{\mathrm{avg}}(\phi^{\mathrm{Q}}_{n,k}, N^{\mathrm{Q}}_{n,k}) = \Theta(n^{-\omega_{\mathrm{Q}}}) \qquad \text{as } n \to \infty.$$

Hence, the FEMQ of degree k is an almost minimal average error algorithm iff $k \geq r + l - 1 + \frac{1}{2}d(p-1)$. □

For $\varepsilon > 0$, we now wish to determine $\text{comp}^{\text{avg}}(\varepsilon, \Lambda^{\text{std}})$, the inherent complexity of computing an ε-approximation in the average case setting using only standard information. We wish to compare this with

$$\text{FEMQ}_k^{\text{avg}}(\varepsilon) = \inf\{\, \text{cost}^{\text{avg}}(\phi_{n,k}^Q, N_{n,k}^Q) : e^{\text{avg}}(\phi_{n,k}^Q, N_{n,k}^Q) \leq \varepsilon \,\},$$

the cost of using the FEMQ to compute an ε-approximation in the average case setting.

Note that our analysis so far has dealt with nonadaptive information. We now remove this restriction. Observe that if Conjecture 7.4.3.1 holds, then the sequence of nth minimal average radii is semiconvex. From Theorem 7.3.4, we know that adaptive information (of possibly-varying cardinality) is no stronger than nonadaptive information. Hence the restriction to nonadaptive information will not change the problem complexity.

As with the "pure" FEM, we may now use the previous results to establish

COROLLARY 7.4.3.2. *Suppose that Conjecture 7.4.3.1 holds.*

(i) *The problem complexity is*

$$\text{comp}^{\text{avg}}(\varepsilon, \Lambda^{\text{std}}) = \Theta(\varepsilon^{-1/\omega_{\text{std}}}) \qquad \text{as } \varepsilon \to 0.$$

(ii) *The average case cost of computing an ε-approximation with the FEMQ of degree k is*

$$\text{FEMQ}_k^{\text{avg}}(\varepsilon) = \Theta(\varepsilon^{-1/\omega_Q}) \qquad \text{as } \varepsilon \to 0.$$

Hence the FEMQ of degree k is an almost optimal complexity algorithm iff $k \geq r + l - 1 + \frac{1}{2}d(p-1)$. □

Again, we note that there is once again a large asymptotic penalty to be paid when we use a FEMQ of too low a degree. That is, suppose that Conjecture 7.4.3.1 holds, and that $k < r + l - 1 + \frac{1}{2}d(p-1)$. Then

$$\frac{\text{FEMQ}_k^{\text{avg}}(\varepsilon)}{\text{comp}^{\text{avg}}(\varepsilon, \Lambda^{\text{std}})} = \Theta(\varepsilon^{-d\delta}) \qquad \text{as } \varepsilon \to 0,$$

where now

$$\delta = \frac{1}{k+1-l} - \frac{1}{r + \frac{1}{2}d(p-1)} > 0.$$

Thus,

$$\lim_{\varepsilon \to 0} \frac{\text{FEMQ}_k^{\text{avg}}(\varepsilon)}{\text{comp}^{\text{avg}}(\varepsilon, \Lambda^{\text{std}})} = \infty.$$

Notes and Remarks

NR 7.4.3:1 It may be difficult to establish Conjecture 7.4.3.1, since lower bounds for minimal radii of restricted information are hard to achieve. For example, consider the multidimensional quadrature problem, the space of problem elements being $L_2([0,1]^d)$ equipped with a Wiener measure. Woźniakowski (1989) has only recently determined minimal radii of standard information for this problem. (The proof involves upper and lower bounds on L_2-discrepancy that were proved twenty years apart!) Another analogous example is given by the multivariate approximation problem discussed in Papageorgiou and Wasilkowski (1986), for which we still do not know optimal standard information.

NR 7.4.3:2 If Conjecture 7.4.3.1 does not hold, we can still obtain bounds on the complexity of our problem by using Theorems 7.4.3.2 and 7.4.3.3, and noting that the lower bounds and upper bounds on the minimal radii form semiconvex sequences. In particular, we find an upper bound

$$\mathrm{comp}^{\mathrm{avg}}(\varepsilon) = O(\varepsilon^{-1/\omega_{\mathrm{std}}}) \qquad \text{as } \varepsilon \to 0,$$

as well as a lower bound

$$\mathrm{comp}^{\mathrm{avg}}(\varepsilon) = \Omega(\varepsilon^{-1/\min\{\omega_{\mathrm{std}}+1/2,\omega^*\}}) \qquad \text{as } \varepsilon \to 0$$

on the average ε-complexity.

7.5   Ill-posed problems

In Chapter 6, we studied ill-posed problems in the worst case setting. We found that ill-posed problems are strongly noncomputable under the absolute error criterion, i.e. there are no algorithms for solving these problems whose error is finite. Faced with this difficulty, we investigated the solution of ill-posed Fredholm problems under the residual error criterion. We found that ill-posed problems become tractable if we use the residual to measure the quality of the solution, since we formally reduce our problem to the approximation of the identity when we change from the absolute to the residual error criterion.

Unfortunately, this solution to our difficulty is not entirely satisfactory. It is a well-known truism of numerical mathematics that a small residual does not necessarily imply a small error. Indeed, if this were not so, then the strong noncomputability under the absolute error criterion would also apply to the residual criterion!

Clearly, error in the average case setting is always less than or equal to that in the worst case setting if the problem elements and error criterion are kept the same. With this in mind, we ask whether it is possible to resurrect the hope of a solution having small absolute error by investigating the average case complexity of ill-posed problems.

This section is divided into two subsections. In the first, we discuss ill-posed problems in the average case setting under the absolute error criterion. Assuming only that the solution operator for our problem is measurable, we show that the average case ε-complexity of any ill-posed problem is finite for any positive ε. Furthermore, we determine optimal information and algorithms.

In the second subsection, we consider ill-posed Fredholm problems in the average case setting under the residual error criterion. We do this mainly for the sake of completeness, formally reducing these problems to the approximation problem in the average case setting. As a result, we find that ill-posed Fredholm problems are always solvable in the average case setting under the residual error criterion.

7.5.1 The absolute error criterion

We now consider ill-posed problems under the absolute error criterion in the average case setting. As in §6.4, we are given a linear unbounded solution operator

$$S: D \subset F_1 \to G.$$

However, we now assume that F_1 and G are both separable infinite-dimensional spaces, with F_1 a Banach space and G a Hilbert space. As before, the domain D is a linear subspace of F_1, and there exist linearly independent $x_1, x_2, \ldots \in D$, having unit norm, such that $\lim_{n \to \infty} \|Sx_n\| = \infty$. Moreover, we choose $\Lambda = \Lambda^*$, i.e. arbitrary continuous linear information is permissible. We assume that μ is a Gaussian measure on F_1 with zero mean and correlation operator C_μ.

We also require the solution operator S to be measurable. This is a very weak requirement; for example, the solution operator for the Fredholm problem of the first kind (the standard example of a linear ill-posed problem) is measurable. Since the domain D of S is a subspace of F_1, the zero-one law implies that $\mu(D) = 0$ or $\mu(D) = 1$; see Shilov and Fan Dyk Tin (1967) for details. Suppose first that $\mu(D) = 0$. Then any algorithm has zero error. This means that the ε-complexity of our problem is zero, and any algorithm is optimal. Hence, we will suppose that $\mu(D) = 1$ in the remainder of this section.

We first prove

THEOREM 7.5.1.1. *Under the conditions above, $\nu = \mu S^{-1}$ is a Gaussian measure.*

PROOF. From Vakhania *et al.* (1987, pp. 149, 152), there exists a separable Hilbert space H with inner product $\langle \cdot, \cdot \rangle$ such that:

(i) H is continuously embedded in F_1.
(ii) $C_\mu(F_1^*)$ is a dense subset of H.
(iii) $\langle C_\mu \lambda_1, C_\mu \lambda_2 \rangle = \lambda_1(C_\mu \lambda_2)$ for any $\lambda_1, \lambda_2 \in F_1^*$.

Since $\mu(D) = 1$, we may use the proof of Theorem 1 in Gattinger (1978) to see that $H \subseteq D$.

Let

$$I = \begin{cases} \{1, \ldots, m\} & \text{if } m = \dim H < \infty, \\ \mathbb{P} & \text{otherwise,} \end{cases}$$

and take a sequence $\{\lambda_k \in F_1^*\}_{k \in I}$ such that $\{C_\mu \lambda_k\}_{k \in I}$ is an orthonormal basis of H. Then for each $n \in I$, we define a continuous linear operator $S_n \colon F_1 \to G$ by

$$S_n f = \sum_{k=1}^{n} \lambda_k(f) S C_\mu \lambda_k \qquad \forall f \in F_1.$$

Suppose first that $\dim H = m$. Then the measure μ is concentrated on $H = C_\mu(F_1^*)$, see Vakhania *et al.* (1987, p. 182), and therefore $S = S_m$ holds μ-a.e. If $\dim H = \infty$, Theorem 3 of Gattinger (1978) says that S_n converges μ-a.e. and that $S = \lim_{n \to \infty} S_n$ holds μ-a.e., implying the weak convergence of the image measures μS_n^{-1} to $\nu = \mu S^{-1}$. Since G is a separable Banach space, the zero-mean Gaussian measures on G form a closed subset (with respect to the weak topology) within the set of all Borel probability measures on G, see Linde (1983, p. 67). Since continuous linear images of Gaussian measures are Gaussian, we conclude that $\nu = \mu S^{-1}$ is also a Gaussian measure.   □

We now look at algorithms using nonadaptive information N. As in §7.3, we assume that

$$N f = \begin{bmatrix} \lambda_1(f) \\ \vdots \\ \lambda_n(f) \end{bmatrix} \qquad \forall f \in D,$$

with

$$\lambda_i(C_\mu \lambda_j) = \delta_{i,j} \qquad (1 \le i, j \le n).$$

We now show that the μ-spline algorithm of §7.3 is well-defined, and is an optimal error algorithm with finite error.

THEOREM 7.5.1.2. *For any nonadaptive information* N, *the* μ-*spline algorithm*

$$\phi^{\mathrm{s}}(N f) = \sum_{j=1}^{n} \lambda_j(f) S C_\mu \lambda_j \qquad \forall f \in D$$

is an optimal error algorithm using N, *and*

$$e^{\mathrm{avg}}(\phi^{\mathrm{s}}, N) = r^{\mathrm{avg}}(N) = \left[\int_D \|S f\|^2 \, \mu(df) - \sum_{j=1}^{n} \|S C_\mu \lambda_j\|^2 \right]^{1/2} < \infty.$$

PROOF. From the proof of Theorem 7.5.1.1, it follows that $C_\mu \lambda_j \in D$ for $1 \le j \le n$. Hence, the spline algorithm is well-defined. It is now easy to see that the proof of Theorem 7.3.1 found on p. 229 of IBC is valid. Finally, since $\nu = \mu S^{-1}$ is Gaussian, its second moment is finite, and so

$$r^{\mathrm{avg}}(N)^2 \le \int_D \|Sf\|^2 \, \mu(df) = \int_G \|g\|^2 \, \nu(dg) < \infty. \qquad \square$$

We now wish to determine the nth minimal average radius and nth optimal information. The a priori measure $\nu = \mu S^{-1}$ on the solution elements is a Gaussian measure with zero mean and correlation operator C_ν. Since C_ν is a self-adjoint compact injection, there is a complete orthonormal basis $\{ u_i \}_{i=1}^\infty$ for G consisting of eigenvectors of C_ν. Hence we have

$$C_\nu u_i = \gamma_i u_i \,,$$

with

$$\gamma_1 \ge \gamma_2 \ge \ldots \ge 0 \qquad \text{and} \qquad \lim_{i \to \infty} \gamma_i = 0 \,.$$

Moreover, since the correlation operator C_ν of $\nu = \mu S^{-1}$ is a trace class operator, we have

$$\int_D \|Sf\|^2 \, \mu(df) = \int_G \|g\|^2 \, \nu(dg) = \operatorname{trace} C_\nu = \sum_{i=1}^\infty \gamma_i < \infty \,.$$

Let $\delta > 0$. Since S is measurable, each linear functional $\langle S\cdot, u_j \rangle$ is measurable, and is hence the limit μ-a.e. of a sequence of bounded linear functionals by Theorem 1 of Gattinger (1978). Since $\nu = \mu S^{-1}$ is Gaussian,

$$\int_D |\langle Sf, u_j \rangle|^2 \, \mu(df) = \int_G |\langle g, u_j \rangle|^2 \, \nu(dg)$$

is finite. It now follows that for each index j, there exists $\lambda_{j,\delta} \in F_1^*$ such that

$$\int_D |\langle Sf, u_j \rangle - \lambda_{j,\delta}(f)|^2 \, \mu(df) < \frac{\delta^2}{n} \,.$$

Then we define our information $N_{n,\delta}$ to be

$$N_{n,\delta} f = \begin{bmatrix} \lambda_{1,\delta}(f) \\ \vdots \\ \lambda_{n,\delta}(f) \end{bmatrix} \qquad \forall f \in D,$$

and our algorithm $\phi_{n,\delta}$ using $N_{n,\delta}$ to be

$$\phi_{n,\delta}(N_{n,\delta} f) = \sum_{j=1}^n \lambda_{j,\delta}(f) u_j \qquad \forall f \in D.$$

We now have

THEOREM 7.5.1.3.

    (i) *The nth minimal average radius is given by*

$$r^{\mathrm{avg}}(n) = \left[\int_D \|Sf\|^2 \, \mu(df) - \sum_{i=1}^n \gamma_i \right]^{1/2} = \left[\sum_{i=n+1}^\infty \gamma_i \right]^{1/2}.$$

    (ii) *For any $\delta > 0$, we have*

$$e^{\mathrm{avg}}(\phi_{n,\delta}, N_{n,\delta}) \le \left[\sum_{i=n+1}^\infty \gamma_i \right]^{1/2} + \delta.$$

    (iii) *In addition, suppose that u_j is in the domain of S^* for $1 \le j \le n$, and set*

$$\lambda_j^*(f) = \gamma_j^{-1/2} \langle Sf, u_j \rangle \qquad \forall f \in D,$$

where we have assumed $\gamma_j > 0$ for simplicity. Then the information N_n^* given by

$$N_n^* f = \begin{bmatrix} \lambda_1^*(f) \\ \vdots \\ \lambda_n^*(f) \end{bmatrix} \qquad \forall f \in D$$

is nth optimal information, and the μ-spline algorithm ϕ_n^* using N_n^* (which is an nth minimal error algorithm) has the form

$$\phi_n^*(N_n^* f) = \sum_{j=1}^n \langle Sf, u_j \rangle u_j \qquad \forall f \in D.$$

PROOF. If the conditions in (iii) hold, then $\lambda_1^*, \dots, \lambda_n^*$ are continuous linear functionals, and the result follows immediately from Theorem 7.3.2. If not, then the proof of Theorem 7.3.2 given in IBC still yields the lower bound

$$r^{\mathrm{avg}}(n) \ge \left[\int_D \|Sf\|^2 \, \mu(df) - \sum_{i=1}^n \gamma_i \right]^{1/2} = \left[\sum_{i=n+1}^\infty \gamma_i \right]^{1/2}.$$

The proof of our theorem will be complete once we establish (ii).

    To show that the desired error bound holds, let $\varepsilon_{j,\delta} = \langle S\cdot, u_j \rangle - \lambda_{j,\delta}$ for each index j. Let

$$I_1 = \int_D \left\| Sf - \sum_{j=1}^n \langle Sf, u_j \rangle \right\|^2 \mu(df)$$

and

$$I_2 = \int_D \left\| \sum_{j=1}^n \varepsilon_{j,\delta}(f) u_j \right\|^2 \mu(df)$$

Using the triangle and Schwarz inequalities, we find that

$$e^{\text{avg}}(\phi_{n,\delta}, N_{n,\delta}) \le I_1^{1/2} + I_2^{1/2}.$$

Since $u_1, u_2, \ldots$ are the orthonormalized eigenvectors of C_ν, we have

$$I_1 = \int_G \left\| g - \sum_{j=1}^n \langle g, u_j \rangle u_j \right\|^2 \nu(dg) = \int_G \left\| \sum_{j=n+1}^\infty \langle g, u_j \rangle u_j \right\|^2 \nu(dg)$$

$$= \sum_{j=n+1}^\infty \int_G \langle g, u_j \rangle^2 \nu(dg) = \sum_{j=n+1}^\infty \langle C_\nu u_j, u_j \rangle = \sum_{j=n+1}^\infty \gamma_j$$

and

$$I_2 = \sum_{j=1}^n \int_D |\varepsilon_{j,\delta}(f)|^2 \mu(df) \le \delta^2.$$

Hence

$$e^{\text{avg}}(\phi_{n,\delta}, N_{n,\delta}) \le \left[\sum_{j=n+1}^\infty \gamma_j \right]^{1/2} + \delta,$$

as required. $\square$

As a result, we see that for any $\delta > 0$, the information $N_{n,\delta}$ is within δ of being nth optimal information of cardinality n, and that the spline algorithm using that information is within δ of being an nth minimal average error algorithm.

Let $\varepsilon \ge 0$. We now determine the ε-complexity of our ill-posed problem in the average case setting under the absolute error criterion. Recall that the average ε-complexity number is the minimal number of (possibly adaptive) information evaluations that are needed to find an ε-approximation.

THEOREM 7.5.1.4.

(i) *The average ε-complexity number satisfies*

$$\lceil m^{\text{avg}}(\varepsilon) \rceil = n^{\text{avg}}(\varepsilon) = \inf \left\{ n \in \mathbb{N} : \sum_{j=n+1}^\infty \gamma_j < \varepsilon^2 \right\},$$

where γ_j is the jth eigenvalue of C_ν.

(ii) *Let $n = n^{\text{avg}}(\varepsilon)$. For $\delta \in [0, \gamma_{n+1}]$ the information $N_{n+1,\delta}$, described in Theorem 7.5.1.3, and the μ-spline algorithm ϕ^s using $N_{n+1,\delta}$ are almost optimal, with*

$$c\, m^{\text{avg}}(\varepsilon) \le \text{comp}^{\text{avg}}(\varepsilon) \le \text{cost}^{\text{avg}}(\phi^s, N_{n,\delta}) \le (c+2)\big(m^{\text{avg}}(\varepsilon)+1\big) - 1.$$

PROOF. This follows almost immediately from Theorem 7.5.1.4, along with the general results in §7.3. The only possible complication arises because the information N_n^* described in Theorem 7.5.1.4 need not be well-defined. If we use information $N_{n+1,\delta}$ with $0 \leq \delta \leq \gamma_{n+1}$, then it is easy to see that

$$e^{\text{avg}}(\phi^s, N_{n+1,\delta}) = r^{\text{avg}}(N_{n+1,\delta}) \leq r^{\text{avg}}(n) \leq \varepsilon.$$

Hence, ϕ^s and $N_{n+1,\delta}$ give an ε-approximation with cost at most $(c+2)(n+1) - 1$. Since $n \leq m^{\text{avg}}(\varepsilon) + 1$, the result follows. □

In summary, we see that the average case ε-complexity is proportional to $c\, m^{\text{avg}}(\varepsilon)$ for large c and small ε.

So far, we have only considered the case of $F = F_1$. That is, we have used the whole space F_1 as our class of problem elements. Next, we turn to the case $F = B_q = qBF_1$ with $q < \infty$. That is, we suppose that the problem elements form a ball of finite radius. In what follows, we shall once again assume that

$$\mu(B_q) \geq \tfrac{1}{2}(\sqrt{21} - 3) \simeq 0.8.$$

A careful reading of the proof of Theorem 7.3.5 (see Section 5.8 in Chapter 6 of IBC) shows that this theorem still holds if $\int_D \|Sf\|^2 \,\mu(df) < \infty$. Hence, we have the following

THEOREM 7.5.1.5. *If S is measurable, then the results of Theorem 7.3.5 still hold. In particular, if*

$$\text{comp}^{\text{avg}}\left(\varepsilon(1 + \eta), \Lambda\right) = \text{comp}^{\text{avg}}(\varepsilon, \Lambda)\left(1 + O(\eta)\right) \qquad \text{as } \eta \to 0,$$

then

$$\text{comp}^{\text{avg}}(\varepsilon, \Lambda, q) \simeq \text{comp}^{\text{avg}}(\varepsilon, \Lambda)$$

for reasonably large q and c. □

Notes and Remarks

NR 7.5.1:1 Much of this material is based on Werschulz (1987a) and on Section 4 in Chapter 7 of IBC. However, the older material required the solution operator to be a closed linear transformation of Hilbert spaces, whereas here we only require that it be a measurable operator from a Banach space to a Hilbert space.

NR 7.5.1:2 The result that μS^{-1} is Gaussian if S is a measurable linear operator defined almost everywhere over a separable Banach space equipped with a Gaussian measure is proved in Kon *et al.* (1990), as well as in a personal communication received from N. N. Vakhania in 1990.

7.5.2 The residual error criterion

We now discuss the solution of ill-posed problems under the residual error criterion in the average case setting. As described in §4.5, we are given a bounded injective problem operator $L: G \to F_1$ with range D dense in F_1. Our class F of problem elements will be either $F = F_1$ or $F = B_q$, i.e. either the whole space or a ball of finite radius q. The solution operator $S: F \cap D \to G$ is given by

$$u = Sf \quad \text{iff} \quad Lu = f \qquad \forall f \in F \cap D.$$

Since we are studying an ill-posed problem, we assume that L is compact. Hence, we are once again studying the Fredholm problem of the first kind, but this time in an average case setting, rather than in the worst case setting.

Let $E: F_1 \to W$ be an embedding. Under the residual error criterion, the error of an algorithm ϕ using information N is given by

$$e^{\text{avg-resid}}(\phi, N) = \left[\int_{F \cap D} \left\| E\big(L\phi(Nf) - f\big) \right\|^2 \mu(df) \right]^{1/2}.$$

The radius of information is given by

$$r^{\text{avg-resid}}(N) = \inf_{\phi \text{ using } N} e^{\text{avg-resid}}(\phi, N),$$

measuring the minimal residual among all algorithms using the information N.

As we did in the worst case setting, we solve our ill-posed problem under the residual criterion by reducing it to the problem of approximating the identity injection E. More precisely, for an algorithm ψ using information N, we let

$$e^{\text{avg}}_{\text{app}}(\psi, N) = \left[\int_F \left\| E\big(\psi(Nf) - f\big) \right\|^2 \mu(df) \right]^{1/2}$$

denote the average case (absolute) error of the algorithm ψ in approximating E, and let

$$r^{\text{avg}}_{\text{app}}(N) = \inf_{\psi \text{ using } N} e^{\text{avg}}_{\text{app}}(\psi, N)$$

be the radius of information N for the approximation problem.

Our main result is the following

THEOREM 7.5.2.1. *For any information* N, *the radii of information are the same for the ill-posed problem and the approximation problem under the residual error criterion, i.e.*

$$r^{\text{avg-resid}}(N) = r_{\text{app}}^{\text{avg}}(N).$$

PROOF. Let ϕ be an algorithm using information N for the Fredholm problem. Then $L\phi$ is an algorithm for the approximation problem. Since D is of full measure in F, we find

$$e^{\text{avg-resid}}(\phi, N)^2 = \int_{F \cap D} \left\| E\big(L\phi(Nf) - f\big) \right\|^2 \mu(df)$$

$$\geq \inf_{\psi \text{ using } N} \int_{F} \left\| E\big(\psi(Nf) - f\big) \right\|^2 \mu(df)$$

$$= r_{\text{app}}^{\text{avg}}(N)^2.$$

Since ϕ is an arbitrary algorithm using N, we see that

$$r^{\text{avg-resid}}(N) = \inf_{\phi \text{ using } N} e^{\text{avg-resid}}(\phi, N) \geq r_{\text{app}}^{\text{avg}}(N).$$

Moreover, since L has dense range, the algorithm $\phi(Nf) = L^{-1}\big(\psi(Nf)\big)$ is well-defined. Hence this lower bound is sharp. $\qquad\square$

Using this result, we can discuss the complexity of ill-posed problems under the residual error criterion. For $\varepsilon > 0$, we let

$$\text{comp}^{\text{avg-resid}}(\varepsilon) = \inf\{\text{cost}^{\text{avg-resid}}(\phi, N) : e^{\text{avg-resid}}(\phi, N) \leq \varepsilon\}$$

denote the complexity of our ill-posed problem under the residual criterion. We also write

$$\text{comp}_{\text{app}}^{\text{avg}}(\varepsilon) = \inf\{\text{cost}_{\text{app}}^{\text{avg}}(\psi, N) : e_{\text{app}}^{\text{avg}}(\psi, N) \leq \varepsilon\}$$

for the complexity of the approximation problem under the absolute criterion. Since for any information operator N, the radii of these problems are the same, we immediately have

COROLLARY 7.5.2.1. *For any* $\varepsilon > 0$,

$$\text{comp}^{\text{avg-resid}}(\varepsilon) = \text{comp}_{\text{app}}^{\text{avg}}(\varepsilon). \qquad\square$$

We now go into more detail. Note that the injection E is bounded. This means that we can immediately apply the results in §7.3 to the operator E, which will allow us to determine optimal information and algorithms. From Theorem 7.3.5, we see that the complexity for the case $F = B_q$ is roughly the same as that for the case $F = F_1$. So, we concentrate on this latter case in the rest of this section.

We first note that the μ-spline algorithm is once again optimal:

COROLLARY 7.5.2.2. *Let N be nonadaptive information, and let $\sigma_N(y)$ be the μ-spline interpolating $y = Nf$. Then the μ-spline algorithm $\phi^s = L^{-1}\sigma_N$ is well-defined, and*

$$e^{\text{avg-resid}}(\phi^s, N) = r^{\text{avg-resid}}(N) = r^{\text{avg}}_{\text{app}}(N).$$

PROOF. From the proof of Theorem 7.5.1.1, we see that $\sigma_N(y) \in \text{range } L$ for any $y \in N(F_1)$, and so the μ-spline algorithm is well-defined. Optimality of the μ-spline algorithm then follows from Theorem 7.3.1. $\square$

We now derive optimal information and minimal residual algorithms. In what follows, we suppose that $\Lambda = \Lambda^*$, i.e. any continuous linear functional is permissible. Furthermore, we treat only the *Hilbert case*, i.e. all spaces involved are separable Hilbert spaces.

Let $n \in \mathbb{N}$, and recall that Λ^*_n denotes the class of all nonadaptive information using n information evaluations from Λ^*. We let

$$r^{\text{avg-resid}}(n) = \inf_{N \in \Lambda^*_n} r^{\text{avg-resid}}(N)$$

and

$$r^{\text{avg}}_{\text{app}}(n) = \inf_{N \in \Lambda^*_n} r^{\text{avg}}_{\text{app}}(N)$$

respectively denote the nth minimal radii for the ill-posed problem under the residual criterion and the approximation problem under the absolute criterion.

Let $\{y_j\}^\infty_{j=1}$ be an orthonormal basis for F_1 consisting of eigenvectors for the correlation operator C_μ, with

$$C_\mu y_j = \beta_j y_j \qquad (j = 1, 2, \dots)$$

and

$$\beta_1 \geq \beta_2 \geq \dots \geq 0 \qquad \text{and} \qquad \sum_{j=1}^\infty \beta_j = \text{trace } C_\mu < \infty.$$

Define information N^*_n by

$$N^*_n f = \begin{bmatrix} \langle f, y_1 \rangle \\ \vdots \\ \langle f, y_n \rangle \end{bmatrix} \qquad \forall f \in F_1$$

and an algorithm ϕ^*_n using N^*_n by

$$\phi^*_n(N^*_n f) = \sum_{j=1}^n \langle f, y_j \rangle L^{-1} y_j \qquad \forall f \in F_1.$$

Using Theorem 7.3.2, we have

THEOREM 7.5.2.2. N_n^* is nth optimal information, and ϕ_n^* is an nth minimal residual algorithm, i.e.

$$e^{\text{avg-resid}}(\phi_n^*, N_n^*) = r^{\text{avg-resid}}(n) = r_{\text{app}}^{\text{avg}}(n) = \left[\sum_{j=n+1}^{\infty} \beta_j \right]^{1/2}. \qquad \square$$

As a result, we see that our problem is convergent, i.e.

$$\lim_{n \to \infty} r^{\text{avg-resid}}(n) = 0.$$

Moreover, for any adaptive information N^{a}, there exists nonadaptive information N^{non} such that

$$\text{card } N^{\text{non}} \le \lceil \text{card } N^{\text{a}} \rceil \quad \text{and} \quad r^{\text{avg-resid}}(N^{\text{non}}) \le r^{\text{avg-resid}}(N^{\text{a}}).$$

Thus adaption is no more powerful than nonadaption.

Having discussed optimal information and minimal residual algorithms, we now consider the complexity

$$\text{comp}^{\text{avg-resid}}(\varepsilon) = \inf\{\text{cost}^{\text{avg-resid}}(\phi, N) : e^{\text{avg-resid}}(\phi, N) \le \varepsilon \}$$

of computing an average case ε-approximation under the residual error criterion. Let

$$n^{\text{avg-resid}}(\varepsilon) = \inf\{\, n \in \mathbb{N} : r^{\text{avg-resid}}(n) \le \varepsilon \,\}$$

denote the minimal number of nonadaptive information evaluations required for an average case ε-approximation under the residual criterion. From Theorem 7.5.1.2, we see that

$$n^{\text{avg-resid}}(\varepsilon) = \inf \left\{\, n \in \mathbb{N} : \left[\sum_{j=n+1}^{\infty} \beta_j \right]^{1/2} \le \varepsilon \right\}.$$

From Theorem 7.3.4, we then have

THEOREM 7.5.2.3.   For any $\varepsilon > 0$,

$$c\big(n^{\text{avg-resid}}(\varepsilon) - 1\big) \le \text{comp}^{\text{avg-resid}}(\varepsilon) \le (c+2)n^{\text{avg-resid}}(\varepsilon) - 1. \qquad \square$$

As an example, we compare the worst and average case settings for an ill-posed Fredholm problem with $F_1 = H^r(\Omega)$ for some $r > 0$ and $W = L_2(\Omega)$, where $\Omega \subset \mathbb{R}^d$. We suppose that $F = B_q$, where q satisfies

$$\mu(B_q) \ge \tfrac{1}{2}(\sqrt{21} - 3) \simeq 0.8.$$

In the worst case setting, we find that

$$\mathrm{comp}^{\mathrm{wor\text{-}resid}}(\varepsilon) = \Theta(\varepsilon^{-d/r}) \qquad \text{as} \, \varepsilon \to 0,$$

see §4.5 and §5.4. Hence, although the complexity is finite, it is exponential in the dimension d of the problem. We now look at the average case setting. Assume that the measure μ is as in §4.1, with $\beta_j = \Theta(j^{-p})$ as $j \to \infty$ for some $p > 1$. As in §4.2, we find that

$$\mathrm{comp}^{\mathrm{avg\text{-}resid}}(\varepsilon) = \Theta(\varepsilon^{-1/\omega^*}) \qquad \text{as} \, \varepsilon \to 0,$$

where

$$\omega^* = \frac{r}{d} + \frac{p-1}{2}.$$

Hence, if p changes slowly with d, then the complexity $\mathrm{comp}^{\mathrm{avg\text{-}resid}}(\varepsilon)$ is no longer exponential in d. In particular, if p is independent of d, then the (pessimistic) upper bound

$$\mathrm{comp}^{\mathrm{avg\text{-}resid}}(\varepsilon) = O(\varepsilon^{-(p-1)/2}) \qquad \text{as} \, \varepsilon \to 0$$

is independent of the dimension d.

Notes and Remarks

NR 7.5.2:1 Much of the material in this subsection is also discussed in Werschulz (1987c).

7.6 The probabilistic setting

We now consider results in the probabilistic setting. This setting blends features of both the worst case and average case settings, in that we measure error by the worst case behavior over the class of problem elements, but are willing to ignore possible bad behavior over sets of small measure δ. Moreover, we measure cost by the worst case.

Our main result will be that results for the probabilistic setting will be similar to those for the average case setting. For shift-invariant problems with δ fixed, the probabilistic complexity will be approximately the same as the average case complexity, and the same algorithms and information will be optimal. In particular, if the finite element method is almost optimal for the average case setting, then we show that it is arbitrarily close to being optimal for the probabilistic setting. For ill-posed problems with arbitrary $\delta \in [0,1]$ and positive ε, information and algorithms that are optimal for the average case setting are also optimal in the probabilistic setting.

7.6.1   General results

Suppose that $S\colon F \to G$ is a solution operator, where G is a normed linear space and F is a set of problem elements. Let μ be a probability measure on F and let $\delta \in [0,1]$. As in §3.5, the probabilistic δ-error of an algorithm ϕ using information N is given by

$$e^{\mathrm{prob}}(\phi, N, \delta) = \inf_{A \subseteq F: \mu(A) \leq \delta} \sup_{f \in F - A} \|Sf - \phi(Nf)\|,$$

and the cost of this algorithm is given by

$$\mathrm{cost}^{\mathrm{prob}}(\phi, N) = \sup_{f \in F} \mathrm{cost}(\phi, N, f),$$

as in the worst case setting.

We deal with the probabilistic setting by reducing it to the average case setting using a different error criterion. Indeed, for $x > 0$, let us define an error functional ER_x on G by

$$\mathrm{ER}_x(g) = \begin{cases} 0 & \text{if } \|g\| \leq x, \\ 1 & \text{otherwise,} \end{cases}$$

and define

$$\begin{aligned} M(\phi, N, x) &= \int_F \mathrm{ER}_x\left(Sf - \phi(Nf)\right) \mu(df) \\ &= \mu(\{f \in F : \|Sf - \phi(Nf)\| > x\}). \end{aligned}$$

That is, $M(\phi, N, x)$ is the average error of ϕ using N, with error measured by the new error criterion ER_x.

Note that

$$e^{\mathrm{prob}}(\phi, N, \delta) = M^{-1}(\phi, N, \delta) = \inf\{\, x \geq 0 : M(\phi, N, x) \leq \delta \,\}.$$

If this infimum is attained, then

$$e^{\mathrm{prob}}(\phi, N, \delta) \leq \varepsilon \quad \text{iff} \quad M(\phi, N, \varepsilon) \leq \delta,$$

showing the relation between the probabilistic error of an algorithm and its average case under the error criterion ER_x.

As always, for given information N, we wish to find the algorithm making best use of N. That is, we wish to find the *probabilistic δ-radius of information*, which is given by

$$r^{\mathrm{prob}}(N, \delta) = \inf_{\phi \text{ using } N} e^{\mathrm{prob}}(\phi, N, \delta).$$

An algorithm ϕ using N for which

$$e^{\mathrm{prob}}(\phi, N, \delta) = r^{\mathrm{prob}}(N, \delta)$$

is said to be an *optimal error algorithm.*

Using the probabilistic radius of information, we can determine bounds on the probabilistic complexity. Let

$$m^{\mathrm{prob}}(\varepsilon, \delta) = \min\{\,\mathrm{card}\,N : N \text{ such that } r^{\mathrm{prob}}(N, \delta) \le \varepsilon\,\}$$

denote the *probabilistic (ε, δ)-cardinality number.* Here, $\mathrm{card}\,N$ denotes the worst case cardinality, i.e. $\mathrm{card}\,N = \sup_{f \in F} n(f)$, where $n(f)$ is the number of functional evaluations used in computing Nf for the problem element $f \in F$. Note that since we are using a worst case cost, it is no restriction to consider only information of fixed cardinality.

We can give a useful characterization of $m^{\mathrm{prob}}(\varepsilon, \delta)$ in terms of induced and conditional measures. For information N, we let $\mu_1 = \mu N^{-1}$ be the *induced measure* generated by μ and N. Then μ_1 is a Gaussian measure on $\mathbb{R}^n$ with mean zero and correlation operator the $n \times n$ identity matrix. For $y = Nf$, we let $\mu_2(\cdot|y)$ be the Gaussian measure on $\mathfrak{B}(F_1)$ whose mean is

$$m_{\mu, y} = \sum_{j=1}^{n} y_j C_\mu \lambda_j$$

and whose correlation operator is

$$C_{\mu, N} \lambda = C_\mu \lambda - \sum_{j=1}^{n} \lambda(C_\mu \lambda_j) \lambda_j C_\mu \qquad \forall \lambda \in F_1^*.$$

As the Appendix to IBC shows, $\{\mu_2(\cdot|y)\}_{y \in N(F_1)}$ is a family of *conditional measures,* i.e.

(i) $\mu_2(N^{-1}y|y) = 1$ for almost all $y \in N(F_1)$,
(ii) $\mu_2(B|\cdot)$ is μ_1-measurable for all $B \in \mathfrak{B}(F_1)$,
(iii) $\mu(B) = \int_{N(F)} \mu_2(B|y)\,\mu_1(dy)$ for all $B \in \mathfrak{B}(F_1)$.

Recall that $\nu = \mu S^{-1}$ is the a priori measure on the set of solution elements. For $y \in N(F)$, we let $\nu(\cdot|y)$ be the corresponding *a posteriori measure,* which is defined by

$$\nu(B|y) = \mu_2\big(S^{-1}(B)|y\big) \qquad \forall B \in \mathfrak{B}(G).$$

If we let

$$M(N, x) = \inf_{\phi \text{ using } N} M(\phi, N, x) = 1 - \int_{N(F)} \sup_{z \in G} \nu\big(B(z, x)|y\big)\,\mu_1(dy),$$

where $B(z, x)$ denotes the ball of radius x with center z, then we find that

$$m^{\text{prob}}(\varepsilon, \delta) = \min\{\, \text{card}\, N : N \text{ such that } M(N, \varepsilon) \le \delta \,\}.$$

Since c is the cost of one information evaluation, we have

$$\text{comp}^{\text{prob}}(\varepsilon, \delta) \ge c\, m^{\text{prob}}(\varepsilon, \delta).$$

As with previous settings, we find that

$$\text{comp}^{\text{prob}}(\varepsilon, \delta) \simeq c\, m^{\text{prob}}(\varepsilon, \delta)$$

if there is an algorithm using information of cardinality $m^{\text{prob}}(\varepsilon, \delta)$ whose error is at most ε and whose combinatory cost is dominated by the informational cost.

We now specialize these results to continuous linear problems, as described in §7.3. Hence, we suppose that S is a continuous linear operator and that the set F of problem elements is a Hilbert space F_1. We also assume that G is a Hilbert space. The measure μ on F_1 is a Gaussian measure with zero mean and positive definite correlation operator C_μ. We also assume that $\Lambda = \Lambda^*$, i.e. any continuous linear functional is permissible as an information operation.

Recall that the probabilistic setting has been reduced to the average case setting under a new error criterion. As shown in Section 6.3 in Chapter 6 of IBC, many of the results in §7.3 also hold when the mean-square error criterion used in that section is replaced by this new error criterion. This observation will allow us to easily determine optimal algorithms and information for our problem in the probabilistic setting.

We first discuss optimal error algorithms for the probabilistic setting. Recall that the μ-spline algorithm was an optimal error algorithm in the average case setting. From Section 5.1 in Chapter 8 of IBC, we have

THEOREM 7.6.1.1. *Let N be nonadaptive information. Then the μ-spline algorithm using N is an optimal error algorithm, and*

$$r^{\text{prob}}(N, \delta) = M^{-1}(N, \delta) = \inf\{\, x \ge 0 : M(N, x) \le \delta \,\}. \qquad \square$$

Hence the same algorithm has optimal error in both the average case and probabilistic case settings.

Next, we need to find nonadaptive information having minimal radius over all information of given cardinality. For any nonnegative integer n, let Λ_n denote nonadaptive information of cardinality at most n, and let

$$r^{\text{prob}}(n, \delta) = \inf_{N \in \Lambda_n} r^{\text{prob}}(N, \delta)$$

denote the nth *minimal probabilistic radius of information*. Information N_n^* for which

$$r^{\text{prob}}(N_n^*, \delta) = r^{\text{prob}}(n, \delta)$$

is said to be nth *optimal information* for the probabilistic setting. We seek to determine such optimal information. Recall that the a priori measure $\nu = \mu S^{-1}$ on the set of solution elements has zero mean and covariance operator $C_\nu = S C_\mu S^*$. Let $\{u_j\}_{j=1}^{\infty}$ be a complete orthonormal basis for G consisting of eigenvectors for C_ν, with

$$C_\nu u_j = \gamma_j u_j \qquad (j = 1, 2, \dots),$$

and

$$\gamma_1 \geq \gamma_2 \geq \dots \geq 0 \qquad \text{and} \qquad \sum_{j=1}^{\infty} \gamma_j = \text{trace}\, C_\nu < \infty.$$

For any index j such that $\gamma_j > 0$, we let

$$\lambda_j^*(f) = \gamma_j^{-1/2} \langle Sf, u_j \rangle.$$

For any index n, we define information N_n^* of cardinality n by

$$N_n^* f = \begin{bmatrix} \lambda_1^*(f) \\ \vdots \\ \lambda_n^*(f) \end{bmatrix} \qquad \forall f \in F_1,$$

and an algorithm ϕ_n^* using N_n^* by

$$\phi_n^*(N_n^* f) = \sum_{j=1}^{n} \lambda_j^*(f) S C_\mu \lambda_j^* = \sum_{j=1}^{n} \langle Sf, u_j \rangle u_j \qquad \forall f \in F_1.$$

Of course, N_n^* is nth optimal information for the average case setting and ϕ_n^* is the μ-spline algorithm using N_n^*. From Section 5.3 in Chapter 8 of IBC, we have

THEOREM 7.6.1.2. *For any index n, the information N_n^* is nth optimal information for the probabilistic setting.* □

Hence the same information is optimal for both the average case and probabilistic case settings.

Note, however, that we do not give tight bounds for the nth minimal probabilistic radius. Such bounds as are known for general S are not tight, namely that

$$\sqrt{\gamma_{n+1}} \psi^{-1}(1 - \delta) \leq r^{\text{prob}}(n, \delta) \leq r^{\text{avg}}(n) \sqrt{2 \ln \frac{5}{\delta}},$$

where

$$\psi(z) = \sqrt{\frac{2}{\pi}} \int_0^z e^{-t^2/2}\, dt,$$

so that

$$\psi^{-1}(1-\delta) = \sqrt{2\ln\frac{1}{\delta}}\,(1 + o(1)) \qquad \text{as } \delta \to 0.$$

See the Notes and Remarks for further discussion.

Notes and Remarks

NR 7.6.1:1 This general material is from Chapter 8 of IBC.

NR 7.6.1:2 The case that F is a ball of finite radius in F_1 may be reduced to the case $F = F_1$, see Section 5.5 in Chapter 8 of IBC.

NR 7.6.1:3 Although tight bounds on $r^{\mathrm{prob}}(n,\delta)$ are not known for general S, Heinrich (1989) has developed tight bounds for a special one-dimensional approximation problem. We suspect that his techniques should extend to include other shift-invariant problems, but this requires further research.

7.6.2   Results for shift-invariant problems

Next, we look at the probabilistic complexity of shift-invariant problems. We show that the results for the probabilistic setting are essentially the same as those for the average case setting, if δ is fixed, independent of ε.

Recall that that our problem is (m,s)-shift-invariant in $\mathbb{R}^d$. The problem elements have smoothness r, and we measure error in the $H^l(\Omega)$-norm for some $l \in [0, \min\{m,s\}]$. The measure μ has zero mean and correlation operator C_μ. The eigenvalues β_j of C_μ tend to zero as j^{-p} for some $p > 1$. The eigenvectors of C_μ are those of $\hat{A} = (A^*A)^{1/2}$, where A and A^* are (m',s')-shift-invariant. For further details, see §7.4.

In what follows, we only consider the case $\Lambda = \Lambda^*$, leaving the case of standard information for future investigation. Recall that

$$\omega^* = \frac{r+s-l}{d} + \frac{p-1}{2}$$

as in §7.4.

THEOREM 7.6.2.1.   *Let $\delta \in (0, \frac{1}{2})$ be fixed. Then the following hold for the probabilistic setting:*

   (i) $r^{\mathrm{prob}}(n,\delta) = \Theta(n^{-\omega^*})$ *as $n \to \infty$.*
   (ii) $\mathrm{comp}^{\mathrm{prob}}(\varepsilon,\delta) = \Theta(\varepsilon^{-1/\omega^*})$ *as $\varepsilon \to 0$.*
   (iii) *If $k \geq s+r+\frac{1}{2}d(p-1)$, then the finite element method of degree k is arbitrarily close to being an almost optimal complexity algorithm for computing an (ε,δ)-approximation with small ε and fixed δ.*

That is, for any nonnegative-valued function $h\colon [0, \varepsilon_0] \to \mathbb{R}$ with $\lim_{\varepsilon \to 0} h(\varepsilon) = 0$, the cost of computing an (ε, δ)-approximation with the FEM of degree k is at most $O\big(\mathrm{comp}^{\mathrm{prob}}(\varepsilon h(\varepsilon), \delta)\big)$.

This means, for instance, that the cost of computing an (ε, δ)-approximation with the FEM of degree k is at most $O(\log \log \ldots \log(\varepsilon^{-1}) \cdot \varepsilon^{-1/\omega^*})$ as $\varepsilon \to 0$.

Before we prove this theorem, we quote the following useful result:

LEMMA 7.6.2.1. *Recall that* $\gamma_1 \geq \gamma_2 \geq \ldots > 0$ *are the eigenvalues of* $C_\nu = SC_\mu S^*$. *Suppose that*

$$\lim_{n \to \infty} \frac{\sum_{k=n+1}^{\infty} \gamma_k^3}{\left(\sum_{k=n+1}^{\infty} \gamma_k^2\right)^{3/2}} = 0$$

and that

$$\lim_{n \to \infty} \frac{\sum_{k=\lceil tn \rceil + 1}^{n} \gamma_k}{\left(\sum_{k=n+1}^{\infty} \gamma_k^2\right)^{1/2}} = \infty$$

for all $t \in (0, 1)$. *Then for any* $\delta \in (0, \frac{1}{2})$,

$$\lim_{\varepsilon \to 0} \frac{m^{\mathrm{prob}}(\varepsilon, \delta)}{m^{\mathrm{avg}}(\varepsilon)} = 1,$$

and hence

$$\lim_{\varepsilon \to 0} \frac{\mathrm{comp}^{\mathrm{prob}}(\varepsilon, \delta)}{\mathrm{comp}^{\mathrm{avg}}(\varepsilon)} = 1.$$

PROOF. See pp. 339–341 of IBC. □

We now proceed to the

PROOF OF THEOREM 7.6.2.1. Recall that $0 < \alpha_1 \leq \alpha_2 \leq \ldots$ are the eigenvalues of the unbounded operator $\hat{A}$, and that $\alpha_j = \Theta(j^{s'/d})$ as $j \to \infty$. From the proof of Theorem 7.4.2.2, we see that

$$\gamma_j = \Theta(\beta_j \alpha_j^{-2(r+s-l)/s'}) = \Theta(j^{-(p+2(r+s-l)/d)}) \qquad \text{as } j \to \infty,$$

from which it is easy to check that the hypotheses of Lemma 7.6.2.1 are satisfied.

Since

$$m^{\mathrm{prob}}(\varepsilon, \delta) \sim m^{\mathrm{avg}}(\varepsilon) = \Theta(\varepsilon^{-1/\omega^*}) \qquad \text{as } \varepsilon \to 0,$$

it immediately follows that

$$r^{\mathrm{prob}}(n, \delta) = \Theta(n^{-\omega^*}) \qquad \text{as } n \to \infty$$

and

$$\mathrm{comp}^{\mathrm{prob}}(\varepsilon,\delta) = \Theta(\varepsilon^{-1/\omega^*}) \qquad \text{as } \varepsilon \to 0.$$

It only remains to prove that the FEM is nearly almost optimal. Let $h\colon [0,\varepsilon_0] \to \mathbb{R}$ be any nonnegative function for which $\lim_{\varepsilon \to 0} h(\varepsilon) = 0$. Since $\delta \in (0,\frac{1}{2})$ is fixed, there exists positive ε_1 such that $(1-5e^{-1/(8h(\varepsilon)^2)})$. $(1-16h^2(\varepsilon)) \geq 1-\delta$ for $\varepsilon \in [0,\varepsilon_1]$. We also let $\nu_{N^{\mathrm{FE}}_{n,k}}$ denote the Gaussian measure with mean zero and correlation operator

$$C_{\nu,N^{\mathrm{FE}}_{n,k}} = C_\nu - \sum_{j=1}^{n} \langle \cdot, SC_\mu\lambda_j\rangle SC_\mu\lambda_j,$$

where

$$N^{\mathrm{FE}}_{n,k}f = \begin{bmatrix} \lambda_1(f) \\ \vdots \\ \lambda_n(f) \end{bmatrix} \qquad \forall f \in F,$$

with μ-orthonormal functionals $\lambda_1,\dots,\lambda_n$. It is straightforward to check that $\mathrm{trace}\, C_{\nu,N^{\mathrm{FE}}_{n,k}} = r^{\mathrm{avg}}(N^{\mathrm{FE}}_{n,k})^2$.

Choose $\varepsilon \in (0,\varepsilon_1]$, and pick $n \in \mathbb{N}$ such that $e^{\mathrm{avg}}(\phi^{\mathrm{FE}}_{n,k}, N^{\mathrm{FE}}_{n,k}) \leq \varepsilon h(\varepsilon)$. Clearly we have

$$\begin{aligned} \mathrm{cost}^{\mathrm{prob}}(\phi^{\mathrm{FE}}_{n,k}, N^{\mathrm{FE}}_{n,k}) = \Theta(n) &= \Theta\big(\mathrm{comp}^{\mathrm{avg}}(\varepsilon h(\varepsilon))\big) \\ &= \Theta\big(\mathrm{comp}^{\mathrm{prob}}(\varepsilon h(\varepsilon))\big) \qquad \text{as } \varepsilon \to 0, \end{aligned}$$

the last equality holding by Lemma 7.6.2.1. We need only show that

$$e^{\mathrm{prob}}(\phi^{\mathrm{FE}}_{n,k}, N^{\mathrm{FE}}_{n,k}, \delta) \leq \varepsilon,$$

i.e. that

$$M(\phi^{\mathrm{FE}}_{n,k}, N^{\mathrm{FE}}_{n,k}, \varepsilon) \leq \delta.$$

Here

$$M(\phi^{\mathrm{FE}}_{n,k}, N^{\mathrm{FE}}_{n,k}, \varepsilon) = 1 - \int_{\mathbb{R}^n} \nu\big(B(\phi^{\mathrm{FE}}_{n,k}(y),\varepsilon)|y\big)\, \mu_1(dy),$$

where $B(z,x)$ denotes the $H^l_E(\Omega)$-ball of radius x and center z (see §7.6.1).

For $y \in \mathbb{R}^n$, let $a(y) = \phi^{\mathrm{FE}}_{n,k}(y) - \phi^s(y)$ and let $Z_n(\varepsilon) = \{\, y \in \mathbb{R}^n : \|a(y)\|_{H^l(\Omega)} \leq \frac{1}{2}\varepsilon \,\}$. Then

$$\begin{aligned} \int_{\mathbb{R}^n} \nu\big(B(\phi^{\mathrm{FE}}_{n,k}(y),\varepsilon)|y\big)\, \mu_1(dy) &= \int_{\mathbb{R}^n} \nu\big(B(a(y),\varepsilon)|0\big)\, \mu_1(dy) \\ &= \int_{\mathbb{R}^n} \nu_{N^{\mathrm{FE}}_{n,k}}\big(B(a(y),\varepsilon)\big)\, \mu_1(dy) \\ &\geq \int_{Z_n(\varepsilon)} \nu_{N^{\mathrm{FE}}_{n,k}}\big(B(a(y),\varepsilon)\big)\, \mu_1(dy), \end{aligned}$$

the equalities holding as on p. 330 of IBC. For any $y \in Z_n(\varepsilon)$, the triangle inequality implies that $B(0, \frac{1}{2}\varepsilon) \subseteq B(a(y), \varepsilon)$. Hence

$$\int_{Z_n(\varepsilon)} \nu_{N_{n,k}^{\mathrm{FE}}} \left(B(a(y), \varepsilon) \right) \mu_1(dy) \geq \int_{Z_n(\varepsilon)} \nu_{N_{n,k}^{\mathrm{FE}}} \left(B(0, \tfrac{1}{2}\varepsilon) \right) \mu_1(dy)$$

$$= \mu_1 \left(Z_n(\varepsilon) \right) \cdot \nu_{N_{n,k}^{\mathrm{FE}}} \left(B(0, \tfrac{1}{2}\varepsilon) \right).$$

From **NR 7.3:4**, we find that

$$\nu_{N_{n,k}^{\mathrm{FE}}} \left(B(0, \tfrac{1}{2}\varepsilon) \right) \geq 1 - 5 \exp \left(\frac{-(\frac{1}{2}\varepsilon)^2}{2 \operatorname{trace} \nu_{N_{n,k}^{\mathrm{FE}}}} \right) = 1 - 5 \exp \left(\frac{-\varepsilon^2}{8 \, r^{\mathrm{avg}}(N_{n,k}^{\mathrm{FE}})^2} \right)$$

$$\geq 1 - 5 e^{-1/(8h(\varepsilon)^2)}.$$

Since ϕ^s is an average case optimal error algorithm using $N_{n,k}^{\mathrm{FE}}$, the triangle inequality and the inequality $(a+b)^2 \leq 2(a^2 + b^2)$ imply that

$$\int_{\mathbb{R}^n} \|a(y)\|_{H^l(\Omega)}^2 \, \mu_1(dy) = \int_{\mathbb{R}^n} \int_{(N_{n,k}^{\mathrm{FE}})^{-1}(y)} \|a(y)\|_{H^l(\Omega)}^2 \, \mu_2(df|y) \, \mu_1(dy)$$

$$\leq 2 \left(e^{\mathrm{avg}}(\phi_{n,k}^{\mathrm{FE}}, N_{n,k}^{\mathrm{FE}})^2 + e^{\mathrm{avg}}(\phi^s, N_{n,k}^{\mathrm{FE}})^2 \right)$$

$$\leq 4 \, e^{\mathrm{avg}}(\phi_{n,k}^{\mathrm{FE}}, N_{n,k}^{\mathrm{FE}})^2$$

$$\leq 4\varepsilon^2 h(\varepsilon)^2.$$

But

$$\int_{\mathbb{R}^n} \|a(y)\|_{H^l(\Omega)}^2 \, \mu_1(dy) \geq \int_{\mathbb{R}^n \setminus Z_n(\varepsilon)} \|a(y)\|_{H^l(\Omega)}^2 \, \mu_1(dy)$$

$$\geq \tfrac{1}{4}\varepsilon^2 \left[1 - \mu_1 \left(Z_n(\varepsilon) \right) \right],$$

and so

$$\mu_1 \left(Z_n(\varepsilon) \right) \geq 1 - 16 h^2(\varepsilon).$$

Putting this all together, we find that

$$\int_{\mathbb{R}^n} \nu \left(B(\phi_{n,k}^{\mathrm{FE}}, \varepsilon) | y \right) \mu_1(dy) \geq \left(1 - 5 e^{-1/(8h(\varepsilon)^2)} \right) \cdot \left(1 - 16 h^2(\varepsilon) \right) \geq 1 - \delta,$$

the second inequality holding because $\varepsilon \in (0, \varepsilon_1]$. Hence $M(\phi_{n,k}^{\mathrm{FE}}, N_{n,k}^{\mathrm{FE}}, \varepsilon) \leq \delta$, and so $e^{\mathrm{prob}}(\phi_{n,k}^{\mathrm{FE}}, N_{n,k}^{\mathrm{FE}}, \delta) \leq \varepsilon$, as required. $\qquad\square$

Of course, the obvious question is to determine whether we can get rid of the factor $h(\varepsilon)$. Since some of the inequalities we have used are quite loose, it seems reasonable to conjecture that we can do so. Moreover, we suspect that the FEM is almost optimal in the probabilistic setting iff it is almost optimal in the average case setting. Hence we pose

CONJECTURE 7.6.2.1. *The following are equivalent:*

(i) $k \geq s + r + \frac{1}{2}d(p-1)$.

(ii) *The finite element method of degree k is an almost optimal complexity algorithm that computes an (ε, δ)-approximation for small ε and fixed δ.*   ⬜

Notes and Remarks

NR 7.6.2:1 The proof that the FEM is close to being almost optimal in the probabilistic setting uses some ideas of H. Woźniakowski (private communication).

7.6.3   Ill-posed problems

We now briefly discuss ill-posed problems in the probabilistic setting. Once again, the solution operator S is a measurable linear transformation, whose domain D is a subspace of a separable Banach space F_1 and whose codomain is a separable Hilbert space G. The measure μ on F_1 is a Gaussian measure with mean zero and positive definite correlation operator C_μ, the domain D of S having full measure.

Recall from §7.5 that the a priori measure $\nu = \mu S^{-1}$ on the solution elements is a Gaussian measure with correlation operator C_ν. The patient reader may check that all the results for linear problems that are found in §7.6.1 still hold, since these proofs (as found in Section 5 of Chapter 8 of IBC) only require that the a priori measure ν be Gaussian. In particular, the following hold:

(i) For any nonadaptive information N, the μ-spline algorithm is an optimal error algorithm in the probabilistic setting.

(ii) For any n, the same information is (almost) optimal for the average and probabilistic settings.

(iii) All estimates in §7.6.1 for the minimal radius, cardinality number, and complexity still hold.

Hence we can always solve ill-posed problems in the probabilistic setting.

8

Complexity in the asymptotic and randomized settings

8.1 Introduction

Recall that our emphasis so far has been on finding an algorithm that computes an ε-approximation over a *set* of problem elements, this computation being done at minimal cost. Moreover, we have only considered information and algorithms that are *deterministic*, not allowing for randomness. We now look at what happens if we measure what happens at a fixed problem element as the algorithms get successively better (the asymptotic setting) or if we allow nondeterministic algorithms (the randomized setting).

We first look at the asymptotic setting. In this setting, we require that a sequence of algorithms approximate the solution at a *particular* problem element, and that this sequence converge as quickly as possible. The main result is that algorithms and information that are optimal for linear problems in the worst case setting are also optimal in the asymptotic setting.

Then, we consider the randomized setting. One idea that has become popular has been to use randomized information and algorithms, a prime example being the Monte Carlo methods used for high-dimensional numerical quadrature. Recall that many of the problems we have studied so far have either been intractable for regions of high dimension or had infinite complexity. Based on the success of Monte Carlo methods for other problems, it is reasonable to ask to what extent randomization can help in the solution of PDE and IE over high-dimensional regions.

We first look at shift-invariant problems (such as elliptic PDE and IE of the second kind) in the randomized setting. Our results are somewhat mixed. Suppose first that we want to approximate the solution everywhere in the region. Then the randomized complexity of these problems is the same as in the deterministic worst case setting. Hence these problems remain intractable, even when we switch to the randomized setting. However, if we only need the solution at a small number of points, then we show that randomization *can* render the problem tractable.

Finally, we consider the randomized complexity of the most important class of ill-posed problems, namely, Fredholm problems of the first kind. We show that these problems have infinite complexity in the randomized setting. Hence randomization does not help us to solve ill-posed problems.

Notes and Remarks

NR 8.1:1 Our results in this chapter are based heavily on work by other authors. The seminal work on complexity in the asymptotic setting was done by G. M. Trojan, in an unpublished manuscript that circulated in 1983, while that for complexity in the randomized setting may be found in Chapter 11 of IBC.

8.2   Asymptotic setting

We now consider the asymptotic setting, in which we look for a sequence of approximations with the fastest possible speed of convergence. The main result is that this fastest speed is given by the sequence of nth minimal worst case radii. Using this result, we shall see that the finite element method (with or without quadrature) is optimal in the asymptotic setting precisely when it is optimal in the worst case setting.

We first define our notion of information and algorithm more precisely for the asymptotic setting. Let F_1 be a Banach space and G a normed linear space. Let $S \colon F_1 \to G$ be a solution operator. We wish to approximate Sf for any $f \in F_1$. Assume that $\Lambda \subseteq \Lambda^*$ is the class of permissible information, i.e. any permissible information operation must be a continuous linear functional.

The *information* we use will have the form

$$\overline{N}f = \begin{bmatrix} \lambda_1(f) \\ \lambda_2(f; y_1) \\ \vdots \\ \lambda_n(f; y_1, \ldots, y_{n-1}) \\ \vdots \end{bmatrix}.$$

Here,

$$y_i = \lambda_i(f; y_1, \ldots, y_{i-1})$$

and

$$\lambda_i(\,\cdot\,; y_1, \ldots, y_{i-1}) \in \Lambda$$

for any index i. Note that we allow information to be adaptive.

Let $N_n f$ denote the first n evaluations in $\overline{N}f$, i.e.

$$N_n f = \begin{bmatrix} \lambda_1(f) \\ \lambda_2(f; y_1) \\ \vdots \\ \lambda_n(f; y_1, \ldots, y_{i-1}) \end{bmatrix}.$$

By an *algorithm*, we shall mean a sequence $\overline{\phi} = \{\phi_n\}_{n \in \mathbb{N}}$ of mappings, where $\phi_n \colon N_n(F_1) \subseteq \mathbb{R}^n \to G$, i.e. ϕ_n is an algorithm using N_n (in the usual sense).

We are now able to define optimality in the asymptotic setting. An algorithm $\overline{\phi}^*$ using information $\overline{N}^*$ is said to be *optimal* if for any algorithm $\overline{\phi}$ using any permissible information $\overline{N}$, and for any positive sequence $\{\delta_n\}_{n \in \mathbb{N}}$ converging to 0,

$$\mathrm{int} \left\{ f \in F_1 : \lim_{n \to \infty} \frac{\|Sf - \phi_n(N_n f)\|}{\delta_n \|Sf - \phi_n^*(N_n^* f)\|} = 0 \right\} = \emptyset.$$

(Here, and in what follows, we use the convention that $0/0 = 1$.) Hence an algorithm is optimal iff it can only be beaten on a set of problem elements whose interior is empty.

We first report results for continuous linear problems, i.e. we assume that S is a bounded linear operator.

Recall that $r(n, \Lambda)$ is the nth minimal radius of nonadaptive continuous linear information from Λ. That is, if Λ_n is the class of nonadaptive continuous linear information from Λ whose cardinality is at most n, then

$$r(n, \Lambda) = \inf_{N \in \Lambda_n} \inf_{\phi \text{ using } N} e(\phi, N).$$

We then have the following

THEOREM 8.2.1. *Let $\overline{N}$ be adaptive continuous linear information, and let $\overline{\phi}$ be any algorithm using $\overline{N}$. Suppose that $\{\delta_n\}_{n \in \mathbb{N}}$ is any positive sequence for which $\lim_{n \to \infty} \delta_n = 0$. Then the interior of the set*

$$\left\{ f \in F_1 : \lim_{n \to \infty} \frac{\|Sf - \phi_n(N_n f)\|}{\delta_n r(n, \Lambda)} = 0 \right\}$$

is empty.

PROOF. See Sections 2.1 and 2.2 in Chapter 10 of IBC. □

Note that the sequence $\{\delta_n\}_{n \in \mathbb{N}}$ can converge to zero arbitrarily slowly. The result of this theorem says that modulo this sequence, no sequence of algorithms can converge faster than the sequence of nth minimal radii. When is this rate of convergence achieved? Suppose that

$$r\left(\lceil \tfrac{1}{4}(n+1) \rceil\right) = \Theta\big(r(n, \Lambda)\big) \qquad \text{as } n \to \infty. \tag{1}$$

Let

$$N_n^{\mathrm{non}} f = \begin{bmatrix} \lambda_{n,1}(f) \\ \lambda_{n,2}(f) \\ \vdots \\ \lambda_{n,n}(f) \end{bmatrix}$$

be nth almost optimal information, i.e.

$$r(N_n^{\text{non}}) = \Theta\big(r(n, \Lambda)\big) \qquad \text{as } n \to \infty.$$

We choose our information $\overline{N^*}$ to be

$$\overline{N^*}f = \begin{bmatrix} N_1^{\text{non}} f \\ N_2^{\text{non}} f \\ N_4^{\text{non}} f \\ \vdots \\ N_{2^k}^{\text{non}} f \\ \vdots \end{bmatrix}.$$

Moreover, let ϕ_n^* be any algorithm using N_n for which

$$\|Sf - \phi_n^*(N_n f)\| \le C\, r(n, \Lambda)\|f\|,$$

with C independent of n and f. Letting $\overline{\phi^*} = \{\phi_n^*\}_{n \in \mathbb{N}}$, it is easy to show

THEOREM 8.2.2.   *Assume that (1) holds. Then the algorithm $\overline{\phi^*}$ using information $\overline{N^*}$ described above is optimal in the asymptotic setting.*

PROOF. See Section 2.2 in Chapter 10 of IBC.   □

Hence if (1) holds, then $\overline{\phi^*}$ has optimal convergence in the asymptotic setting. Note that although adaptive information is allowed, $\overline{\phi^*}$ uses non-adaptive information $\overline{N^*}$. Hence, adaption is no stronger than nonadaption in the asymptotic setting for continuous linear problems.

We now look at the asymptotic setting for the problems we have studied so far, namely, shift-invariant problems, ill-posed problems, and ordinary differential equations.

Recall that shift-invariant problems include the solution of elliptic PDE, elliptic systems, and Fredholm problems of the second kind. For shift-invariant problems, we may use Theorems 7.4.2.1 and 7.4.2.1 to see that condition (1) always holds. Hence the same algorithms and information that were optimal in the worst case setting are also optimal in the asymptotic setting. In particular, the conditions that we found to be necessary and sufficient for the finite element method (with or without quadrature) to be optimal in the worst case setting are also necessary and sufficient for its asymptotic optimality.

For ill-posed problems, the results of these theorems do not directly apply, since continuity of S is one of the initial hypotheses. Indeed, this continuity figures heavily in the proof of Theorem 8.2.1. As we show in NR 8.2:3, it is possible to have a sequence of algorithms that converges to the solution of an ill-posed problem; however, this convergence may be arbitrarily slow. Based on this example, we conjecture that no sequence of algorithms for an ill-posed problem can "have a rate of convergence," in the following sense:

CONJECTURE 8.2.1. *Let the solution operator* $S \colon D \subset F_1 \to G$ *be an unbounded, closed, densely defined linear transformation of Banach spaces, and let* $\overline{\phi} = \{\phi_n\}_{n \in \mathbb{N}}$ *be an algorithm using information* $\overline{N} = \{N_n\}_{n \in \mathbb{N}}$ *for its approximate solution. For any sequence* $\{\delta_n\}_{n \in \mathbb{N}}$ *converging to zero, the set*

$$\left\{ f \in F_1 : \lim_{n \to \infty} \frac{\|Sf - \phi_n(N_n f)\|}{\delta_n} < \infty \right\}$$

has empty interior. 🛛

We now consider the initial value problem for ordinary differential equations. This is a nonlinear problem, and hence Theorems 8.2.1 and 8.2.2 do not apply. However, Kacewicz (1988) has derived a generalization of Theorems 8.2.1 and 8.2.2 that hold for nonlinear problems satisfying certain special conditions. Furthermore, these conditions hold for ODE initial value problems. Hence, the same algorithms that are optimal in the worst case setting for ODE are also optimal in the asymptotic setting.

Notes and Remarks

NR 8.2:1 The results in this section are based mainly on Chapter 10 of IBC, which should be consulted for proofs and further details.

NR 8.2:2 The initial work on optimal algorithms for linear problems in the asymptotic setting was done in 1983 by G. M. Trojan in an unpublished manuscript.

NR 8.2:3 Let $F_1 = G$ be a Hilbert space with orthonormal basis $e_1, e_2, \ldots$. Let $\{\alpha_j\}_{j \in \mathbb{N}}$ be any increasing positive sequence for which $\lim_{j \to \infty} \alpha_j = \infty$. Let

$$D = \left\{ f \in F_1 : \sum_{j=1}^{\infty} \alpha_j^2 \langle f, e_j \rangle^2 < \infty \right\},$$

and define the closed unbounded operator $S \colon D \subset F_1 \to G$ by

$$Sf = \sum_{j=1}^{\infty} \alpha_j \langle f, e_j \rangle e_j \qquad \forall f \in D.$$

For $f \in D$, choose information

$$\overline{N}f = \begin{bmatrix} \langle f, e_1 \rangle \\ \langle f, e_2 \rangle \\ \vdots \end{bmatrix},$$

so that

$$N_n f = \begin{bmatrix} \langle f, e_1 \rangle \\ \vdots \\ \langle f, e_n \rangle \end{bmatrix}.$$

Define an algorithm $\overline{\phi}$ using $\overline{N}$ by

$$\phi_n(N_n f) = \sum_{j=1}^{n} \alpha_j \langle f, e_j \rangle e_j \qquad \forall f \in D.$$

It is easy to check that $\lim_{n \to \infty} \|Sf - \phi_n(N_n f)\| = 0$ for $f \in D$, i.e. $\overline{\phi}$ is convergent. However, the convergence can be arbitrarily slow.

NR 8.2:4 In this section, we have said that an algorithm was deemed optimal iff its rate of convergence could not be beaten except on a set having empty interior, i.e. on a "small" set. We can also say that a set is "small" if its measure is zero. This leads us to consider an "asymptotic average case setting." In this new setting, we say that an algorithm is optimal iff its rate of convergence can only be beaten on a set of measure zero. Section 4 in Chapter 10 of IBC studies continuous linear problems defined over a Banach space with Gaussian measure μ. Their main result is that the sequence of μ-spline algorithms is optimal in the asymptotic average case setting. Moreover, it is easy to check that these results also hold for ill-posed problems whose solution operator is bounded on the average.

Recall that in Chapter 7, we found conditions for the FEM and FEMQ to be optimal for shift-invariant problems in the average case setting. It is not difficult to show that when these conditions hold, these methods are also optimal in the asymptotic average case setting.

8.3   Randomized setting

We now consider the randomized setting. Whereas in the previous settings the information and algorithm were deterministic, we allow a random choice of the information and algorithm in the randomized setting. By allowing randomized information, we hope that the problems that were intractable in the worst case setting may be made tractable. We first consider shift-invariant problems. Suppose that we want to approximate the solution everywhere. It turns out that the randomized and deterministic complexities are roughly the same, and so randomization does not help make our problem tractable. However, if we only want the approximate values at a few points, randomization makes these problems tractable. Finally, we consider ill-posed problems. In particular, we show that the Fredholm problem of the first kind in the Hilbert case has infinite complexity.

We first give a formal description of randomized information and algorithms. As always, we wish to approximate a solution operator $S \colon F \to G$, with Λ the class of permissible information operations. Suppose that ρ is a probability measure on a space T. To approximate Sf, we first randomly select $t \in T$. Based on this random t, we then choose information $N_t f$ and an algorithm ϕ_t. Then $\phi_t(N_t f)$ is our approximation to Sf. Thus a *randomized algorithm* is given by (ϕ, N, T, ρ).

Note that we make our random selection only once, at the beginning of the algorithm. By analogy with game theory, we could call this a "mixed strategy" randomized setting. One can also define randomized information that makes a random choice of each information functional as the evaluations proceed. Although this would appear to be a more general notion of randomization, Nemirovsky and Yudin (1983) showed that this is not so; there exists a Polish space T such that the mixed strategy methods coincide with the randomized methods in the more general sense. Hence we restrict our attention to mixed strategy methods in what follows.

We now define complexity in the randomized setting. The *error* of a randomized algorithm (ϕ, N, T, ρ) is given by

$$e^{\mathrm{ran}}(\phi, N, T, \rho) = \left(\sup_{f \in F} \int_T \|Sf - \phi_t(N_t f)\|^2 \, \rho(dt) \right)^{1/2},$$

and the *cost* of (ϕ, N, T, ρ) is given by

$$\mathrm{cost}^{\mathrm{ran}}(\phi, N, T, \rho) = \sup_{f \in F} \int_T \mathrm{cost}(\phi_t, N_t) \, \rho(dt).$$

Then for any $\varepsilon > 0$, the *ε-complexity with randomization* is defined in the usual way, i.e.

$$\mathrm{comp}^{\mathrm{ran}}(\varepsilon, \Lambda) = \inf \{ \, \mathrm{cost}^{\mathrm{ran}}(\phi, N, T, \rho) : e^{\mathrm{ran}}(\phi, N, T, \rho) \leq \varepsilon \, \},$$

the infimum being over all randomized algorithms (ϕ, N, T, ρ). Moreover, a randomized algorithm $(\phi^*, N^*, T^*, \rho^*)$ is *optimal* if

$$e^{\mathrm{ran}}(\phi^*, N^*, T^*, \rho^*) \leq \varepsilon$$

and

$$\mathrm{cost}^{\mathrm{ran}}(\phi^*, N^*, T^*, \rho^*) = \mathrm{comp}^{\mathrm{ran}}(\varepsilon).$$

Note that allowing randomization can sometimes make an intractable problem tractable. Indeed, consider the problem of approximating the integral $\int_{I_d} f(x) \, dx$, where $I_d = [0, 1]^d$ is the d-dimensional unit cube. Suppose that our problem elements have r continuous derivatives, the rth derivatives being uniformly bounded on the unit cube I_d. The deterministic complexity is $\Theta(\varepsilon^{-d/r})$, which grows exponentially in d, and so this problem is intractable in the deterministic worst case setting. However, the randomized complexity is $\Theta(\varepsilon^{-2d/(2r+d)})$ as $\varepsilon \to 0$, see Bakhvalov (1959). Note that the randomized complexity of this problem is $O(\varepsilon^{-2})$ as $\varepsilon \to 0$, which is independent of the dimension. Hence, this problem is tractable in the randomized setting.

We now investigate whether randomization can also break intractability for the problems we have studied in this book. Recall that ODE are *not* intractable, since there is no exponential dependence on the dimension. Hence we restrict our attention to shift-invariant problems and ill-posed problems.

Before we proceed further, we state two useful lemmas. To do this, we need to define the nth *minimal radius $r^{\mathrm{ran}}(n, \Lambda)$ of randomized information* to be the minimal randomized error over all algorithms using randomized information of cardinality at most n from Λ. As usual, we let $r(n, \Lambda)$ denote the nth minimal radius in the worst case deterministic setting, omitting the reference to Λ if $\Lambda = \Lambda^*$, merely writing $r^{\text{ran}}(n)$.

The first lemma covers the case $\Lambda = \Lambda^*$, i.e. any continuous linear information is permissible.

LEMMA 8.3.1. *Let* $S: F_1 \to G$ *be a continuous linear transformation of Hilbert spaces with pure point spectrum. If* $\Lambda = \Lambda^*$, *then*

$$r^{\text{ran}}(n) \geq \frac{1}{\sqrt{2}} r(2n - 1).$$

PROOF. See Novak (1990). □

Our second lemma holds for arbitrary $\Lambda \subseteq \Lambda^*$, i.e. we allow a restricted class of information.

LEMMA 8.3.2. *Let* $S: D \subseteq F_1 \to G$ *be a measurable linear transformation of a separable Banach space into a separable Hilbert space, and let* F *be the unit ball of* F_1. *For any zero mean Gaussian measure* μ *on* F_1, *let* $m^{\text{avg}}(\varepsilon, \Lambda, \mu)$ *denote the average case* ε*-cardinality number with respect to the measure* μ. *Then*

$$\text{comp}^{\text{ran}}(\varepsilon, \Lambda) \geq c\big(m^{\text{avg}}(\varepsilon, \Lambda, \mu) - \tfrac{1}{2}\big),$$

where c *is the cost of evaluating one information functional.*

PROOF. See pp. 419–420 of IBC. Note that the statement of this result in IBC assumes that S is continuous. However it is easy to check that the proof is valid, since S measurable and μ Gaussian imply that $\int_D \|Sf\|^2 \, \mu(df)$ is finite. □

We now apply these lemmas. First, we discuss shift-invariant problems.

THEOREM 8.3.1. *For any shift-invariant problem,*

$$\text{comp}^{\text{ran}}(\varepsilon, \Lambda^*) = \Theta\big(\text{comp}(\varepsilon, \Lambda^*)\big) \qquad \text{as } \varepsilon \to 0.$$

PROOF. Clearly we have

$$\text{comp}^{\text{ran}}(\varepsilon) = O\big(\text{comp}(\varepsilon)\big) \qquad \text{as } \varepsilon \to 0.$$

We need only show that

$$\text{comp}^{\text{ran}}(\varepsilon) = \Omega\big(\text{comp}(\varepsilon)\big) \qquad \text{as } \varepsilon \to 0.$$

For a shift-invariant problem, we have $r(n) = \Theta(n^{-\alpha})$ as $n \to \infty$ for some $\alpha \geq 0$. Thus

$$\mathrm{comp}(\varepsilon) = \Theta(\varepsilon^{-1/\alpha}) \qquad \text{as } \varepsilon \to 0.$$

Using Lemma 8.3.1, we have

$$r^{\mathrm{ran}}(n) \geq \frac{1}{\sqrt{2}} r(2n - 1, \Lambda^*) = \Theta\big((2n - 1)^{-\alpha}\big) = \Theta(n^{-\alpha})$$

as $n \to \infty$. Hence

$$\mathrm{comp}^{\mathrm{ran}}(\varepsilon) = \Theta(\varepsilon^{-1/\alpha}) \qquad \text{as } \varepsilon \to 0,$$

proving the theorem. $\qquad\qquad\qquad\qquad\qquad\qquad\qquad\qquad\qquad\qquad$ □

Hence if we want randomization to help break intractability, we need to change our problem somehow. One possibility is to relax our requirement that our algorithm approximate the solution everywhere in the region Ω over which the problem is defined. Suppose we need the solution only at a fixed set of points $x_1, \ldots, x_m \in \Omega$. Since we have a Green's function representation for the solution of the shift-invariant problems we have studied in this book, we may write

$$(Sf)(x_j) = \int_\Omega g(x_j, x) f(x) \, dx \qquad (1 \leq j \leq m).$$

Hence if the Green's function g is known, our problem can be reduced to the approximation of several weighted integrals. Using the results in Section 4.2 in Chapter 11 of IBC, it is easy to show

THEOREM 8.3.2.   *Consider the problem of evaluating the solution of a shift-invariant problem at a fixed finite set of points. Suppose that standard information is permissible and that F is the unit ball of $H^r(\Omega)$. Then*

$$\mathrm{comp}^{\mathrm{ran}}(\varepsilon, \Lambda^{\mathrm{std}}) = \Theta(\varepsilon^{-2d/(2r+d)}) \qquad \text{as } \varepsilon \to 0. \qquad\qquad □$$

In particular, consider the special case of $r = 0$, i.e. L_2 data. Then $\mathrm{comp}^{\mathrm{ran}}(\varepsilon, \Lambda^{\mathrm{std}}) = \Theta(\varepsilon^{-2})$ as $\varepsilon \to 0$, no matter what the dimension d is. Furthermore, weighted Monte Carlo quadrature is almost optimal.

Having dealt with shift-invariant problems, we now turn to ill-posed problems. In particular, we consider the Fredholm problem of the first kind in the Hilbert case. Here, we are given a injective compact transformation $L: G \to F_1$ of Hilbert spaces, with dense range $D = L(G)$. Our solution operator $S: D \subset F_1 \to G$ is defined by

$$u = Sf \quad \text{iff} \quad Lu = f \qquad \forall f \in D.$$

We now show that the complexity of this problem is infinite.

THEOREM 8.3.3. *For the Fredholm problem of the first kind,*

$$\mathrm{comp}^{\mathrm{ran}}(\varepsilon, \Lambda^*) = \infty \qquad \forall\, \varepsilon \in (0,1).$$

PROOF. Let σ_i^{-2} be the ith-largest eigenvalue of L^*L, corresponding to the eigenvector u_i, with $u_1, u_2, \ldots$ an orthonormal basis of G. Set $e_i = \sigma_i L u_i$, so that $e_1, e_2, \ldots$ is an orthonormal basis of F_1. For $p > 0$, let μ be a zero mean Gaussian measure on F_1 whose correlation operator C_μ is given by

$$C_\mu f = \sum_{i=1}^{\infty} i^{-(1+2/p)} \sigma_i^{-2} \langle f, e_i \rangle e_i \qquad \forall\, f \in F_1.$$

Then the a priori measure $\nu = \mu L$ on the solution elements is a zero mean Gaussian measure with correlation operator

$$C_\nu g = \sum_{i=1}^{\infty} i^{-(1+2/p)} \langle g, u_i \rangle u_i \qquad \forall\, g \in G.$$

As in §7.5.1, we find that

$$\lceil m^{\mathrm{avg}}(\varepsilon, \Lambda^*, \mu) \rceil = \min\left\{ k : \sum_{i=k+1}^{\infty} i^{-(1+2/p)} \leq \varepsilon^2 \right\}$$

$$\geq \min\{ k : (2/p)^{-1/2}(k+1)^{-2/p} \leq \varepsilon^2 \}$$

$$= \left\lceil \varepsilon^{-p}\,(2/p)^{-p/2} \right\rceil - 1.$$

From Lemma 8.3.2, we thus have

$$\mathrm{comp}^{\mathrm{ran}}(\varepsilon, \Lambda^*) \geq c\big(m^{\mathrm{avg}}(\varepsilon, \Lambda^*, \mu) - \tfrac{1}{2}\big)$$

$$\geq c\big(\varepsilon^{-p}(\tfrac{1}{2}p)^{p/2} - \tfrac{3}{2}\big).$$

Remembering that $\varepsilon < 1$, we let $p \to \infty$, finding that $\mathrm{comp}^{\mathrm{ran}}(\varepsilon, \Lambda^*) = \infty$, as required. □

Notes and Remarks

NR 8.3:1 The results in this section are based mainly on Chapter 11 of IBC, which should be consulted for proofs and further details.

NR 8.3:2 Note that we used an L_2 norm to define randomized error. Other norms (especially, the L_1 norm) could have been used as well.

NR 8.3:3 The randomized setting described above is a "worst case randomized setting," since error and cost are defined in a worst case sense. We can also use an "average case randomized setting," in which error and cost are defined as expected values. However, there is no advantage in doing this, since the average case complexity is essentially the same in both the deterministic and randomized settings. For further details, see Section 3 in Chapter 10 of IBC.

NR 8.3:4 Of course, in many practical situations, we do not know the Green's function. For these cases, Theorem 8.3.2 gives a lower bound on the complexity. Finding algorithms that achieve this bound is still an open research problem. For additional discussion, see, for example, Mascagni (1990) and the references quoted therein.

Appendix

A.1 Introduction

To make this book reasonably self-contained, we summarize basic technical material in this Appendix. In the first section, we define and describe Sobolev spaces. In the second section, we give basic results concerning weakly coercive linear equations in Banach and Hilbert spaces.

For further reference about Sobolev spaces, the reader should consult Adams (1975), Babuška and Aziz (1972), Ciarlet (1978), or Oden and Reddy (1976). Moreover, Babuška and Aziz (1972) and Oden and Reddy (1976) contain further discussion of weakly coercive equations.

A.2 Sobolev spaces

In this section, we describe Sobolev spaces $W^{r,p}(\Omega)$ over a domain $\Omega \subset \mathbb{R}^d$ with Lipschitz continuous boundary. These are spaces of functions whose rth derivatives belong to $L_p(\Omega)$. Hence, these functions have smoothness r in a Lebesgue sense.

A.2.1 Definition

We first introduce the standard multi-index notation in $\mathbb{R}^d$. A vector $\alpha = (\alpha_1, \ldots, \alpha_d) \in \mathbb{N}^d$ of non-negative integers is said to be a *multi-index* of *order* $|\alpha| = \alpha_1 + \cdots + \alpha_d$. For $x = (x_1, \ldots, x_d) \in \mathbb{R}^d$, we let $x^\alpha = x_1^{\alpha_1} \ldots x_d^{\alpha_d}$ Then for $k \in \mathbb{N}$ and $K \subseteq \mathbb{R}^d$,

$$P_k(K) = \left\{ \sum_{|\alpha| \leq k} a_\alpha x^\alpha : a_\alpha \in \mathbb{R} \text{ for } |\alpha| \leq k, \, x \in \Omega \right\}$$

denotes the space of polynomials of degree at most k over K For smooth functions v defined on a subset of $\mathbb{R}^d$, we let

$$D^\alpha v = \partial_1^{\alpha_1} \ldots \partial_d^{\alpha_d} v,$$

where ∂_i denotes the partial derivative in the ith coordinate direction x_i.

Recall that $L_p(\Omega)$ is the space of all functions v on Ω such that

$$\|v\|_{L_p(\Omega)} = \begin{cases} \left[\int_\Omega |v(x)|^{1/p}\, dx \right]^{1/p} & \text{for } p < \infty \\ \operatorname{ess\,sup}_{x \in \Omega} |v(x)| & \text{for } p = \infty \end{cases}$$

is finite. Here, and in what follows, we use Lebesgue integrals exclusively (see §7.2).

We describe (weak) derivatives of functions in $L_p(\Omega)$. Recall that $C^\infty(\Omega)$ is the space of all functions that are infinitely differentiable in Ω, and that $C_0^\infty(\Omega)$ is the space of all elements of $C^\infty(\Omega)$ having compact *support*, i.e. that are nonzero only on a compact subset of Ω. For $v \in L_p(\Omega)$ and a multi-index α, we let $D^\alpha v$ denote the *weak derivative* defined by

$$\int_\Omega D^\alpha v(x) \cdot w(x)\, dx = (-1)^{|\alpha|} \int_\Omega v(x) \cdot D^\alpha w(x)\, dx \qquad \forall\, w \in C_0^\infty(\Omega).$$

This formula is motivated by an integration by parts.

Next, we define Sobolev spaces $W^{r,p}(\Omega)$, where $p \in [1, \infty]$ and $r \in \mathbb{R}$. We do this in stages. First, we will assume that $r \in \mathbb{N}$ is a non-negative integer. Then

$$W^{r,p}(\Omega) = \{\, v \in L_p(\Omega) : D^\alpha v \in L_p(\Omega) \text{ for all } \alpha \text{ with } |\alpha| \leq r \,\}.$$

Define the norm $\| \cdot \|_{W^{r,p}(\Omega)}$ by

$$\|v\|_{W^{r,p}(\Omega)} = \begin{cases} \left[\sum_{|\alpha| \leq r} \|D^\alpha v\|_{L_p(\Omega)} \right]^{1/p} & \text{if } p < \infty \\ \sup_{|\alpha| \leq r} \|D^\alpha v\|_{L_\infty(\Omega)} & \text{if } p = \infty \end{cases}$$

Then $W^{r,p}(\Omega)$ is a Banach space under the norm $\| \cdot \|_{W^{r,p}(\Omega)}$. Note that we can also write

$$\|v\|_{W^{r,p}(\Omega)} = \begin{cases} \left[\sum_{j=0}^{r} |v|_{W^{j,p}(\Omega)} \right]^{1/p} & \text{if } p < \infty \\ \sup_{0 \leq j \leq r} |v|_{W^{j,\infty}(\Omega)} & \text{if } p = \infty \end{cases}$$

where the seminorm $| \cdot |_{W^{r,p}(\Omega)}$ is defined by

$$|v|_{W^{r,p}(\Omega)} = \begin{cases} \left[\sum_{|\alpha| = r} \|D^\alpha v\|_{L_p(\Omega)}^{1/p} \right]^{1/p} & \text{if } p < \infty \\ \sup_{|\alpha| = r} \|D^\alpha v\|_{L_\infty(\Omega)} & \text{if } p = \infty. \end{cases}$$

There are other characterizations of the Sobolev space $W^{r,p}(\Omega)$. In particular:

(i) $W^{r,p}(\Omega)$ is the completion of $C^\infty(\Omega)$ in the $W^{r,p}(\Omega)$-norm, and

(ii) $W^{r,p}(\Omega)$ is the closure in $L_p(\Omega)$ of the set of all functions whose $W^{r,p}(\Omega)$-norm is finite.

The case $p = 2$ is especially important. We generally write $H^r(\Omega) = W^{r,2}(\Omega)$. The spaces $H^r(\Omega)$ are Hilbert spaces, the inner product being given by

$$\langle v, w \rangle_{H^r(\Omega)} = \sum_{|\alpha| \leq r} \int_\Omega v(x)w(x)\,dx \qquad \forall\, v, w \in H^r(\Omega).$$

Sobolev spaces $W^{r,p}(\Omega)$ with negative r are defined by duality, i.e.

$$\|v\|_{W^{r,p}(\Omega)} = \sup_{\substack{w \in C_0^\infty(\Omega) \\ w \neq 0}} \frac{\int_\Omega v(x)w(x)\,dx}{\|w\|_{W^{-r,p'}(\Omega)}},$$

where $1/p + 1/p' = 1$. Hence for $r < 0$, the space $W^{r,p}(\Omega)$ is the dual of the space $W_0^{-r,p'}(\Omega)$.

Finally, we discuss the definition of Sobolev spaces $W^{r,p}(\Omega)$ with $r \in \mathbb{R}$. These spaces are defined by various techniques that allow us to define (for $0 < \theta < 1$) intermediate spaces X_θ that are "between" two known Banach spaces X_0 and X_1. The interested reader may consult standard references such as Butzer and Berens (1967), Löfstrom (1970), or Peetre (1963) for further details on the construction. If $r \in (r_0, r_1)$, these techniques may be used to construct the space $W^{r,p}(\Omega)$ as an intermediate space that interpolates the known spaces $W^{r_0,p}(\Omega)$ and $W^{r_1,p}(\Omega)$, with interpolating parameter $\theta = r/(r_1 - r_0)$.

This concludes the discussion of the spaces $W^{r,p}(\Omega)$. We can also define Sobolev spaces $W^{r,p}(\partial\Omega)$, defined on the boundary $\partial\Omega$ of Ω. Furthermore, we can also define vector-valued Sobolev spaces, i.e. Sobolev spaces of $\mathbb{R}^d$-valued functions, in the obvious way.

A.2.2 Some useful results and inequalities

We now state some results and inequalities about Sobolev spaces that are used throughout this book. Proofs may be found in the standard references mentioned previously.

First, we state a basic result about linear operators and interpolation spaces that will be useful in computing errors of algorithms defined over Sobolev spaces of non-integer order.

LEMMA A.2.2.1 (INTERPOLATION LEMMA). *Let* $T\colon X_i \to Y_i$ *be a bounded linear operator with norm* M_i *(for* $i = 0$ *and* $i = 1$*). For* $0 < \theta < 1$, *suppose that* X_θ *interpolates* X_0 *and* X_1 *and that* Y_θ *interpolates* Y_0 *and* Y_1. *Then* $T\colon X_\theta \to Y_\theta$ *is a bounded linear operator with norm* M, *where* $M \leq M_0^\theta M_1^{1-\theta}$. $\qquad\square$

Our next result deals with the Sobolev space $W_0^{r,p}(\Omega)$:

LEMMA A.2.2.2 (POINCARÉ-FRIEDRICH INEQUALITY). *Let $r \geq 0$. Then there exists a constant $C > 0$ such that*

$$\|v\|_{W^{r,p}(\Omega)} \leq C|v|_{W^{r,p}(\Omega)} \qquad \forall v \in W_0^{r,p}(\Omega).$$

Hence the Sobolev r-seminorm is a norm on $W_0^{r,p}(\Omega)$, equivalent to the original Sobolev r-norm. □

We can find out how much classical smoothness the Sobolev spaces have by appealing to results of Kondrasov, Rellich, and Sobolev:

LEMMA A.2.2.3 (COMPACT EMBEDDING THEOREMS).
  (i) *Let $rp < d$. Then $W^{r,p}(\Omega)$ is compactly embedded in $L_q(\Omega)$ for all $q \in [1, p^*)$, where $p^* = pd/(d - r)$.*
 (ii) *Let $rp = d$. Then $W^{r,p}(\Omega)$ is compactly embedded in $L_q(\Omega)$ for all $q \geq 1$.*
(iii) *Let $rp > d$. Then $W^{r,p}(\Omega)$ is compactly embedded in $C(\Omega)$.* □

We finally mention a result concerning quotient spaces. Let $k \in \mathbb{N}$ and $p \in [1, \infty]$, and let

$$[v] = \{\, v + p : p \in P_k(\Omega) \,\}$$

denote the *equivalence class* of $v \in W^{k+1,p}(\Omega)$. Then the *quotient space*

$$W^{k+1,p}(\Omega)/P_k(\Omega) = \{\, [v] : v \in W^{k+1,p}(\Omega) \,\},$$

is a Banach space under the *quotient norm*

$$\|[v]\|_{W^{k+1,p}(\Omega)/P_k(\Omega)} = \inf_{p \in P_k(\Omega)} \|v + p\|_{W^{k+1,p}(\Omega)}$$

$$\forall [v] \in W^{k+1,p}(\Omega)/P_k(\Omega).$$

A somewhat less well-known (but useful) result is that $|\cdot|_{W^{k+1,p}(\Omega)}$ is a norm over $W^{k+1,p}(\Omega)/P_k(\Omega)$, equivalent to the quotient norm:

LEMMA A.2.2.4 (Ciarlet 1978, p. 115). *For any smooth region $\Omega \subseteq \mathbb{R}^d$ and for any $k \in \mathbb{N}$ and $p \in [1, \infty]$, there exists a constant $C = C(\Omega, k, p) > 0$ such that*

$$\inf_{p \in P_k(\Omega)} \|v + p\|_{W^{k+1,p}(\Omega)} \leq C(\Omega)|v|_{W^{k+1,p}(\Omega)} \qquad \forall v \in W^{k+1,p}(\Omega).$$

Hence

$$|[v]|_{W^{k+1,p}(\Omega)} \leq \|[v]\|_{W^{k+1,p}(\Omega)/P_k(\Omega)} \leq C(\Omega)|[v]|_{W^{k+1,p}(\Omega)}$$

for any $[v] \in W^{k+1,p}(\Omega)/P_k(\Omega)$. $\qquad\qquad\qquad\qquad\qquad\qquad$ $\square$

A.2.3 Interpolation by polynomials

We now discuss the approximation of functions belonging to a Sobolev space by polynomials of degree k.

We first define a *reference element* $\hat{K}$ to be a fixed polyhedron in $\mathbb{R}^d$. Associated with $\hat{K}$ is a set $\hat{\Psi} = \{\hat{\psi}_1, \dots, \hat{\psi}_J\}$ of linear functionals on $P_k(\hat{K})$, called *degrees of freedom*. We assume that $\hat{\Psi}$ is $P_k(\hat{K})$-*unisolvent*, i.e. that for any $\hat{\alpha}_1, \dots, \hat{\alpha}_J \in \mathbb{R}$, there exists a unique $\hat{p} \in P_k(\hat{K})$ such that

$$\hat{\psi}_i(\hat{p}) = \hat{\alpha}_i \qquad (1 \leq i \leq J).$$

Hence, we may choose a basis $\{\hat{p}_1, \dots, \hat{p}_J\}$ for $P_k(\hat{K})$ satisfying

$$\hat{\psi}_i(\hat{p}_j) = \delta_{i,j} \qquad (1 \leq i, j \leq J).$$

For functions $\hat{v} \colon \hat{K} \to \mathbb{R}$, we define a $P_k(\hat{K})$-*interpolant* $\hat{\Pi}\hat{v}$ by

$$(\hat{\Pi}\hat{v})(\hat{x}) = \sum_{j=1}^{J} \hat{\psi}(\hat{v})\hat{p}_j(\hat{x}) \qquad \forall\, \hat{x} \in \hat{K}.$$

Since

$$\hat{\Pi}\hat{p} = \hat{p} \qquad \forall\, \hat{p} \in P_k(\hat{K}),$$

we can use Lemma A.2.2.4 to prove the following basic result on approximation over the reference element:

LEMMA A.2.3.1 (Ciarlet 1978, p. 122). *Suppose that*

(i) $W^{k+1,p}(\hat{K})$ *is continuously embedded in* $W^{m,q}(\hat{K})$, *and*
(ii) $\hat{\psi}_1, \dots, \hat{\psi}_J$ *are bounded linear functionals on* $W^{k+1,p}(\hat{K})$.

Then there exists a positive constant C such that

$$|\hat{v} - \hat{\Pi}\hat{v}|_{W^{m,q}(\hat{K})} \leq C|\hat{v}|_{W^{k+1,p}(\hat{K})} \qquad \forall\, \hat{v} \in W^{k+1,p}(\hat{K}). \qquad \square$$

We will also have occasion to use the following result from p. 192 of Ciarlet (1978), which also follows from Lemma A.2.2.4:

LEMMA A.2.3.2 (BRAMBLE-HILBERT LEMMA). *Let $\hat{f}$ be a continuous linear functional on* $W^{k+1,p}(\hat{K})$. *Suppose that*

$$\hat{f}(\hat{p}) = 0 \qquad \forall\, \hat{p} \in P_k(\hat{K}).$$

Then there exists a positive constant C such that

$$|\hat{f}(\hat{v})| \leq C\|\hat{f}\|^*_{W^{k+1,p}(\hat{K})}|\hat{v}|_{W^{k+1,p}(\hat{K})} \qquad \forall \hat{v} \in W^{k+1,p}(\hat{K}),$$

where $\|\cdot\|^*_{W^{k+1,p}(\hat{K})}$ denotes the norm in the dual space of $W^{k+1,p}(\hat{K})$.   □

Next, we move from polynomials over the reference element $\hat{K}$ to polynomials over a (presumably small) element K. Let K be the affine image under F_K of $\hat{K}$, so that there exists an invertible matrix $B_K \in \mathbb{R}^{d \times d}$ and a vector $b_K \in \mathbb{R}^d$ such that the affine mapping $F_K : \hat{K} \to K$ defined by

$$F_K \hat{x} = B_K x + b_K \qquad \forall \hat{x} \in \hat{K}$$

is a bijection of $\hat{K}$ onto K. Note that the bijection F_K of $\hat{K}$ onto K induces a bijection between functions $\hat{v} : \hat{K} \to \mathbb{R}$ and $v : K \to \mathbb{R}$, namely

$$v(x) = \hat{v}(\hat{x}) \quad \text{with} \quad x = F_K(\hat{x}) \qquad \forall \hat{x} \in \hat{K}.$$

In particular, let us define polynomials $p_{1,K}, \ldots, p_{J,K} \in P_k(K)$ by

$$p_{j,K}(x) = \hat{p}_j(\hat{x}) = \hat{p}_j(F_K^{-1}x) \qquad \forall x \in K$$

and a set $\Psi_K = \{\psi_{1,K}, \ldots, \psi_{J,K}\}$ of linear functionals on $P_k(K)$ by

$$\psi_{i,K}(p) = \hat{\psi}(\hat{p}) = \hat{\psi}(p \circ F_K^{-1}) \qquad \forall p \in P_k(K).$$

Then $\{p_{1,K}, \ldots, p_{J,K}\}$ is a basis for $P_k(K)$, and Ψ_K satisfies

$$\psi_{i,K}(p_{j,K}) = \delta_{i,j} \qquad (1 \leq i, j \leq J).$$

We say that the elements of Ψ_K are the (local) *degrees of freedom* over K. For a function $v : K \to \mathbb{R}$, we define the $P_k(K)$-*interpolant* $\Pi_K v$ by

$$(\Pi_K v)(x) = (\hat{\Pi}\hat{v})(\hat{x}) = \sum_{j=1}^{J} \psi_{j,K}(v) p_{j,K}(x) \qquad \forall x \in K.$$

The following basic result on local approximation over K holds because

$$\Pi_K p = p \qquad \forall p \in P_k(\hat{K}).$$

Introducing the geometric quantities

$$h_K = \text{diam } K$$

and

$$\rho_K = \sup\{\text{diam } B : \text{spheres } B \supseteq K\}$$

we have

LEMMA A.2.3.3 (Ciarlet 1978, pp. 118–122).

(i) *Let $p \in [1, \infty]$ and $j \in \mathbb{N}$. There exists a constant $C > 0$ such that for $v \in W^{j,p}(K)$,*

$$|\hat{v}|_{W^{j,p}(\hat{K})} \leq C h_K^j (\operatorname{vol} K)^{-1/p} |v|_{W^{j,p}(K)}$$

and

$$|v|_{W^{j,p}(K)} \leq C \rho_K^{-j} (\operatorname{vol} K)^{1/p} |\hat{v}|_{W^{j,p}(\hat{K})}.$$

(ii) *Let $p, q \in [1, \infty]$ and $j \in \{0, 1, \ldots, k+1\}$. Suppose that $W^{k+1,p}(K)$ is continuously embedded in $W^{j,q}(K)$, and that $\psi_1, \ldots, \psi_J$ are bounded linear functionals on $W^{k+1,p}(K)$. Then there exists a constant $C > 0$ such that for $v \in W^{k+1,p}(K)$,*

$$|v - \Pi_n v|_{W^{j,q}(K)} \leq C (\operatorname{vol} K)^{(1/q)-(1/p)} \frac{h_K^{k+1}}{\rho_K^j} |v|_{W^{k+1,p}(K)}. \qquad \square$$

Using these results, it is easy to prove the following inequality, which shows how to relate two seminorms of a polynomial over K:

LEMMA A.2.3.4 (INVERSE INEQUALITY, Ciarlet 1978, p. 140 ff.). *If l and m are nonnegative integers with $l \leq m$, and if $q, r \in [1, \infty]$, then there is a positive constant C, such that*

$$|v|_{W^{m,q}(K)} \leq C (\operatorname{vol} K)^{1/q - 1/r} \frac{h_K^l}{\rho_K^m} |v|_{W^{l,r}(\Omega)} \qquad \forall v \in P_k(K). \qquad \square$$

A.3 Weakly coercive linear equations

Let V and W be Banach spaces, with respective duals V^* and W^*. Let $B : V \times W \to \mathbb{R}$ be a bilinear form. For $f \in W^*$, we wish to determine whether there exists a unique $u \in V$ such that

$$B(u, w) = f(w) \qquad \forall w \in W, \tag{1}$$

We say that B is *weakly coercive* if the following hold:

(i) B is *bounded*, i.e. there exists a finite positive M such that

$$|B(v, w)| \leq M \|v\|_V \|w\|_W \qquad \forall v \in V, w \in W.$$

(ii) B satisfies the *inf-sup conditions*, i.e. there exists a finite positive γ such that

$$\inf_{\substack{v \in V \\ \|v\|_V \leq 1}} \sup_{\substack{w \in W \\ \|w\|_W \leq 1}} |B(v, w)| \geq \gamma.$$

(iii) B satisfies the *adjoint inf-sup conditions*, i.e.

$$\sup_{v \in V} |B(v, w)| > 0 \qquad \forall \text{ nonzero } w \in W.$$

Although we do not have a general criterion for solvability of (1) when V and W are Banach spaces, we do have the following result (see Babuška and Aziz 1972 p. 112; Oden and Reddy 1976, p. 310):

LEMMA A.3.1 (GENERALIZED LAX-MILGRAM LEMMA). *Let V and W be Hilbert spaces, and let B be weakly coercive. Then for any $f \in W^*$, there exists a unique $u \in V$ such that (1) holds.* □

We now wish to approximate the solution of (1). Let $\{V_n\}_{n=1}^{\infty}$ and $\{W_n\}_{n=1}^{\infty}$ be sequences of finite dimensional subspaces of V and W, with $\dim V_n = \dim W_n = n$ for each $n \in \mathbb{P}$. For each index n, let $B_n : V_n \times W_n \to \mathbb{R}$ be a bilinear form that approximates B, and let $f_n \in W_n^*$ be a linear functional approximating f. Our approximation of (1) is as follows: for $f \in W^*$ and $n \in \mathbb{P}$, we wish to find $u_n \in V_n$ such that

$$B_n(u_n, w_n) = f_n(w_n) \qquad \forall \, w_n \in W_n. \tag{2}$$

Note that there are three sources of approximation in computing the approximation u_n to u:

(i) The spaces V and W are (respectively) replaced by the finite-dimensional spaces V_n and W_n.

(ii) The bilinear form B is replaced by the bilinear form B_n.

(iii) The linear functional f is replaced by the linear functional f_n.

It *is* possible to have $B_n = B$ and $f_n = f$, as in the "pure" finite element method. However, if V and W are infinite-dimensional, then $V_n \neq V$ and $W_n \neq W$ for any $n \in \mathbb{P}$.

The following result (see Babuška and Aziz 1972, p. 290 *ff.*; Ciarlet 1978, p. 186) gives us conditions for solvability, as well as an error estimate:

LEMMA A.3.2 (STRANG'S LEMMA). *Suppose that the bilinear form B is uniformly weakly coercive, i.e. there is a positive constant γ' and a positive integer n_0 such that if $n \geq n_0$, then*

$$\forall \, v_n \in V_n, \exists \, w_n \in W_n : |B(v, w)| \geq \gamma' \|v_n\|_V \|w_n\|_W.$$

Suppose further that there exists a non-negative sequence $\{\delta_n\}_{n=1}^{\infty}$ with $\lim_{n \to \infty} \delta_n = 0$, such that for any $n \in \mathbb{P}$,

$$|B(v_n, w_n) - B_n(v_n, w_n)| \leq \delta_n \|v_n\|_V \|w_n\|_W \qquad \forall \, v_n, w_n \in V_n.$$

Then there exists $n_1 \in \mathbb{P}$ such that for any $n \geq n_1$ and $f \in W^*$, there is a unique $u_n \in V_n$ such that (2) holds. Moreover, there exists a positive constant C, independent of f and n, such that if u is a solution to (1), then

$$
\|u - u_n\|_V \leq C \left[\inf_{v_n \in V_n} \|u - v_n\|_V \right.
$$
$$
\left. + \sup_{w_n \in W_n} \left(\frac{|B(v_n, w_n) - B_n(v_n, w_n)|}{\|w_n\|_W} + \frac{|f(w_n) - f_n(w_n)|}{\|w_n\|_W} \right) \right].
$$

PROOF. Since $\delta_n \to 0$, there exists $n_2 \in \mathbb{N}$ such that $\delta_n \leq \frac{1}{2}\gamma'$ for all $n \geq n_2$. Let $n_1 = \max\{n_0, n_2\}$. We claim that if $n \geq n_1$, then for any $v_n \in V_n$, there exists $w_n \in W_n$ such that

$$
|B_n(v_n, w_n)| \geq \tfrac{1}{2}\gamma' \|v_n\|_V \|w_n\|_W. \tag{3}
$$

Indeed, let $n \geq n_1$ and $v_n \in V_n$. Since B is uniformly weakly coercive, there exists $w_n \in V_n$ such that

$$
|B(v_n, w_n)| \geq \gamma' \|v_n\|_V \|w_n\|_W.
$$

Hence

$$
|B_n(v_n, w_n)| \geq |B(v_n, w_n)| - |B(v_n, w_n) - B_n(v_n, w_n)|
$$
$$
\geq (\gamma' - \delta_n)\|v_n\|_V \|w_n\|_W \geq \tfrac{1}{2}\gamma' \|v_n\|_V \|w_n\|_W,
$$

as claimed.

Let $f \in W^*$ and $n \in \mathbb{P}$. Since V_n and W_n are finite-dimensional, the problem (2) may be rewritten as an $n \times n$ linear system $Ma = b$ of equations. Since B_n satisfies (3), we see that $\ker M = \{0\}$. Hence M is invertible, so that there exists a unique solution u_n to (2).

We now turn to the error bound. Let $v_n \in V_n$. Since $u_n - v_n \in V_n$, we may use (1), (2), and (3) to see that there exists $w_n \in W_n$ such that

$$
\tfrac{1}{2}\gamma' \|u_n - v_n\|_V \|w_n\|_W \leq |B_n(u_n - v_n, w_n)|
$$
$$
\leq |f(w_n) - B_n(u_n, w_n)| + |f(w_n) - B(v_n, w_n)|
$$
$$
+ |B(v_n, w_n) - B_n(v_n, w_n)|
$$
$$
= |f(w_n) - f_n(w_n)| + |B(u - v_n, w_n)|
$$
$$
+ |B(v_n, w_n) - B_n(v_n, w_n)|
$$
$$
\leq |f(w_n) - f_n(w_n)| + M\|u - v_n\|_V \|w_n\|_W
$$
$$
+ |B(v_n, w_n) - B_n(v_n, w_n)|.
$$

Hence

$$\|u_n - v_n\|_V \le \frac{2\max\{M,1\}}{\gamma'}\left(\|u - v_n\|_V + \frac{|B(v_n, w_n) - B_n(v_n, w_n)|}{\|w_n\|_W} + \frac{|f(w_n) - f_n(w_n)|}{\|w_n\|_W}\right).$$

Let $C = 1 + 2\max\{M,1\}/\gamma'$. Then

$$\|u - u_n\|_V \le \|u - v_n\|_V + \|u_n - v_n\|_V$$
$$\le C\left[\|u - v_n\|_V + \frac{|B(v_n, w_n) - B_n(v_n, w_n)|}{\|w_n\|_W} + \frac{|f(w_n) - f_n(w_n)|}{\|w_n\|_W}\right]$$
$$\le C\left[\|u - v_n\|_V + \sup_{w_n \in W_n}\left(\frac{|B(v_n, w_n) - B_n(v_n, w_n)|}{\|w_n\|_W} + \frac{|f(w_n) - f_n(w_n)|}{\|w_n\|_W}\right)\right].$$

Since $v_n \in V_n$ is arbitrary, we may take the infimum over all $v_n \in V_n$ to get our desired conclusion. □

Suppose now that V is embedded in a Banach space H, and that we want estimates of the error in the H-norm $\|\cdot\|_H$, rather than in the V-norm $\|\cdot\|_V$. We assume that for any $g \in H^*$, there exists a unique $\tilde{S}g \in W$ such that

$$B(v, \tilde{S}g) = g(v) \qquad \forall v \in V. \tag{4}$$

For example, suppose that $W = V$, V is a Hilbert space, and B is symmetric and weakly coercive; then (4) holds by the Generalized Lax-Milgram Lemma.

Under these hypotheses, the following result (based on Ciarlet 1978, p. 203) shows how we can get an estimate of the H-norm error, once we know the V-norm error.

LEMMA A.3.3 (AUBIN-CIARLET-NITSCHE DUALITY ARGUMENT). *Let (4) have a unique solution for any $g \in H^*$. Then*

$$\|u - u_n\|_H \le \sup_{g \in H^*} \frac{1}{\|g\|_{H^*}} \inf_{w_n \in W_n}\left[M\|u - u_n\|_V\|\tilde{S}g - w_n\|_W + |B(u_n, w_n) - B_n(u_n, w_n)| + |f(w_n) - f_n(w_n)|\right].$$

PROOF. Let $g \in H^*$. Then (4) implies that

$$g(u - u_n) = B(u - u_n, \tilde{S}g).$$

Let $w_n \in V_n$, so that

$$|B(u - u_n, w_n)| \le |B(u_n, w_n) - B_n(u_n, w_n)| + |B(u, w_n) - B_n(u_n, w_n)|$$
$$= |B(u_n, w_n) - B_n(u_n, w_n)| + |f(w_n) - B_n(u_n, w_n)|.$$

Hence

$$|g(u - u_n)| \le |B(u - u_n, \tilde{S}g - w_n)| + |B(u - u_n, w_n)|$$
$$\le M\|u - u_n\|_V \|\tilde{S}g - w_n\|_W + |B(u_n, w_n) - B_n(u_n, w_n)|$$
$$+ |f(v) - f_n(w_n)|.$$

Since $w_n \in W_n$ is arbitrary, we may take the infimum over all $w_n \in W_n$ in this inequality. Finally, since

$$\|u - u_n\|_H = \sup_{g \in H} \frac{|g(u - u_n)|}{\|g\|_{H^\bullet}},$$

we get the desired conclusion.   □

Bibliography

ADAMS, R. A.
(1975) "Sobolev spaces." Academic Press, New York.

AGMON, S.
(1965) "Lectures on elliptic boundary value problems." Van Nostrand, Princeton.

AGMON, S., DOUGLIS, A., NIRENBERG, L.
(1969) Estimates near the boundary for higher-order elliptic partial differential equations satisfying general boundary conditions, *Communications on Pure and Applied Mathematics* **12**, 623–727.

ANDERSSEN, R. S., DEHOOG, F. R., AND LUKAS, M. A.
(1980) "The application and numerical solution of integral equations," Monographs and Textbooks on Mechanics of Solids and Fluids (Mechanics: Analysis), Vol. 6. Slijthoff & Noordhoof, Alphen aan den Rijn.

ARAUJO, A. AND GINÉ, E.
(1980) "The central limit theorem for real and Banach valued random variables." Wiley, New York.

ATKINSON, K. E.
(1976) "A survey of numerical methods for the solution of Fredholm equations of the second kind." Society for Industrial and Applied Mathematics, Philadelphia.

AZIZ, A. K. AND LEVENTHAL, S.
(1978) Finite element approximation for first-order systems, *SIAM Journal on Numerical Analysis* **15**, 1103–1111.

AZIZ, A. K. AND WERSCHULZ, A. G.
(1980) On the numerical solution of Helmholtz equation by the finite element method, *SIAM Journal on Numerical Analysis* **17**, 681–686.

AZIZ, A. K., KELLOGG, R. B., AND STEPHENS, A. B.
(1985) Least squares methods for elliptic systems, *Mathematics of Computation* **44**, 53–70.

BABUŠKA, I.
(1987) Information-based numerical practice, *Journal of Complexity* **3**, 331–346.

(1988) Recent progress in the p and h,p-versions of the finite element method, *in* "Numerical analysis 1987" (D. F. Griffiths and G. A. Watson, eds.), pp. 1–17. Wiley, New York.

BABUŠKA, I. AND AZIZ, A. K.
(1972) Survey lectures on the mathematical foundations of the finite element method, *in* "The mathematical foundations of the finite element method with applications to partial differential equations" (A. K. Aziz, ed.), pp. 3–359. Academic Press, New York.

BABUŠKA, I. AND GUO, B. Q.

(1988) The *h-p* version of the finite element method for domains with curved boundaries, *SIAM Journal on Numerical Analysis* **25**, 837–861.

BAKER, C. T. H.

(1977) "The Numerical treatment of integral equations." Clarendon, Oxford.

BAKER, C. T. H. AND MILLER, G. F.

(1982) "Treatment of integral equations by numerical methods." Academic Press, New York.

BAKER, G. A.

(1973) Simplified proofs of error estimates for the least squares method for Dirichlet's problem, *Mathematics of Computation* **27**, 229–235.

BAKHVALOV, N. S.

(1959) On approximate calculation of integrals (in Russian), *Vestnik MGV, Ser. Mat. Mekh. Astron. Fiz. Khim.* **4**, 3–18.

(1977) "Numerical methods." Mir, Moscow.

BEREZANSKII, JU. M.

(1968) "Expansions in eigenfunctions of self-adjoint operators," Translations of Mathematical Monographs, Vol. 17. American Mathematical Society, Providence, Rhode Island.

BÖRGERS, CH. AND WIDLUND, O. B.

(1986) Finite element capacitance matrix methods, Technical Report 261, Courant Institute of Mathematical Sciences, New York, 1986.

BORELL, C.

(1975) The Brunn-Minkowski inequality in Gauss space, *Invent. Math.* **30**, 207–216.

(1976) Gaussian Radon measures on locally convex spaces, *Mathematica Scandinavica* **38**, 265–285.

BRAMBLE, J. H. AND NITSCHE, J.

(1973) A generalized Ritz-least-squares method for Dirichlet problems, *SIAM Journal on Numerical Analysis* **10**, 81–93.

BRAMBLE, J. H. AND SCHATZ, A. H.

(1971) Least squares methods for 2*m*th order elliptic boundary-value problems, *Mathematics of Computation* **25**, 1–32.

BUTCHER, J. C.

(1987) "The numerical analysis of ordinary differential equations: Runge Kutta and general linear methods." Wiley, New York.

BUTZER, P. L. AND BERENS, H.

(1967) "Semi-groups of operators and approximation." Springer-Verlag, Berlin.

CHOW, C. S. AND TSITSIKLIS, J. N.

(1989*a*) The information-based complexity of dynamic programming, to appear in *Journal of Complexity* .

(1989*b*) An optimal multigrid algorithm for continuous state discrete time stochastic control, Research Report, Deptartment of Electrical Engineering and Computer Science, Massachusetts Institute of Technology, Cambridge, 1989.

CIARLET, P. G.
(1978)   "The finite element method for elliptic problems." North-Holland, Amsterdam.

CIARLET, P. G. AND RAVIART, P. A.
(1972)   Interpolation theory over curved elements, *Computer Methods in Applied Mechanics and Engineering* **1**, 217–249.

DEBNATH, L. AND THOMAS, J. G.
(1976)   On finite Laplace transformation with applications, *Zeitschrift für angewandte Mathematik und Mechanik* **56**, 559–565.

DELLWO, D. R.
(1988)   Accelerated refinement with applications to integral equations, *SIAM Journal on Numerical Analysis* **25**, 1327–1339.

DELVES, L. M. AND WALSH, J.
(1974)   "Numerical solution of integral equations." Clarendon, Oxford.

DOUGLASS, J. JR., DUPONT, T., AND WAHLBIN, L.
(1975)   The stability in L^q of the L^2 projection into finite element function spaces, *Numerische Mathematik* **23**, 193–197.

DRYJA, M.
(1981)   A finite element-capacitance method for elliptic problems on regions partitioned into subregions, *Numerische Mathematik* **44**, 153–168.

DUGUNDJI, J.
(1966)   "Topology." Allyn and Bacon, Boston.

DUNFORD, N. AND SCHWARTZ, J. T.
(1963)   "Linear operators–Part I: General theory." Wiley-Interscience, New York.

DUNN, H. S.
(1967)   A generalization of the Laplace transform, *Proceedings of the Cambridge Philosophical Society* **63**, 155–160.

EDWARDS, R. E.
(1965)   "Functional analysis." Holt, New York.

EMELYANOV, K. V. AND ILIN, A. M.
(1967)   Number of arithmetic operations necessary for the approximate solution of Fredholm integral equations, *USSR Computational Mathematics and Mathematical Physics* **7**, 259–267.

FIX, G. J, AND GURTIN, M. E.
(1977)   On patched variational principles with application to elliptic and mixed elliptic-hyperbolic problems, *Numerische Mathematik* **28**, 259–271.

FIX, G. J. AND STRANG, G.
(1969)   Fourier analysis of the finite element method in Ritz-Galerkin theory, *Studies in Applied Mathematics* **48**, 265–273.

FIX, G. J., GUNZBERGER, M. AND NICOLAIDES, R. A.
(1981)   On mixed finite element methods for first order elliptic systems, *Numerische Mathematik* **37**, 29–48.

FREDERIKKSON, B. AND MACKERLE, I.
(1984)   "Structural mechanics: Finite element computer programs," 4th ed. Advanced Engineering Corporation, Linkoping.

FRIED, I.
(1973) Boundary and interior approximation errors in the finite-element method, *Journal of Applied Mechanics* **40**, 1113–1117.

FRIEDMAN, A.
(1970) "Foundations of modern analysis." Holt, Rinehart & Winston, New York.

FRIEDMAN, M. B. AND DELLWO, D. R.
(1982) Accelerated projection methods, *Journal of Computational Physics* **45**, 113–136.

GAREY, M. R. AND JOHNSON, D. S.
(1979) "Computers and intractability." Freeman, San Francisco.

GATTINGER, M.
(1978) Representation theorems for operators and measures on abstract Wiener spaces, *in* "Measure Theory Applications to Stochastic Analysis, Proceedings" (G. Kallianpur, D. Kölzow, eds.), pp. 239–249. Lecture Notes in Mathematics **695**, Springer-Verlag, Berlin.

GEHATIA, M. AND WIFF, D. R.
(1970) Solution of Fujita's equation for equilibration sedimentation by applying Tikhonov's regularizing functions, *Journal of Polymer Science Part A-2* **8**, 2039–2049.

GEL'FAND, I. M. AND VILENKIN, N. YA.
(1964) "Generalized functions. Volume 4: Applications of harmonic analysis." Academic Press, New York.

GIHMAN, I. I. AND SKOROHOD, A. V.
(1974) "The theory of stochastic processes I," Grundlehren der mathematischen Wissenschaften **210**. Springer-Verlag, Berlin.

GTOA See Traub and Woźniakowski (1980).

GUI, W. Z.
(1988) Hierarchical elements, local mappings, and the h-p version of the finite element method (I), *Journal of Computational Mathematics* **6**, 54–67.

HACKBUSH, W.
(1985) "Multi-grid methods and applications." Springer-Verlag, Berlin.

HADAMARD, J.
(1952) "Lectures on the Cauchy problem in linear partial differential Equations." Dover, New York.

HEINRICH, S.
(1985) On the optimal error of degenerate kernel methods, *Journal of Integral Equations* **9**, 251–266.

(1989) Probabilistic complexity analysis for linear problems in bounded domains.

HENRICI, P.
(1962) "Discrete variables methods in ordinary differential equations." Wiley, New York.

HORMANDER, L.
(1983) "Analysis of linear partial differential operators." Springer-Verlag, Berlin.

HUERTA, I. P.
(1986)   Adaption helps for some nonconvex classes, *Journal of Complexity* **2**, 333–352.

IBC   See Traub, Wasilkowski, and Woźniakowski (1988).

IKEBE, Y.
(1972)   The Galerkin method for the numerical solution of Fredholm integral equations of the second kind, *SIAM Review* **14**, 465–491.

JEROME, J. W.
(1968)   Asymptotic estimates of the L_2 n-width, *Journal of Mathematical Analysis and Applications* **22**, 449–464.

KACEWICZ, B. Z.
(1984)   How to increase the order to get minimal-error algorithms for systems of ODE, *Numerische Mathematik* **45**, 93–104.

(1988)   Minimal asymptotic error of algorithms for solving ODE, *Journal of Complexity*, 373–389.

KACEWICZ, B. Z. AND WASILKOWSKI, G. W.
(1986)   How powerful is continuous nonlinear information for linear problems?, *Journal of Complexity* **2**, 306–316.

KNUTH, D. E.
(1976)   Big omicron and big omega and big theta, Association for Computing Machinery, *SIGACT News*.

KON, M. A. AND NOVAK, E.
(1989)   On the adaptive and continuous information problems, to appear in *Journal of Complexity* **5**.

KON, M. A., RITTER, K. H., AND WERSCHULZ, A. G.
(1990)   All ill-posed problems are solvable on the average for Gaussian measures, Research Report, Computer Science Department, Columbia University, New York.

KUNG, H. T.
(1976)   The complexity of obtaining starting points for solving operator equations by Newton's method, *in* "Analytic computational complexity" J. F. Traub, ed., pp. 35–57. Academic Press, New York.

KUO, H.-H.
(1975)   "Gaussian measures in Banach spaces," Lecture Notes in Mathematics **463**. Springer-Verlag, Berlin.

LAVRENTJEV, M. M.
(1955)   On the Cauchy problem for the Laplace equation, (in Russian), *Dokl. Akad. Nauk SSSR* **127**, 205–206.

LEE, D., PAVLIDIS, T., AND WASILKOWSKI, G. W.
(1987)   A note on the trade-off between sampling and quantization in signal processing, *Journal of Complexity* **3**, 359–371.

LEYK, Z.
(1990)   Galerkin method and optimal error algorithms in Hilbert spaces, *Numerische Mathematik* **57**, 1–14.

LINDE, W.
(1983) "Infinitely Divisible and Stable Measures on Banach Spaces." Teubner Verlagsgesellschaft, Leipzig.

LIONS, J. L. AND MAGENES, E.
(1972) "Non-homogeneous boundary value problems and applications." Springer-Verlag, Berlin.

LÖFSTROM, J.
(1970) Besov spaces in theory of approximation, *Ann. Mat. Pura. Appl.* **85**, 93–184.

MASCAGNI, M.
(1990) A Monte Carlo method based on Wiener integration for solving elliptic boundary value problems, to appear in *SIAM Journal on Scientific and Statistical Computing* .

MARTI, J. T.
(1983) Numerical solution of Fujita's equation, *in* "Improperly posed problems and their numerical treatment" (G. Hämmerlin and K.-H. Hoffmann, eds.), pp. 179–187. Birkhäuser-Verlag, Basel.

MOROZOV, M. A.
(1985) "Methods for solving incorrectly posed problems." Springer-Verlag, Berlin.

NEČAS, J.
(1967) "Les méthods directes en théorie des equations elliptiques." Masson, Paris.

NEDELKOV, I. P.
(1972) Improper problems in computational physics, *Computer Physics Communications*, 157–164.

NEMIROVSKY, A. S. AND YUDIN, D. B.
(1983) "Problem complexity and method efficiency in optimization." Wiley-Interscience, New York.

NOOR, A. K.
(1985) Books and monographs on finite element technology, *Finite Elements in Analysis and Design* **1**, 101–111.

NOVAK, E.
(1990) Optimal randomized methods in Hilbert spaces, in preparation.

(1975) "Deterministic and stochastic error bounds in numerical analysis," Lecture Notes in Mathematics **1349**. Springer-Verlag, Berlin.

ODEN, J. T. AND CAREY, G. F.
(1983) "Finite elements: Mathematical aspects." Prentice-Hall, Englewood Cliffs.

ODEN, J. T. AND REDDY, J. N.
(1976) "An introduction to the mathematical theory of finite elements." Wiley-Interscience, New York.

PACKEL, E. W.
(1986) Linear problems (with extended range) have linear optimal algorithms, *Aequationes Mathematicae* **30**, 18–25.

PACKEL, E. W. AND TRAUB, J. F.
(1987) Information-based complexity, *Nature* **328**, 29–33.

PACKEL, E. W. AND WOŹNIAKOWSKI, H.
(1987)   Recent developments in information-based complexity, *Bulletin of the American Mathematical Society* **17**, 9–36.

PAPGEORGIOU, A. AND WASILKOWSKI, G. W.
(1986)   Average complexity of multivariate problems, Research Report, Department of Computer Science, Columbia University, New York.

PARTHASARATHY, K. R.
(1967)   "Probability measures on metric spaces." Academic Press, New York.

PEETRE, J.
(1963)   "A theory of interpolation of normed spaces." Lecture Notes, Brazilia.

PEREVERZEV, S. V.
(1989)   Complexity of the Fredholm problem of the second kind, *in* "International symposium on optimal algorithms" Bulgarian Academy of Sciences.

PINKUS, A.
(1985)   "n-Widths in approximation theory." Springer-Verlag, Berlin.

POGGIO, T., TORRE, V., AND KOCH, C.
(1985)   Computational vision and regularization theory, *Nature* **317**, 314–319.

POUR-EL, M. B. AND RICHARDS, I.
(1983)   Noncomputability in analysis and physics: a complete determination of the class of noncomputable linear operators, *Advances in Mathematics* **48**, 44–74.

TE RIELE, H. J. J.
(1979)   "Colloquium numerical treatment of integral equations," Mathematics Center Syllabus **41**. Mathematisch Centrum, Amsterdam.

RIESZ, F. AND SZ.-NAGY, B.
(1955)   "Functional analysis." Ungar, New York.

ROĬTBERG, J. AND ŠEFTEL, Z.
(1969)   A theorem on homeomorphisms for elliptic systems, *Mathematics of the USSR–Sbornik* **7**, 439–465.

SCHECHTER, M.
(1971)   "Principles of functional analysis." Academic Press, New York.

SCHIPPERS, H.
(1979)   Multigrid techniques for the solution of Fredholm integral equations of the second kind, *in* "Colloquium numerical treatment of integral equations" (H. J. J. te Riele, ed.), pp. 27–46. Mathematisch Centrum **41**, Amsterdam.

SCHOCK, E.
(1982)   Galerkin-like methods for equations of the second kind, *Journal of Integral Equations* **4**, 361–364.

SCHULTZ, M.
(1973)   "Spline analysis." Prentice-Hall, Englewood Cliffs.

SHILOV, G. E. AND FAN DYK TIN
(1967)   "Integral, measure, and derivative on linear spaces." Nauka, Moscow.

SHILOV, G. E. AND GUREVICH, B. L.
(1966)   "Integral, measure, and derivative: A unified approach." Prentice-Hall, Englewood Cliffs.

SKOROHOD, A. V.
(1974) "Integration in Hilbert space." Springer-Verlag, Berlin.

STRANG, G. AND FIX, G. J.
(1973*a*) "An analysis of the finite element method." Prentice-Hall, Englewood Cliffs.

(1973*b*) A Fourier analysis of the finite element variational method, *in* "Constructive Aspects of Functional Analysis, Part II", pp. 793–840. CIME, Rome.

SWARTZTRAUBER, P. N.
(1977) The methods of cyclic reduction, Fourier analysis and the FACR algorithm for discrete solution of Poisson's equation on a rectangle, *SIAM Review* **19**, 490–501.

TIKHOMIROV, V. M.
(1976) "Some Problems in Approximation Theory," in Russian. Moscow State University, Moscow.

TIKHONOV, A. N.
(1963) Solution of incorrectly formulated problems and the regularization method, *Soviet Mathematics Doklady* **4**, 1036–1038.

TIKHONOV, A. N. AND ARSENIN, V. Y.
(1977) "Solutions of ill-posed problems." V. H. Winston and Sons, Washington, DC.

TRAUB, J. F. AND WOŹNIAKOWSKI, H.
(1980) "A general theory of optimal algorithms." Academic Press, New York.

(1984) Information and computation, *in* "Advances in Computers **23**" (M. C. Yovits, ed.), pp. 35–92. Academic Press, New York.

(1990) Information-based complexity: new questions for mathematicians, to appear in *Mathematical Intelligencer* .

TRAUB, J. F., WASILKOWSKI, G. W., AND WOŹNIAKOWSKI, H.
(1988) "Information-based complexity." Academic Press, New York.

TSITSIKLIS, J. N.
(1989) On the control of discrete-event dynamical systems, *Mathematics of Control, Signals, and Systems* **2**, 95–107.

TWOMEY, S.
(1977) "Introduction to the mathematics of inversion in remote sensing and indirect measurement," Developments in Geomathematics **3**. Elsevier Scientific Publ., Amsterdam.

VAKHANIA, N. N.
(1981) "Probability distributions on linear spaces." North-Holland, New York.

VAKHANIA, N. N., TARIELADZE, V. I. AND CHOBANYAN, S. A.
(1987) "Probability Distributions on Banach Spaces." Reidel Publishing Company, Dordrecht.

WAHBA, G.
(1984) Cross validated spline methods for the estimation of multivariate functions from data on functionals, *in* "Statistics: An appraisal, proceedings 50th anniversary conference Iowa State Statistics Laboratory" H. A. David and H. T. David, eds., pp. 205–235. Iowa State Univ. Press.

(1985)   A comparison of GCV and GML for choosing the smoothing parameter in the generalized spline smoothing problem, *The Annals of Statistics* **13**, 1378–1402.

WASILKOWSKI, G. W.
(1986)   Information of varying cardinality, *Journal of Complexity* **2**, 204–228.

WASILKOWSKI, G. W. AND WOŹNIAKOWSKI, H.
(1984)   Can adaption help on the average?, *Numerische Mathematik* **44**, 169–190.

WEIDMAN, J.
(1980)   "Linear operators in Hilbert spaces," Graduate Texts in Mathematics **68**. Springer-Verlag, Berline.

WENDLAND, W.
(1979)   "Elliptic systems in the plane." Pitman, London.

WERSCHULZ, A. G.
(1980)   Computational complexity of one-step methods systems of ordinary differential equations, *Mathematics of Computation* **34**, 155–174.

(1982)   Optimal error properties of finite element methods for second order elliptic Dirichlet problems, *Mathematics of Computation* **38**, 401–413.

(1984)   Does increased regularity lower complexity?, *Mathematics of Computation* **42**, 66–93.

(1985*a*) What is the complexity of elliptic systems?, *Journal of Approximation Theory* **45**, 69–84.

(1985*b*) What is the complexity of the Fredholm problem of the second kind?, *Journal of Integral Equations* **9**, 213–241.

(1986)   Complexity of indefinite elliptic problems, *Mathematics of Computation* **46**, 457–477.

(1987*a*) What is the complexity of ill-posed problems?, *Numerical Functional Analysis and Optimization* **9**, 945–967.

(1987*b*) Finite elements are not always optimal, *Advances in Applied Mathematics* **8**, 354–375.

(1987*c*) An information-based approach to ill-posed problems, *Journal of Complexity* **3**, 270–301.

(1989)   Average case complexity of elliptic partial differential equations, *Journal of Complexity* **5**, 306–330.

(1990)   Optimal residual algorithms for linear operator equations, to appear in *Journal of Complexity* **6**.

WOŹNIAKOWSKI, H.
(1986) Information-based complexity, *in* "Annual Review of Computer Science", pp. 319–380. Annual Reviews, Inc., William Kaufman, Inc., Palo Alto.

(1989)   Optimal average case multivariate quadrature, Research Report, Department of Computer Science, Columbia University, New York.

ŽENIŠEK, A.
(1972)   Hermite interpolation on simplexes in the finite element method, *in* "Proceedings EquaDiff 3", pp. 271–277. J. E. Purkyně University, Brno.

Author index

Subject index